CURRENT ISSUES OF SCIENCE AND RESEARCH IN THE GLOBAL WORLD

PROCEEDINGS OF THE INTERNATIONAL CONFERENCE ON CURRENT ISSUES OF SCIENCE AND RESEARCH IN THE GLOBAL WORLD, VIENNA, AUSTRIA, 27–28 MAY 2014

Current Issues of Science and Research in the Global World

Editors

Vlasta Kunova

Faculty of Law, Pan-European University, Bratislava, Slovakia

Martin Dolinsky

Faculty of Economics and Business, Pan-European University, Bratislava, Slovakia

CRC Press
Taylor & Francis Group
Boca Raton   London   New York   Leiden

CRC Press is an imprint of the
Taylor & Francis Group, an **informa** business

A BALKEMA BOOK

CRC Press/Balkema is an imprint of the Taylor & Francis Group, an informa business

© 2015 Taylor & Francis Group, London, UK

Typeset by V Publishing Solutions Pvt Ltd., Chennai, India
Printed and bound in Great Britain by CPI Group (UK) Ltd, Croydon, CR0 4YY

Published by: CRC Press/Balkema
　　　　　　 P.O. Box 11320, 2301 EH Leiden, The Netherlands
　　　　　　 e-mail: Pub.NL@taylorandfrancis.com
　　　　　　 www.crcpress.com – www.taylorandfrancis.com

ISBN: 978-1-138-02739-8 (Hbk)
ISBN: 978-1-317-52510-3 (eBook PDF)

Current Issues of Science and Research in the Global World – Kunova & Dolinsky (Eds)
© 2015 Taylor & Francis Group, London, ISBN 978-1-138-02739-8

Table of contents

Faculty of economics and business, section current topics of economic theory and practice in international business

Faculty of massmedia, section media-science-culture

Faculty of informatics, section new trends in information technology applications

Preface

The present volume of "Current Issues of Science and Research in the Global World 2014" contains papers presented at the International Conference in Vienna, which was organized by the Pan-European University (PEU).

The phenomenon of globalization is now recognized as a key factor in the post-modern civilization and touches every citizen of our planet. Issues of globalization appear currently in nearly all disciplines and research areas. Interest in globalization, its manifestations and trends are reinforced by the fact that communication and transport systems, transfer of capital and information, population migration and the creation of "social networking"—as a result of the digital revolution—are rapidly developing.

The Conference's research aims were the analysis, synthesis and comparison of the current impacts of globalization on those scientific disciplines which are in the centre of Pan-European University's academic orientation. Thus scientific teams and individuals from the fields of law, economics, media sciences, informatics and psychology defined their scholarly opinions and conclusions. Interdisciplinary research on complex issues such as globalization may ultimately provide innovative and creative solutions. The great challenge for our researchers was the opportunity to compare and discuss directly their views with their foreign colleagues. The choice to organize a conference at the Technical University of Vienna symbolized the international dimension of their field of research.

Law

In the past, the reception of Roman law constituted the unifying factor of law in the legal history of Continental Europe. The processes of harmonization and unification were well known in public and private international law. Law as a discipline has caused a sharp increase in demand for the creation of the so-called "*ius unum*" since the late 19th century. After World War II these efforts culminated in the creation of an entirely new legal system, unprecedented in history, namely that of European Union law. It was the result of the process of European integration which united gradually 28 European States under the roof of one legal system which presently replaces—or at least decisively influences—some 80 per cent of the Member States internal law, and which affects the day-to-day life of more than 500 million citizens of the European Union. The legal instrument for that is the method of harmonization, by which the national law of the Member States is adapted on the basis of certain framework laws, so-called directives, which are enacted on the European level. Harmonization pays respect to the national traditions and peculiarities of the various national legal systems, ranging from common law traditions in Ireland and the United Kingdom to the continental systems. It serves several purposes. For instance, harmonization is currently used to remove trade barriers not only within the EU internal market but also towards third countries all over the world on the basis of numerous cooperation agreements. It is also to establish standards for legal systems which have been transformed after the fall of totalitarian systems. Thus such aspects of globalization are now considered to be one of the most important topics of European and international jurisprudence.

Economics

One of the most important areas, that have experienced a great increase in risk factors related to the process of globalization, is the world economy. Researchers point to the increasing risk of financial and economic crises on a global scale, widening gap between wealth and poverty and the associated growth of social tension, negative impacts of migration and uneven population demographics. Although a close team of scientists from the field of economics cannot find answers and solutions to all emerging problems, it is certainly necessary to participate in the global debate of theorists in this field of research.

Media Sciences

There are two key issues of globalization of culture, internationalism, multiculturalism, regionalism, individualism and questions on the relationship between homogenous global culture on the one hand and national identity as a collective ideology of the 19th century on the other hand for media sciences. The current role of the media is to interpret existing partial answers to these questions and learning from a wide range of people.

Informatics

Globalization brings the need for processing and transmission of unprecedented amounts of information. New trends in information and communication technology in a global world incredibly improved communication and can boost knowledge of people. A great tool for the transfer of information is information encoding. The current situation shows that it is necessary to build a suitable visualization to this huge influx of information so that people are able to take and use it.

Psychology

Similarly, scientists from the field of psychology largely discuss the phenomenon of so-called "Bicultural identity" and its impact on human personality.

To sum up: The "International Conference: Current Issues of Science and Research in the Global World 2014" is the result of an international collaboration of scientists from the Pan-European University (PEU) and of leading international scholars from other renowned institutions. Our scientific outcomes constitute an essential contribution to the international debate on globalization. It is expected that research on this topic will be further developed.

We regret very much that we have had to refuse more than 20 percent of contributions to select the highest quality scientific contributions to the abovementioned topics. We are deeply grateful to authors, participants, reviewers, the International Scientific Committee, Scientific and Managing editor, student helpers and administrative assistants, for contributing to the success of this conference which, as mentioned in the beginning, was jointly endorsed by PEU in Bratislava and the *Studienzentrum Hohe Warte* in Vienna.

Univ. Prof. Dr. Dr. h.c. Peter Fischer

Organization

ORGANIZING COMMITTEE

Ing. Zuzana Godárová
Mag. Sonja Losert
Mag. Bettina Bartl
Ing. Ján Doboš
Dr. Antonín Doležal
Ing. Martin Dolinský, PhD.
Mgr. Dagmar Hocmanová
Barbora Illéšová
Dr. Andrej Karpát, PhD.
Ing. Mária Lukáčová
Mgr. Marek Ševčík
Mgr. Jana Šuchová
Ing. Tatiana Tretinová, PhD.

SCIENTIFIC COMMITTEE

Prof. Brian A. Barsky, *Berkeley, University of California, USA*
Prof. Eva Cihelková, *The University of Economics, Prague, Czech Republic*
Dr. Silvester Czanner, *Manchester Metropolitan University, UK*
Prof. Jiří Fárek, *Faculty of Economics, Technical University Liberec, Czech Republic*
Prof. Peter Fischer, *Professor Emeritus, Faculty of Law, University of Vienna, Austria*
Prof. Eva Gajdošová, *Pan-European University, Bratislava, Slovakia*
Prof. Bernd Glazinski, *Director Institut für angewandte Managementforschung, Rheinische Fachhochschule Köln, Germany*
Prof. Ruslan Grinberg, *Institute of Economics, Russian Academy of Science, Russia*
Assoc. Prof. Ľudovít Hajduk, *Pan-European University, Bratislava, Slovakia*
Prof. Květoň Holcr, *Pan-European University, Bratislava, Slovakia*
Prof. Kajetana Hontyová, *Pan-European University, Bratislava, Slovakia*
Prof. Jaroslav Ivor, *Pan-European University, Bratislava, Slovakia*
Prof. Jiří Jelínek, *Faculty of Law, Charles University in Prague, Czech Republic*
Prof. Jozef Klimko, *Pan-European University, Bratislava, Slovakia*
Assoc. Prof. Jitka Kloudová, *Pan-European University, Bratislava, Slovakia*
Prof. Teodor Kollárik, *Pan-European University, Bratislava, Slovakia*
Prof. Sergey Korkonosenko, *Saint Petersburg State University, Faculty of Journalism and Mass Communication, Russia*
Prof. Aleksandr P. Korochensky, *Belgorod State University, Faculty of Journalism, Russia*
Assoc. Prof. Vlasta Kunová, *Pan-European University, Bratislava, Slovakia*
Prof. Jozef Leikert, *Pan-European University, Bratislava, Slovakia*
Prof. Peter Linnert, *Sales Manager Akademie, Vienna, Austria*
Assoc. Prof. Elena Lisá, *Pan-European University, Bratislava, Slovakia*
Prof. Igor Lvovich, *Pan-European University, Bratislava, Slovakia*
Prof. Konrad Obermann, *Manheim Institute of Public Health, University of Heidelberg, Germany*

List of reviewers

Assoc. Prof. Andrej Ferko, *Comenius University in Bratislava, Slovakia*
Assoc. Prof. Anton Heretik, *Comenius University in Bratislava, Slovakia*
Assoc. Prof. Daniela Škutová, *Matej Bel University in Banská Bystrica, Slovakia*
Prof. Eva Muchová, *University of Economics in Bratislava, Slovakia*
Assoc. Prof. Ivan Sarmány-Schuller, *Slovak Academy of Sciences, Slovakia*
Assoc. Prof. Mária Bohdalová, *Comenius University in Bratislava, Slovakia*
Prof. Marta Orviská, *Matej Bel University in Banská Bystrica, Slovakia*
Assoc. Prof. Milan Šimek, *Brno University of Technology, Czech Republic*
Prof. Pavel Šturma, *Pan-European University in Bratislava, Slovakia*
Prof. Dr. hc. Peter Fischer, *Universität Wien, Austria*
Assist. Prof. Yuriy Preobrazhenskiy, *Voronezh Institute of High Technologies, Russia*
Prof. Zdeněk Molnár, *University of Economics in Prague, Czech Republic*

List of reviewers

Introductory articles

Current Issues of Science and Research in the Global World – Kunova & Dolinsky (Eds)
© *2015 Taylor & Francis Group, London, ISBN 978-1-138-02739-8*

Requirements on the staff of an application oriented research organization

Werner Purgathofer
Vienna University of Technology, Vienna, Austria
VRVis Research Center, Vienna, Austria

ABSTRACT: The VRVis Research Center in Vienna is the largest technology transfer institution in the area of Visual Computing in Austria. The requirements of the funding body FFG include the publication of scientific research results in first class peer reviewed media, and the active cooperation with co-funding companies. As a consequence the requirements on the staff of VRVis are manifold: they have to communicate with real users, use real data, know about software and hardware, understand the market, do professional documentation, initiate new projects and write funding proposals for these, be part of the scientific community and publish and review papers, manage several projects in parallel and obey strict deadlines for their projects and some more. Such staff is barely available and must be trained on the job.

1 INTRODUCTION

Visual Computing helps for many applications, often it provides visual solutions to data intensive problems. However, academic research often solves problems suspected by the researchers, rather than problems that appear in industry. The VRVis Research Center is set up to solve real problems from the real world, which is mostly less spectacular in the pure scientific sense, but it leads to practically useable results. This sort of research is not less demanding, however seen as second class research by many academics. For the researchers themselves it is rather more difficult than academic research, because the goal cannot be changed to avoid complications. In addition, VRVis researchers need to have numerous other "non academic qualifications".

2 THE VRVis RESEARCH CENTER

The VRVis Competence Center (VRVis 2014) was founded in 2000 funded by the Austrian Kplus program, and since 2010 it is funded by the Austrian COMET program as a K1 center. Its mission is to perform research and development in Visual Computing and bridge the gap between science and industry with translational research. VRVis is located in the TechGate building in Vienna, Austria, and has an annual budget of around 5 million Euros from which mainly around 65 researchers are paid.

The VRVis Research Center is set up as a non-profit-making limited company which is owned by universities and many companies. Currently VRVis is organized in three areas, which are Rendering, Visualization, and Visual Analytics. Each area performs some five to ten projects, most of which are multi-firm projects, that means that more than one company contributes to its budget.

3 THE UNDERSTANDING OF INNOVATION

In theory there is a common understanding how innovation happens. Scientists and researchers investigate the unexplored world and produce publically accessible publications and patents. This process is mostly done at universities and public research institutions. Then developers in companies make use of these findings and results to produce solutions for the real world, stable products for the world market. In large companies often additional research is added, the results of which are mostly protected by patents immediately, or they are kept secret altogether.

In practice this innovation chain rarely happens that way. The research results from academics have assumed ideal conditions, non-realistic constraints and flawless input data, so that an additional step is necessary to adapt such results to realistic requirements. This step is larger than many scientists are aware of, and it justifies specialized translational research institutions such as VRVis, AIT, Fraunhofer, and others. In close cooperation with developers from companies, these institutions have the competence to adapt pure research results to the requirements of the real world, and the people working there need special skills to perform these tasks.

4 REQUIREMENTS ON TRANSLATIONAL RESEARCH STAFF

In addition to the necessity to perform first class project work, i.e. to meet the defined project goals, researchers at a translational research institution such as VRVis need to be able to cope with a variety of additional challenges. These are partly due to the real life conditions, and partly because the overall requirements exceed those of pure research.

4.1 *Communication with real users*

Dealing with people from outside the field of computer science uncovers the use of different languages for the same things, and the complexity of some aspects that seem simple for the computer scientist. Often the user describes a problem and her/his envisaged solution in the idealistic world of her/his realm. The project leader misunderstands parts, and adds new ideas based on her/his knowledge of technical possibilities. The programmer interprets these requirements based on her/his personal experience and attitudes and produces something the user believes is wrong. Only during this last discussion it turns out that the first description by the user was incomplete, the interpretation by the project leader went in the wrong direction and the creativity of the programmer (and possible errors in the code) produced an unusable result. Coping with this situation is a challenge in its own. In addition some users misunderstand the options and possibilities, and also all users (and scientists, of course) are different characters, that don't always fit together.

4.2 *Use of real data*

Idealized test data are always created optimally so that the foreseen variations of an algorithm work fine. Data from the real world is erroneous, incomplete, inexact, includes many exceptions to the defined rules, and, above all that, is often huge, much larger than the algorithm was intended for. Especially the adaptation of functioning and published state-of-the-art algorithms to extreme data sizes is a main research topic for VRVis researchers.

4.3 *Hardware and market knowledge*

Researchers operating under realistic conditions need to be experts with all the various hardware they use. Many fast algorithms have to be implemented with hardware dependent components to ensure low level optimization. Today not only the large amount of companies creates many different interfaces, but also the immense variety of special hardware components. To make optimal use of these, it is also necessary to keep an overview over new product developments, hardware trends and announced products. Visiting technical exhibitions and fairs, reading computer journals and watching announcements in electronic media are necessary and expected activities.

4.4 *Project initiation and proposal writing*

New projects don't come by themselves. VRVis researchers are also experts in generating new projects. They need to be aware of all funding opportunities and know how to write a successful funding proposal. Based on intensive contacts with many companies they must also communicate the functioning of a translational research institution to these. For the proposals they need to be aware of the state-of-the-art of their science field and of the intended application field. They have to be trained in project planning, including time and budget plans. And they have to be experienced and consequent to collect the proposal parts the company partners have to deliver in time!

4.5 *Scientific publishing*

Evaluators of any research institution are often pure academic researchers. Their main evaluation criterion is then scientific output, i.e. the number and quality of produced scientific publications. Therefore, VRVis researchers have to fully understand this aspect and be able to write scientific publications that will be accepted at high quality media. This includes knowing where to publish, knowing the scientific state-of-the-art, and giving good talks at conferences. But it is also a challenge to get the o.k. from companies to publish at all, and to judge any conflicts such a publication might have with patents or other intellectual property rights involved.

4.6 *Professional documentation*

As opposed to usual scientific results which are produced for one time demonstration of the correctness of a result, software or system components intended for practical use need to be accompanied by understandable and complete documentation and user manuals. In addition, companies involved in a project often require training sessions for the people who either will use the results or who are involved in strategic decisions at the company. Thus it is necessary that the programmers and project leader maintain the user view of their products—not always easy when you are deep into the details.

4.7 *Multi-tasking under time pressure*

Projects for the real world also have real deadlines. University scientists are used to extendible deadlines, flexible result progress, and no concrete consequences in case they miss these. Only submission deadlines at conferences generate some stress, but again, no severe consequences other than postponing the result to the next available publication option occur. In contrast, when cooperating with companies there are project contracts, fixed deadlines and there is little compromise accepted for the results. This increased time stress is accompanied by nervous roommates, noise, telephone calls, emails, reviews, guests, continued education, project initiations, and many more disturbing factors. It needs special concentration to be able to work under such conditions.

5 CONCLUSIONS

Translational research is a challenge in its own, often underestimated by academic scientific staff. Besides excellence in computer science involved researchers need to have various valuable skills not taught in usual university courses. Thus the value of a research organization depends even more on its well selected employees and the relevant skills learned on the job than on any other aspect.

ACKNOWLEDGEMENT

The VRVis Research Center is funded by the COMET program of the Austrian FFG (The Austrian Research Promotion Agency) as a K1 Center from 2010 to 2016.

REFERENCE

VRVis, 2014. *www.vrvis.at*

Current Issues of Science and Research in the Global World – Kunova & Dolinsky (Eds)
© *2015 Taylor & Francis Group, London, ISBN 978-1-138-02739-8*

Globalization and Indian youth: Findings from Moral Foundations Theory

W. Renner
Pan-European University, Bratislava, Slovakia

ABSTRACT: The present research assumed that individual differences in personal values would go along with globalization. Study 1 confirmed the hypotheses of Moral Foundations Theory (MFT) by showing that Indian youth (N = 336) with a comparatively low degree of globalization put significantly more emphasis than European youth (N = 163) on values related to authority, in-group loyalty, and spiritual purity, as measured by questionnaires. Study 2, however, asking the Indian sub-sample for their decisions in moral dilemmas, found that their degree of globalization predicted "utilitarian" decisions, i.e., more globalized participants tended to lose sight of their ethical values in return for financial gain. Whereas Study 1 points to the fact that Indian youth still differ from European youth with respect to their ethical positions, Study 2 indicates that Indian society is changing and only part of Indian youth is willing to adhere to their cultural principles when some short-term benefit is at stake.

1 INTRODUCTION

1.1 *Tradition vs. globalization: The example of India*

Traditional India is a collectivist culture emphasizing the well-being of the group and traditional as well as spiritual values (Basham, 1981). Shweder (2008) developed his Cultural Psychology by comparing the ethics of India to those of the United States. Rather than relying on scientific progress, many Indians believe that the real "age of truth" has existed in ancient times and talk of this era highly respectfully (Shweder, 2008).

In contrast, among young people in India, there is the tendency to follow the influence of globalization: English is replacing many local dialects and almost everybody has access to worldwide satellite TV, mobile phones, the Internet, and social media. Besides other health risks like increased smoking and other substance use, the consumption of fast food is increasing by 40% per year in India (Jensen, 2011).

From these considerations it may well be stated that Indian youth is living at the crossroad of traditional culture and globalization, posing an interesting field for studying the possible effects of globalization on moral or ethical decision making.

1.2 *Moral Foundations Theory*

Referring to Shweders Cultural Psychology Moral Foundations Theory (MFT) postulated five moral or ethical principles as follows:

1. "Harm/care: Concerns for the suffering of others […].
2. Fairness/reciprocity: Concerns about unfair treatment […].
3. Ingroup/loyalty: Concerns related to obligations of group membership […].
4. Authority/respect: Concerns related to social order […].
5. Purity/sanctity: Concerns about physical and spiritual contagion […]" (Haidt & Kesebir, 2010, p. 822).

Western middleclass respondents put high emphasis on MFT foundations 1 and 2, while neglecting 3, 4, and 5. Conversely, people from collectivist societies like Indians, are expected to put equal emphasis on all five foundations. In a large online study, these expectations were confirmed, but effect sizes in the sense of Cohen (1992) were surprisingly small (Graham et al., 2011). These small effects may have resulted from the online administration of the questionnaire, thus excluding less globalized individuals.

2 THE PRESENT RESEARCH

2.1 *Research question and hypotheses*

2.1.1 *Study 1*
I examined cultural differences of value related ethics and relationships between globalization and ethical or moral decisions. In accordance with previous results by Graham et al. (2011), I hypothesized that, according to MFT, European Students would put less emphasis on Moral Foundations 3, 4, and 5 as compared to Indian students. In contrast to the (Graham et al.'s, 2011) findings obtained online, I expected large effect sizes (Cohen, 1992) in the less globalized sample contacted by paper questionnaires.

2.1.2 *Study 2*
I examined the concomitants of globalization in the field of moral and/or ethical decision making: globalization should predict a utilitarian, as opposed to a deontological style of decision making in Indian participants. Utilitarianism prefers decisions following a rationalist paradigm (e.g., advocating euthanasia in certain cases of terminal diseases going along with severe suffering). Deontologist decisions follow supreme or divine laws transcending rationalism (e.g., assuming that under no circumstances whatever, it can be right to kill).

2.2 *Participants*

2.2.1 *Study 1*
N = 336 students (226 of them female, mean age 21.9 years, s = 3.1) from Pondicherry University, India with various study majors participated. The European sample comprised N = 163 students (131 of them female, mean age 25.7 years, s = 8.4) mostly of German or Austrian descent at Alpen-Adria-Universität Klagenfurt (Austria), also comprising various majors.

2.2.2 *Study 2*
In Study 2 only the Indian sub-sample as described in 2.2.1 participated.

2.3 *Questionnaires*

2.3.1 *Study 1*
The questionnaires were administered in a paper-pencil form (Table 1). The Moral Foundations Questionnaire (MFQ, "relevance" and "judgment" sections) measures the five moral foundations proposed by MFT (Graham et al., 2011). E.g., the "Harm/care" foundation was assessed by "Whether or not someone suffered emotionally" (p. 385) in the Relevance part and by "One of the worst things a person could do is hurt a defenseless animal" (p. 385) in the Judgment section; the Purity/sanctity foundation was measured by "Whether or not someone acted in a way that God would approve of" in the Relevance section and by "Chastity is an important and valuable virtue" in the Judgment section. In total, both sections comprised 15 items each, i.e., three items for each Moral Foundation respectively.

"Need for Cognition" is an individual's tendency to prefer rational tasks. "Faith in Intuition" is characterized by trusting one's "gut" feelings and "Idealism" indicates a tendency to avoid harm to others.

Table 1.   Questionnaires.

Name of questionnaire	Nr. of items	Answer format	Source
1. Need for Cognition (NFC): Sub-scale of Rational Experiential Questionnaire, REI	10	1 ... 5	(Epstein, Pacini, DenesRaj & Heier, 1996)
2. Faith in Intuition (FI): Subscale of REI	10	1 ... 5	(Epstein et al., 1996)
3. Idealism: Subscale of Ethics Positions Questionnaire (EPQ)	10	1 ... 9	(Forsyth, 1980)
4. Social Desirability: Impression Management Subscale of the Balanced Inventory of Desirable Responding	20	1 ... 7	(Paulhus, 1991)
5. Conservatism: Single item as used by Graham et al. (2011): "very liberal" to "very conservative"	1	1 ... 7	(Graham et al., 2011)
6. Intrinsic religiosity	8	0 ... 4	(Gorsuch & McPherson, 1989)
7. Globalization (adapted from original source)	12	0 ... 7	(Redfern & Crawford, 2010)
8. Moral foundations questionnaire	30	0 ... 5	(Graham et al., 2011)

2.3.2   *Study 2*

Participants received 15 moral scenarios, three for each moral foundation, asking them, how they would decide ("+2" indicating saying "certainly yes", "–2" "certainly no" to the utilitarian decision. Thus, the mean scores indicated the personal degree of utilitarianism. For example, a scenario for Purity/sanctity was:

SOUL. At university, you are participating in an experiment on emotion. [...] you have signed a contract and agreed to follow the instructions [...]. The program now informs you that, while your skin conductance will be measured as an indicator of your level of arousal, you will have to sign a piece of paper that says: "I hereby sell my soul, after my death, to whoever has this piece of paper". [...] If you disagree to sign the paper, you will be accused of having broken the contract with severe academic consequences. While considering what you should do, terrifying folk tales from your childhood about people who have sold their souls for money come to your mind where these people have suffered eternal damnation and agony. Will you overcome your uneasy feelings, and sign the paper?

NO   –2   –1   +1   +2   YES

3   RESULTS

3.1   *Study 1: Endorsement of the five moral foundations by Indian vs. Austrian students*

Figure 1 shows the results for the Relevance section of the MFQ, Figure 2 shows the results for the Judgment section of the MFQ.

As the assumption of normality distribution was not fulfilled, group differences were tested for significance by Mann-Whitney U-Tests. For the Relevance section of the MFQ all the differences were statistically highly significant. In the case of the Judgment section, for the Harm/care foundation, the differences were not significant, whereas for all the remaining foundations, differences were statistically significant. Details are given in Table 2 and Figures 1 and 2.

In line with the first hypothesis, Indian students scored significantly higher than Austrian ones on the Moral Foundations 3 (Authority/respect), 4 (Ingroup/loyalty) and 5 (Purity/

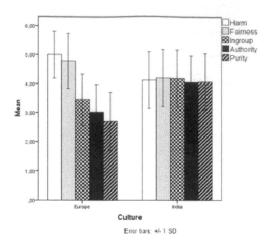

Figure 1.   Differences between Indian and European students on the Relevance Section of the MFQ.

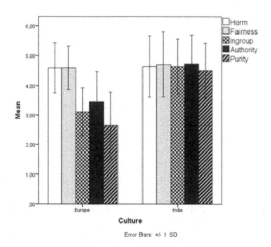

Figure 2.   Differences between Indian and European students on the Judgment Section of the MFQ.

Table 2.   Results of significance tests, effect sizes (Cohen's d) and Confidential Limits (CL).

Moral foundation	Relevance section				Judgment section			
	Z	p	d	CL 95%	Z	p	d	CL 95%
Harm/care	−9.603	0.000	0.95	0.75 ... 1.14	1.152	0.249	−0.05	−0.24 ... 0.14
Fairness/reciprocity	−6.474	0.000	0.60	0.41 ... 0.79	−2.524	0.012	−0.12	−0.31 ... 0.06
Ingroup/loyalty	7.892	0.000	−0.76	−0.95 ... −0.57	−13.991	0.000	1.71	−1.93 ... 1.50
Authority/respect	−10.203	0.000	−1.13	−1.33 ... −0.93	−11.480	0.000	−1.18	−3.18 ... 0.82
Purity/sanctity	−12.147	0.000	−1.39	−1.59 ... −1.18	−14.110	0.000	−1.86	−2.08 ... −1.64

sanctity) on both sections of the MFQ. When accounting for multiple testing by the Bonferroni correction, these differences still remain highly significant ($p < 0.01$).

According to Bortz & Döring (2005) large effects in the sense of Cohen are defined as exceeding $d = 0.8$. Thus, with the only exception of "Ingroup/loyalty" for the Relevance section of the MFQ ($d = -0.76$), in accordance with my hypothesis, large effects in the sense of

Cohen (1992) were achieved with respect to the differences of Austrian and Indian Students on Moral Foundations 3, 4 and 5.

3.2 *Study 2: Globalization as a predictor of utilitarianism in Indian students*

In the next step the role of globalization as compared to other possible predictors of utilitarianism in Indian students was examined. In Table 3 the descriptive results for the Indian subsample have been summarized (for the range of scales see the column "answer format" in Table 1).

On utilitarianism, as measured by the 15 scenarios, the Indian sub-sample had a mean of –0.53 (s = 0.92), with +2 indicating the maximum possible degree and –2 indicating the least possible degree of utilitarianism. An Exploratory Principal Components Analysis with Varimax rotation was computed in order to examine the dimensionality of the results. The scree plot shown in Figure 3 clearly suggests a one-dimensional solution, which explained 37.9% of the variance. Alternatively a Confirmatory Factor Analysis (CFA) might have been considered, but this method tends to reject valid models even in medium sample sizes (Gatignon, 2010). Thus, I preferred an EFA to a CFA approach.

Accordingly, when the results for all 15 scenarios were combined to one single scale, satisfactory internal consistency was achieved (Cronbach's α = 0.873). Thus, for subsequent

Table 3. Descriptive results of Indian sub-sample.

Scale	Mean	SD
1. Need for cognition	3.51	0.67
2. Faith in intuition	3.46	0.59
3. Idealism	6.65	1.26
4. Social desirability	4.22	0.62
5. Conservatism	4.09	1.86
6. Intrinsic religiosity	3.05	0.63
7. Globalization	4.17	1.16

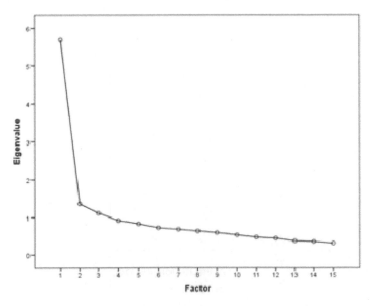

Figure 3. Exploratory principal components analysis of moral decisions by the indian sub-sample: scree plot.

analyses, the mean score obtained from the 15 moral decisions was used as a global indicator of utilitarianism.

The scores summarized in Table 3 were entered into a multiple linear regression model (stepwise method) to predict the personal degree of utilitarianism. The final regression model explained 34% of the variance ($R^2 = 0.343$, corrected $R^2 = 0.328$) of the dependent variable. The significant predictors are summarized in Table 4.

From Table 4 it can be seen that the most powerful predictor of utilitarian decisions was high political conservatism, followed by a low Need for Cognition and a high degree of globalization and low Idealism. It is also important to note that a response stile biased by social desirability predict lower self reported utilitarianism.

As globalization has been shown to be an important predictor of utilitarian decisions in Indian youth, the second central hypothesis of this study was confirmed. In order to examine this result in more detail, the Indian sub-sample was divided into four quartiles, with Quartile 1 scoring lowest and Quartile 4 scoring highest on the globalization scales. Mean Utilitarianism for the four quartiles is shown in Figure 4. From this bar chart it can be seen that respondents scoring below the median on Globalization only very rarely considered a utilitarian as opposed to a deontological decision. Respondents in the Quartile 4 of

Table 4.   Multiple linear regression—predictors of utilitarian decisions.

	Non-standardized coefficients		Standardized coefficients		
	Regression coefficient	Standard error	Beta	T	Sig.
Constant	1.398	0.539		2.595	0.010
Need for cognition	−0.391	0.089	−0.71	−4.398	0.000
Conservatism	0.153	0.031	0.294	4.998	0.000
Social desirability	−0.244	0.097	−0.152	−2.522	0.012
Globalization	0.126	0.049	0.149	2.588	0.010
Idealism	−0.097	0.046	−0.122	−2.133	0.034

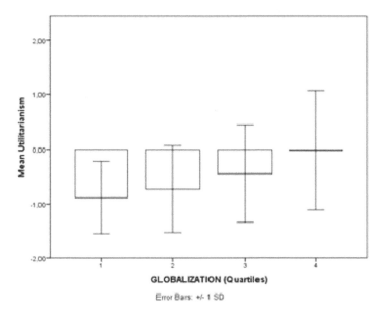

Figure 4.   Mean utilitarianism in Indian respondents with low (Quartile 1), intermediate (Quartiles 2 and 3) and high (Quartile 4) degrees of globalization.

12

Globalization, on the other hand, had an average score near zero, implying that utilitarian decisions were frequent among them.

4 CONCLUSIONS

A limitation of the present study pertains to the restricted student sample.

Study 1 confirmed the expectation of MFT, according to which Indian respondents scored significantly higher than Europeans on moral foundations pertaining to group identity, authority, and religious purity and dignity. The results of Study 1 also suggest that for Indian respondents, all aspects of morality or ethics are equally important.

The latter conclusion is also supported by Study 2: although the 15 moral scenarios had been developed along the lines of five different moral foundations, they were perceived as uni-dimensional by the Indian respondents. More globalized individuals tended to solve the moral scenarios in a utilitarian rather than a deontological way, i.e., they tended to ignore moral principles in favour of personal advantage.

This result suggests that globalization should not be seen as a phenomenon affecting all people to the same extent. Even within the Indian students' sample, differences with respect to the personal degree of globalization were found which predicted various types of decision making in the light of utilitarian as opposed to deontological alternatives. Future research thus should aim at the possibility of training cultural identity in youth and emerging adults.

REFERENCES

Basham, A.L. 2008. *The wonder that was India.* Calcutta: Fontana Books.
Bortz, J. & Döring, N. 2005. *Forschungsmethoden und Evaluation* [Research methods and evaluation]. Berlin (Germany): Springer.
Cohen, J. (1992). A power primer. *Psychological Bulletin,* 112, 155–159.
Epstein, S., Pacini, R., DenesRaj, V., & Heier, H. 1996: Individual differences in intuitive-experiential and analytical-rational thinking styles. *Journal of Personality and Social Psychology,* 71: 390–405.
Forsyth, D.R. 1980. A taxonomy of ethical ideologies. *Journal of Personality and Social Psychology,* 39: 175–184.
Gatignon, H. (2010). *Confirmatory factor analysis in statistical analysis of management data.* DOI: 10.1007/978-1-4419-1270-1_4.
Gorsuch, R.L. & McPherson, S.E. 1989.: Intrinsic/extrinsic measurement: I/E revised and single-item scales. *Journal for the Scientific Study of Religion,* 28, 348–354.
Graham, J., Nosek, B.A., Haidt, J., Iyer, R., Koleva, S., & Ditto, P.H. 2011. Mapping the moral domain. *Journal of Personality and Social Psychology,* 101: 366–385.
Haidt, J. & Kesebir, S. 2010. Morality. In S.T. Fiske, D.T. Gilbert, & G. Lindzey (eds), *Handbook of social psychology* (5th Ed.): 797–832. Hoboken, NJ: Wiley.
Jensen, L.A. 2011. Navigating local and global worlds: Opportunities and risks for adolescent cultural identity development. *Psychological Studies,* 56: 62–70.
Paulhus, D.L. 1991. Measurement and control of response biases. In J.P. Robinson et al. (eds) *Measures of personality and social psychological attitudes:* 17–59. San Diego: Academic Press, 1991. ISBN: 978-0-12-590241-0.
Redfern, K. & Crawford, J. 2010. Regional differences in business ethics in the People's Republic of China: A multi-dimensional approach to the effects of modernisation. *Asia Pacific Journal of Management,* 27: 215–235.
Shweder, R.A. 2008. The cultural psychology of suffering: The many meanings of health in Orissa, India (and elsewhere). *Ethos,* 36, 60–77, 2008.

Current Issues of Science and Research in the Global World – Kunova & Dolinsky (Eds)
© *2015 Taylor & Francis Group, London, ISBN 978-1-138-02739-8*

The design and trademark in the European Union from the perspective of public law

R. Kaššák
Faculty of Law, Pan-European University, Bratislava, Slovakia

ABSTRACT: In the following text some of the basic facts relating to intellectual property law will be addressed, not providing however, a complete analysis of IP laws within the EU legal framework. The author of the study focuses mainly on the area of the Community Trade Mark, International trade mark designating the European Union, Procedural Perspectives and The Registration Procedure of the Community Trade Mark and also the Community Design, Procedural Perspectives and The Registration Procedure of the Registered Community Design.

1   INTRODUCTION

In the following text some of the basic facts relating to intellectual property law will be addressed, not providing however, a complete analysis of IP laws within the EU legal framework.

The law of intellectual property usually provides some **exclusive rights** to an entity. These rights have some specific features. In the following sections, we will analyse some of these features:

a. These exclusive rights are usually **negative in nature**. If a right is categorised as negative in nature, it gives the holder of these rights the ability to exclude other persons (third parties) from **infringing on** his/her monopoly. The most important of these instruments is the power of the **holder** to start an action at a court of law if one violates his right(s) connected to his/her IP product. The existence of these rights, makes the person interested in using the work of the right holder, apply for his/her permission to do so.
b. Exclusive rights are derived from various intellectual property laws and can be **transferred** or **licensed** (in general).
c. The exclusive rights are usually **awarded** for a limited period of time.[1]

A trademark is probably one of the most visible forms of intellectual property. In different forms, it is displayed to the public at large. Typical symbols of fast-food restaurants, symbols used by an international producer of sports clothing and equipment, slogans of international companies and sentences defining a product (or a service) in TV advertisements etc. are referred to as trademark. It is actually **any commercial indicator (name, symbol, word, device, signs etc)**, a distinctive sign, used in order to identify certain goods/services produced/ provided by a specific person or company.

One of the simplest definitions of a trademark encompasses that it is **everything that identifies a product and distinguishes it from other products (in order to identify its source)**.

Nevertheless, it is important to mention that something could be categorised as a trademark (servicemark) if it is unique enough and if it does not become generic (which means

---

[1]Please refer to e.g.: POTASCH, P. In: Kummerová, S. et al.: *Legal English—Fundamental Terms and Topics*. Bratislava: Paneurópska vysoká škola/Žilina: Eurokódex s.r.o., 2010, ISBN 9788089447343.

that the trademark is no longer used to describe a particular product/service but has become a general term describing a certain product/service).[2]

Office for Harmonisation in the Internal Market

The Office for Harmonisation in the Internal Market (OHIM) is the official trade marks and designs office of the European Union. The Office register the Community Trade Mark (CTM) and Registered Community Design (RCD), which are essential components of the European single market. These registrations provide trade mark and design protection throughout the European Union. The Office also works in close partnership with national IP offices (intellectual property offices) in the EU member states, with international offices, and the European Commission on a wide range of issues affecting the owners and users of intellectual property rights.[3] The seat of the Office is in Alicante, Spain.

2   COMMUNITY TRADE MARK

As suggested above, a trade mark is a sign which serves to distinguish the goods and services of one organisation from those of another. The following part of the text discusses the suggested topic from the point of substantive and procedural law by looking at the basic elements of both perspectives.

2.1   *Community trade mark—basic substantive elements*

Before discussing the selected issues in detail, some time will have to be devoted to the methods of acquiring protection (in the form of trademark) in the EU. In general, there are two forms that may be considered:

1. **A Community trade mark:** an exclusive right that protects distinctive signs, valid across the EU, registered directly with OHIM in accordance with the conditions specified in the CTM Regulations. The proprietor of the Community trade mark is required to make genuine use of the trade mark for the five years following its registration, must not interrupt the use of the Community trade mark for a period of longer than 5 years. If not, they may be subject to sanctions for non-use. However, such sanctions would logically not apply, if the proprietor had proper reasons for non-use.
2. **An international trade mark designating the European Union (EU):** likewise an exclusive right but administered by the International Bureau of the World Intellectual Property Organization (WIPO) in Geneva according to the Madrid Protocol. WIPO processes the application and then sends it to OHIM for examination according to the conditions specified in the CTM (Community Trade Mark) Regulations. This has the same effect as applying directly for a Community trade mark.[4]

Trade marks are words, logos, devices or other distinctive features which can be represented graphically. They can consist of, for example, the shape of goods, their packaging, sounds and even smells.[5] The Community trade mark we will be referring to in this text, may consist of (according to Article 4 of the Regulation) any signs capable of being represented graphically (particularly words, personal names, designs, letters, numerals, the shape of goods or of their packaging) **provided** that such signs are capable of distinguishing the goods or services of one undertaking from those of other undertakings.

---

[2]Please refer to e.g.: POTASCH, P. In: Kummerová, S. et al.: *Legal English—Fundamental Terms and Topics*. Bratislava: Paneurópska vysoká škola/Žilina: Eurokódex s.r.o., 2010, ISBN 9788089447343.
[3]For more information on this topic, please see: http://oami.europa.eu
[4]For more information on this topic, please see: http://oami.europa.eu
[5]For more information on this topic, please see: http://oami.europa.eu

Within the EU a special system of trade marks has been established in the past and recently, the basic legal framework provided in the matter is Council Regulation (EC) No 207/2009 of 26 February 2009 on the Community trade mark ("the Regulation" or "the CTM Regulation").

According to Article 5 of the Regulation, any natural or legal person, including authorities established under public law, may obtain a Community trade mark. It is interesting to note that the Regulation does not provide for any special criteria that would distinguish the Community trade mark from other trade marks (as to their nature), the Regulation simply defines that a trademark registered according to the Regulation becomes a Community trade mark thus the only way to establish a Community trade mark is via registration. Unlike other (standard) trade mark which can be used and established simply by prior use (and by meeting some conditions prescribed by law), Community trade mark being a special category may be established only through registration. There is no other way of establishing a trademark according to Article 6 of the Regulation.

There is no need to go into scientific reasoning to understand that one decides to register a trade mark in order to have a commercial benefit. By now it is also clear that trade mark as one of the forms of intellectual property rights and that it is exclusive in nature. The fact that it is exclusive in nature means that it guarantees certain rights only to the proprietor. At the same time, trade mark is also negative in nature which means that once duly registered it may prohibit third parties from using the following for commercial purposes:

– any sign which is identical with the Community trade mark in relation to goods or services which are identical with those for which the Community trade mark is registered;
– any sign where there exists a likelihood of confusion with another trade mark;
– any sign which is identical with, or similar to, the Community trade mark in relation to goods or services which are not similar to those for which the Community trade mark is registered, where use of that sign takes advantage of the repute and distinctive character of the trade mark.

Another principle that must be stressed at this point is the fact that the Community trade mark is treated **as if it was a national trade mark** (i.e. a trade mark established according to the national laws) of the proprietor or his/her national law. This applicable law can be identified from the information in the register with the main rule being that it is the law of the Member State in which the proprietor has his seat or domicil (at the moment of registration) that will govern the status of the Community trade mark. There are some exceptions to this rule but also alternative ways of stating the applicable national laws, are declared by the Regulation.[6]

This is interesting also from the perspective that it is the national courts that provide the proprietor with judicial protection in case his rights relating to the Community trade mark are infringed. To this end, a landmark case must be analysed (briefly). The Court of Justice of the European Community (ECJ) declared an important judgment in a preliminary ruling case (C-235/09) regarding the scope of prohibition against infringement or threatened infringement of a Community Trade Mark registration (CTM). The ECJ declared that the effect of such a prohibition in one member state normally extends to the entire area of the European Union. In the relevant case, Chronopost SA is the owner of the French and Community trade marks 'WEBSHIPPING', applied for in 2000 and registered in respect of, inter alia, services relating to logistics and data transmission, collecting and distributing mail, and express mail management. In spite of that registration, DHL Express France SAS (the successor to DHL International) used the same word in order to designate an express mail management service accessible principally via the Internet. By judgment of 15 March 2006, the Tribunal de grande instance de Paris (Regional Court, Paris, France)—which heard the case as a Community trade mark court—found that DHL Express France had infringed the French trade mark WEBSHIPPING, although it did not adjudicate upon the infringement of the Community

---

[6]For more information on the issue, please refer to Article 16, alt. Article 14 of the Regulation.

trade mark. The Cour d'appel de Paris (Court of Appeal, Paris), on an appeal by Chronopost, upheld that decision on 9 November 2007, and prohibited DHL, subject to a periodic penalty payment in the event of infringement of the prohibition, from continuing to use the signs 'WEBSHIPPING' and 'WEB SHIPPING'. However, it did not allow Chronopost's claim to extend the effects of the prohibition to the entire area of the European Union, and it thus restricted the effects of the prohibition to French territory only. DHL brought an appeal in cassation. That appeal was dismissed, but since Chronopost had brought a cross-appeal against restricting the prohibition and the periodic penalty payment territorially, the Cour de cassation (Court of Cassation) held that it was necessary to refer the matter to the Court of Justice. The relevant matter in the case, concerned the interpretation of the rule on sanctions in the CTM Regulation (Article 98 (1) of the previous Regulation No. 40/94, Article 102 (1) of the present Regulation No. 207/2009). According to the Article concerned, a CTM court shall issue an order prohibiting the defendant from proceeding with the acts that infringed or would infringe the CTM where it finds that the defendant has infringed or threatened to infringe a CTM, unless there are special reasons for not doing so. It shall also take such measures in accordance with its national law. The Court ruled, that the regulation must be interpreted as meaning that **a prohibition issued by a national court, hearing a case as a Community trade mark court, extends, as a rule, to the entire area of the European Union.**[7]

It is clear from the above, that the Community trade mark is of **unitary nature**—it produces the same effects throughout the whole European Union. The unitary nature of the Community trade mark at its very heart is strongly connected to the basic principles of the internal market. For clarity, it needs to be stressed that the fact the Community trademark is unitary means that even if the applicant wanted to, the Office (OHIM) is not able and empowered to limit the application of the Community trademark to a certain geographical region.

2.2 *Community trade mark—procedural perspectives—the registration procedure*

The Community Trade Mark Application

Applicants may file an application for a Community trade mark with—according to Article 25 of the Regulation—one of the following institutions:

– the Office for Harmonisation in the Internal Market
– the central industrial property office of a Member State;
– the Benelux Office for Intellectual Property.[8]

The Regulation provides no rules on which institution is the one to serve the application on thus this is entirely on the applicant to decide. In case the application is not filed with OHIM, the applicant may have to bear additional costs usually in the form handling fees to be paid directly to the national authority (if the application is filed with the national authority).

Although the applications may be filed with all those institutions mentioned above, in case they are filed either with b) or c), they have a duty to forward the application within two weeks (at most) to the Office. In other words, the industrial offices of Member States or the Benelux Office for IP do not decide on the merits of the case, they simply forward the file to the Office. Applications for Community trade marks shall be filed in one of the official languages of the European Union. At the same time, the applicant must indicate a second language which shall be one of the official a languages of the Office (German, English, Spanish, French and Italian) the use of which he accepts as a possible language of proceedings.

In line with Article 26 of the Regulation, the application must contain the following information:

1. a request for the registration of a Community trade mark;
2. information identifying the applicant;

---

[7]Judgment in case C-235/09, DHL Express France SAS v Chronopost SA. Court of Justice of the European Union. Press release No. 35/11.
[8]Refer to Article 25 of the Regulation.

3. a list of the goods or services in respect of which the registration is requested;
4. a representation of the trade mark.

In addition to this, the applicants must also pay a filing fee.

Examination procedure
On receipt of the CTM application, the OHIM starts the examination procedure. The Office inspects the file (the application) continuously until decided on the merits or otherwise settled. In case the Office identifies a deficiency during the examination process, in line with the administrative procedures of most national laws, it serves an objection letter on the applicant. The aim of this letter is to give the applicant the chance to remedy the deficiences (and to be able to proceed with the case). If the deficiency is not remedied, the application will be provisionally refused or, if the deficiency concerns a priority or seniority claim, the claim will be refused. These refusals can be appealed before the Boards of Appeal of the OHIM.

If no problems arise during the examination procedure, or once the objections are successfully dealt with, the trade mark is published after the Office has received the translations into all official languages of the EU from the Translation Centre. The OHIM does not inform the applicant of the publication date and Bulletin reference. If the CTM application is refused, it will not be published.

Publication of the application
Once the CTM application is accepted, it will be published in Part A of the Community Trade Marks Bulletin. Publication takes place as soon as the national office and OHIM search reports have been issued to the applicant. The publication of the application in Part A of the Bulletin opens the three month period for filing an opposition. After the publication of a Community trade mark application, third party observations can be filed (Article 40 CTM regulation) which may relate to the existence of an absolute ground for refusal (Article 7 CTM regulation). Observations must be in written form. The third party will not become a party to the proceedings before the OHIM. The observations will be transmitted to the applicant. The OHIM will then consider whether the observations are well-founded, i.e. whether an absolute ground for refusal exists. If so, an objection will be sent to the applicant by the OHIM. The third party will be informed of the action taken, namely whether or not the observations give rise to an objection.

In addition, publication of the CTM application in Part A of the Community Trade Marks Bulletin permits file inspection to be requested.

Searches
The Community search report is drawn up from OHIM's database and lists any identical or similar earlier Community trade marks (including International Registrations designating the EU). When the new application has been published, proprietors of the earlier trade marks or trade mark applications cited in the report are informed by letter about the new application. This is called a "surveillance letter".

Community search reports and surveillance letters are sent as a service that is included in the application procedure and, as such, are covered by the basic application fee. When the national search option is selected by the applicant, OHIM requests the production of national search reports from the relevant participating national offices, who remain responsible for the content of these reports. The results of the search reports as well as the surveillance letters are for information only, i.e. the citation of a given trade mark should not be considered to constitute a finding that a conflict in fact exists. Such a conclusion can only be reached if and when an opposition is filed and decided upon. The purpose of the search reports is to give the applicants the possibility of withdrawing their applications after analysing the report contents.

Opposition
'Opposition' is a procedure that takes place before OHIM when a third party requests the Office to reject a Community Trade Mark Application (CTMA). Once published there is a

three-month opposition period. If no oppositions are filed during this period, the application proceeds to registration. In general terms, an opponent must have rights in an earlier trade mark or other form of trade sign. The grounds on which an opposition may be made (called 'relative grounds for refusal') are indicated in Article 8 of the Regulation. For an opposition to be successful, the trade mark applied for must be found to be incompatible with such rights. Any proceedings start with a period during which parties can negotiate an agreement, the so-called "cooling-off" period.

When an opposition is filed, the proceedings will include an exchange of observations from both the opponent and the applicant (the 'parties'). After considering these observations, and if agreement has not been reached between the parties, the Opposition Division of OHIM will decide either to reject the contested application totally or in part. If the opposition is not well founded it will be rejected. If the CTMA is not totally rejected, and provided there are no other oppositions pending, it will proceed to registration. The decision of the Opposition Division is subject to appeal by any of the parties. The appeal is decided by OHIM's Boards of Appeal.

A further appeal can be made to the General Court and ultimately to the Court of Justice of the European Union.

Registration of the Community Trade Mark

An application will be registered when the following conditions have been met:

- the examination of the trade mark has raised no objections or the objections raised have been waived; and
- either no opposition has been filed, or any oppositions filed have been rejected.

Publication of the Community Trade Mark

The registered trade mark will be published in Part B of the Community Trade Marks Bulletin, and OHIM will send the applicant a link in order to download a certificate in PDF format.

The rights conferred by a Community trade mark prevail against third parties from the date of publication of registration of the trade mark. Community trade marks are registered for a period of **10 years** from the date of filing of the application. It can be renewed indefinitely for further periods of ten years.

Appeal proceedings in case of non-registration

Notice of appeal may be filed in writing within two months after the date of notification of the detrimental decision. It shall have suspensive effect. If the appeal is declared admissible, it is transferred to the Board of Appeal which examines it and rules.

Decisions taken by the Board of Appeal may also be subject to appeal before the Court of Justice of the EU in cases of:

- lack of competence;
- infringement of an essential procedural requirement;
- infringement of the Treaty, of the Regulation or of any rule relating to their application;

The action may be brought by any party to the proceedings within two months of the date of notification of the decision of the Board of Appeal. The Office has a duty to take account of final decisions made by Court of Justice of the EU.

3   COMMUNITY DESIGN

The basic source of law on Community Designs is the Council Regulation (EC) n°6/2002 of 12 December 2001 on Community Designs (also referred to as "CD Regulation") and its implementing regulation Commission Regulation (EC) n°2245/2002 of 21 October 2002 implementing Council Regulation (EC) No 6/2002 on Community designs. Just like in the case of Community Trade Mark, in the case of Community Designs we will also introduce the reader to the substantive and procedural elements of the above legal institute.

3.1 *Community design—basic substantive elements*

Categories of Community design and the way they are established could be outlined as follows:

1. **A Registered Community Design** (RCD) is an exclusive right that covers the outward appearance of a product or part of it. An RCD initially has a life of five years from the filing date and can be renewed in blocks of five years up to a maximum of 25 years. Applicants may market a design for up to 12 months before filing for an RCD without destroying its novelty. This, however, means that in order for a design to be registered as a RCD it must be applied for within 12 months from disclosure.
2. **An Unregistered Community Design** (UCD) is defined by the CD Regulation in the same way as the RCD but protects a design for a period of three years from the date on which the design was first made available to the public within the territory of the European Union. An unregistered Community design confers on its holder a right to prevent copying. RCDs and UCDs have to meet the same conditions to be protected.

    In reference to the above information, please bear in mind that unlike in the case of Community trade marks, in which such trademarks are created only via registration, in the case of Community designs, the situation is different. The negative side of this "simplicity" is the fact, that the owner of an unregistered Community design, may—in some cases—have problems proving the existence and relevance of the design and of the legal protection.
3. **Since 1 January 2008, it has been possible to designate the European Community in an international application for an industrial design** filed with the International Bureau of the World Intellectual Property Organization (WIPO) in Geneva. WIPO registers the international application and sends it to OHIM. It will have the same effect as applying directly for a Registered Community Design.[9]

3.2 *Registered community design—procedural perspectives—the registration procedure*

The minimum requirements to register a design are as follows:
One needs to complete an application form with:

- name and address of the applicant
- indication of the first and second language (for the purposes of proceedings)
- at least one visual representation of the design
- indication of the type of product designed
- signature
- payment of fees.

For a multiple application, the applicant can file as many designs as he/she wishes, the only condition being that the products to which the design is applied, belong to the same class of product.

Claiming priority
Priority may be claimed on the basis of a previous (first) application(s) of a design or utility model (but not on the basis of a patent application) filed in or for a State which is party to the Paris Convention or a member of the World Trade Organisation (WTO). Priority can only be claimed where the application for a Community design is filed within six months from the date of filing of the first application(s).

---

[9]For more information on this topic, please see: http://oami.europa.eu

In order to avoid the loss of one's right to priority, it is highly recommended that the priority is claimed in the application. The claim of priority needs to be supported by a priority document which is recommended to be enclosed with the application.

Examination procedure

Applications are checked mainly for formalities (name, address, language, signature, priority date(s), fees, description, designer and indication of product/classification; issues of public policy and morality, and if the application qualifies as a design).

If the application does not meet the formal requirements, an objection will be raised by OHIM. This is usually called a "deficiency letter" and will contain a time limit for the applicant's reply. This may lead to the amendment of the application or to its refusal in the case of non-compliance with the requirement. If the deficiency letter is not replied to within two months, this could lead to the refusal of the application, the loss of the claim of the priority right etc.

There is no substantive examination (e.g. no search for novelty), except to verify that the application is for a design and that the design is not contrary to public policy or morality.

If examination reveals no problems, the design is then registered and published immediately. The registration date will be the date of filing of the application. Once the design has been registered, it will be published in the Community Designs Bulletin. OHIM will then issue the registration certificate.[10]

The publication of the registration is thus the last phase of the procedure

Before concluding this chapter, it is relevant to add that the existing system of protection provided by EU law and institutions to the products of the mind may lead to various problems. This is proved by a set of cases out of which we have decided to present the followings: *Volvo v. Eric Veng* (C-238/87), European Court of Justice (ECJ)—Volvo sued Eric Veng in the UK for infringement of their registered design No 968895 of the front wing of the Volvo Series 200 cars, in the UK High Court. Volvo refused to license any other parties to make wings. The High Court referred to the ECJ the question whether a refusal by a car manufacturer to grant licences for spare parts (even in return for reasonable royalties) could be an abuse of a dominant position. The ECJ held that the subject matter of a registered design was the right to exclude third parties, so merely refusing to license was not an abuse. However, conduct might be an abuse if it involved: the arbitrary refusal to supply spare parts to independent repairers, the fixing of prices for spare parts at an unfair level or a decision no longer to produce spare parts for a particular model even though many cars of that model are still in circulation, provided that such conduct is liable to affect trade between Member States.[11]

REFERENCES

Judgment in case C-238/87, Volvo v. Eric Veng, European Court of Justice (ECJ), http://design-law. wikispaces.com/Volvo+v+Veng.

http://oami.europa.eu

Potasch, P. 2010. In: Kummerová, S. et al.: *Legal English—Fundamental Terms and Topics*. Bratislava: Paneurópska vysoká škola/Žilina: Eurokódex s.r.o., 2010, ISBN 9788089447343. Judgment in case C-235/09, DHL Express France SAS v Chronopost SA. Court of Justice of the European Union. Press release No. 35/11.

---

[10]For more information on this topic, please see: http://oami.europa.eu
[11]Volvo v. Eric Veng (C-238/87), European Court of Justice (ECJ), http://design-law.wikispaces.com/ Volvo+v+Veng

Risks of the global economy

J. Sipko
Faculty of Economics and Business, Pan-European University, Bratislava, Slovakia

ABSTRACT: The main goal of the paper is to analyze the main risks facing the world economy on the way towards global recovery. Based on comprehensive data analysis, the author points out the main risks that the world economy faces such as growing inequality, indebtedness, global economic imbalances, instability of the international monetary system, unfinished reform of the international financial system, huge capital flows, etc. The conclusion of the article is that in order to bring the world economy to solid, stable and sustainable economic growth, adopting critical measures by all players of the global economy would be needed. In addition, to stabilize the world economy, international cooperation is essential.

1 INTRODUCTION

The mortgage crisis in the USA caused the global financial crisis and a global recession. Since the global recession was severe, it caused the debt crisis. Despite the mortgage crisis outbreak in 2007, the outlook for the global economy is connected with several risks that undermined the recovery of the global economy, but in particular, the economy of the eurozone countries.

The main goal of this paper is to define the critical risks for the world economy as well as for the eurozone's economy. Although some crucial steps have been adopted and implemented in fostering the global economy, the following risks still exist. Even though the outlook for the global economy is more promising in comparison with previous years, in order to put the global economy on a sustainable path, diminishing the essential global risks is necessary. At this stage of development of the global economy, the following risks exist: uneven global economic growth, the growing role of financial derivatives, persistence of global economic imbalances, unsustainable fiscal outlook for some industrial as well as for some emerging and developing countries, international capital flows, restructuring of the financial sector, including cleaning of bad loans.

Although the global financial crisis started in the USA, the epicentrum of the debt crisis is the eurozone. Since there is limited space for analysis of all risks that the world economy is facing, the paper will analyze only some of them. Although the latest global economic outlook is promising, there is a certain level of uncertainties.

1.1 Global economic outlook

In comparison with previous years, the most recent global economic outlook presented is more promising. As Table 1 (industrial countries, including Slovakia) clearly demonstrates, the global economic outlook for 2014 is improving in comparison with 2013. However, 3.6% of the global GDP is slightly over the limit of 3.0% (GDP of below 3% is considered a limit for the global recession). The economic growth for eurozone is expected to be higher in comparison with that of USA and Japan. In Slovakia, the economic growth is the highest among the developed countries.

The positive development of the global economic outlook was supported by the higher level of economic growth in some emerging market economies, but in particular, in China, India and in some developing countries. Table 2 clearly shows that in comparison with 2010,

Table 1.   The global economic outlook, industrial countries.

	2008	2009	2010	2012	2013	2014
World GDP	2.7	−0.4	5.2	3.2	3.0	3.6
Industrial countries	0.1	−2.8	3.0	1.4	1.3	2.2
USA	−0.3	−3.1	2.0	2.8	1.9	2.8
Eurozone	0.4	−4.4	2.0	−0.7	−0.5	1.2
Germany	0.8	−5.1	3.9	0.9	0.5	1.7
EU	0.5	−4.2	2.0	−0.2	0.0	2.0
Japan	−1.0	−5.5	4.7	1.4	1.5	1.4
Slovakia	5.8	−4.9	4.4	1.8	0.9	2.4

Source: World Economic Outlook, April 2014.

Table 2.   The global economic outlook for industrial countries.

Regions/countries	2008	2009	2010	2012	2013	2014
EMs + DCs	5.9	3.1	7.6	5.1	5.3	6.2
Sub-Sah. Africa	5.6	2.7	7.5	5.0	4.7	4.9
CEEurope	3.1	−3.6	4.6	1.6	2.2	3.8
Brazil	5.2	−0.3	7.5	1.0	2.3	1.8
RF	5.2	−7.8	4.5	3.4	1.3	1.3
India	3.9	8.5	10.3	4.7	4.4	5.4
China	9.6	9.2	10.4	7.7	7.7	7.5
South Africa	3.6	−1.5	3.1	2.5	1.9	2.3
Singapore	1.9	−0.6	15.1	1.9	4.1	3.6

Source: World Economic Outlook, April 2014.

there has been a significant reduction in the economic growth of some emerging economies, namely in the BRICS countries (Brazil, Russian Federation, India, China and South Africa). If this trend continues, there will be real risks for the global economy. There is a positive development in some emerging economies and in some developing countries. Singapore is a country that has seen the highest reduction of economic growth between 2010 and 2012.

The latest development confirms that there is a sign of positive growth in global economic GDP in 2014 in comparison with 2013. In some developed countries, the economic outlook has been improving; however, in some emerging market economies, especially in the BRICS countries, there is a downward trend. If this trend continues, there will be a relatively high risk for the global economy as a whole. Here, there are some negative developments related to the relatively low level of domestic demand (mostly in BRICS countries). Regarding the economic outlook for the Russian Federation, due to the geopolitical risk and the close ties with Ukraine's economy, the Russian economic growth might be in negative territory in 2014. The global GDP is always influenced by the level of global imbalances.

1.2   Global economic imbalances

One of the most important issue related to the global economy are the global economic imbalances that significantly contributed to the global financial crisis. To assess the global economic imbalances, the position on the current account deficit is crucial. The global imbalances are considered as the most disputable and well-known of the global current economic problem, which possibly explains the causes of the global financial crisis.

The global financial imbalances were quite massive even before the outbreak of the global financial crisis (2008). Then, there has been some pressure through the global policies to tackle them on a different level, but one notion that we can seem to agree on is that the

US current account deficit is one of the significant factors to fuel the recent financial crisis. Considering global imbalances in retrospect and prospect, about 5% of GDP was the rate for imbalances of G7 economies, which sounds like a small portion, but it was not indeed since the economic powerhouse which are the seven industrialized nations represent more than half of net global wealth—close to 66% of net global wealth ($223 trillion), according to Credit Suisse Global Wealth Report September 2012. Thus, global imbalances definitely cause many policymakers to have a hard time due to their effort of correcting imbalances and restructuring the international monetary system.

To better understand the present development of the external accounts in Table 3, various countries such as those with a high level of export (China—CHN, Germany—DE, Japan—JPN, Singapore—SNG, and Russian Federation—RF) have been selected. In addition, in the table, included is the eurozone as the whole group of countries that adopted and implemented the single currencies. There are some specific countries such as Iceland—ISL, Ireland, Ukraine—UKR and USA. The latter, i.e. the USA, is a cornerstone problem of the global imbalances. There are some specific cases that have been suffering.

From the current account deficit there is in particular, Iceland. This country was one of the countries with a relatively very high level of current account deficit before the outbreak of the global financial crisis. The good news is that due to the stabilization and adjustment program provided by the International Monetary Fund and based on the principle of conditionality, Iceland's authorities have been able to significantly reduce its current account deficit since 2008. A similar development, even better, has been recorded in Ireland. It is a textbook example of implementing a stabilization and adjustment program for its economy. The current account deficit in 2008 was at a level of 5.7% and since then, it has been reduced and even this country has reached the current account surplus count of 2.7%.

The most difficult case is Ukraine. As Table 3 shows, the country has come to an unprecedented and complicated situation concerning the external account. Since 2009, the current account deficit has increased and in 2013, it reached 9.2%. The present civil war in the southeastern part of Ukraine, including mismanagement of the economy, could lead to the deterioration of the current account in 2014. This might be considered a risk for the region itself and for the Slovak economy.

Historically, the main problem of the global imbalances is related to both China and the USA. In the USA, there has been a relatively very high current account deficit in comparison with the Chinese high level of current account surplus. The latest development between both has confirmed that imbalances between China and the USA are narrowing and for 2014, there is an expectation that the current account deficit will be at a level of 2.2% in comparison with the Chinese current account surplus of 2.2%. So, there is a very good trend in reducing the global economic imbalances.

Although the main imbalances have been reduced, still some exist. To correct the global imbalances, further concentrated efforts should be adopted to reduce the still high level of global imbalances. The global financial crisis was closely related to a growing level of financial derivatives.

Table 3. Current account deficit in some selected economies.

Year	ISL	UKR	USA	IRL	EUR	JPN	CHN	RF	DE	SNG
2008	−28.4	−7.1	−4.7	−5.7	−0.7	3.3	9.3	6.3	6.2	13.9
2009	−11.6	−1.1	−2.6	−2.3	0.3	2.9	4.9	4.1	5.9	17.2
2010	−8.5	−2.2	−3.0	1.1	0.6	3.7	4.0	4.4	6.4	25.3
2011	−5.6	−6.3	−2.9	1.1	0.8	2.0	1.9	5.1	6.8	23.2
2012	−5.0	−8.1	−2.7	1.8	2.0	1.0	2.3	3.6	7.4	17.4
2013	−0.4	−9.2	−2.3	2.7	2.9	0.7	2.1	1.6	7.5	18.4
2014	−0.8	–	−2.2	3.8	2.9	1.2	2.2	2.9	7.3	17.7

Source: World Economic Outlook, April 2014.

1.3 Financial derivatives and the global economic growth

Although the latest development of derivatives is significant, so far there haven't been many articles comparing the growth of financial derivatives and the growth of the real economy. Therefore, it is important to compare the growth of derivatives with the growth of the real economy to be able to explain the increased systemic risk of the financial system. Based on official data published by the International Bank for Settlements, total derivatives have been significantly growing since the abolishment of the Glass-Steagall Act.

In order to better understand the real economic growth for the last 50 years, it is necessary to compare nominal and real world GDP. According to Figure 1, there is a historical development of both world nominal and world real GDP. The development of both nominal and real GDP is correlated since the beginning of the 60 s until the beginning of the last decade. At the end of 2000, world nominal GDP accounted for USD 31.8 trillion. For the same period of time, world real GDP accounted for USD 32.5 trillion. However, since the beginning of 2001, there has been a significant growth in world nominal GDP. One explanation of this might be that the world nominal GDP is growing faster due to the huge increase of the derivatives markets. Nonetheless, the increasing size of derivatives markets is not a part of the real economy.

The development of the global economic growth both nominal and real is described since 1960 in Figure 1. This figure clearly shows that economic growth was on an upward trend, but increased mildly during the last 50 years. The development of both nominal and real economic growth was in strong correlation. Nominal economic growth before the outbreak of the global financial crisis (2007) reached USD 62.173 trillion.

A completely different situation arises when comparing derivatives with interest rates. After the breakdown of the Bretton Woods system, there has been a gradual increase in currency and interest rates. One of the most critically important financial derivatives which have significantly contributed to the global financial crisis were credit default swaps. This type of financial derivative was introduced at almost the same period of time as the abolishment of the Glass-Steagall Act.

The volume of credit default swaps was USD 62.173 trillion at the end of 2007. In reality, credit default swaps at the end of 2007 have the same volume as the global gross

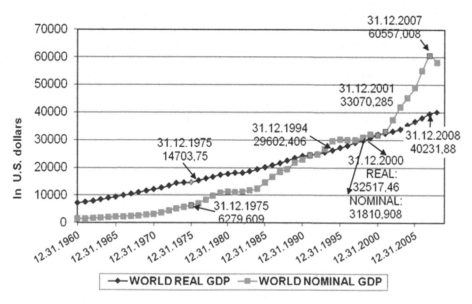

Figure 1.   Comparison of nominal and real world GDP.
Source: BIS data, author's calculations.

national product. That means that this kind of derivatives, which are non-transparent, reached the same volume as the annual global gross domestic product. However, since the end of 2007, there was a decline of credit default swaps and at the end of 2013, they reached a volume of USD 21.020 trillion. As a result of the global financial crisis, the global recession and global debt crisis, but in particular, debt crisis in the eurozone were induced.

2 EUROZONE FINANCIAL CRISIS

2.1 *Eurozone debt crisis*

The global financial crisis led to the global recession which caused the severe debt crisis in eurozone countries. Some eurozone countries, mostly now well-known as a PIIGS (Portugal, Ireland Italy, Greece, and Spain), came to the point where the fiscal stance is not sustainable. Fiscal unsustainability was a result of poor structural policies in these countries, including loss of competitiveness of PIIGS countries.

For fiscal sustainability, reducing the fiscal deficit is crucial. The development of the fiscal balance of some eurozone countries such as Estonia, Finland, Germany, Greece, Ireland, Italy, Portugal, Slovak Republic and Spain is shown in Figure 2.

A part of this comparable analysis is Switzerland, a non-member country of eurozone. Figure 2 shows the historical development of the overall fiscal balance since 2006, e.g., before the outbreak of the global financial crisis, during the crisis and the global recession until 2017. Before the outbreak of the global financial crisis in 2006, there were only a few countries in the eurozone that reached a fiscal surplus, namely Estonia (3.2% of GDP), Finland (4.0% of GDP) and Spain (2.0% of GDP).

Implementing the generally adopted fiscal rules within the eurozone countries during the good years is essential. However, some countries such as Greece and Portugal have permanently broken the rules mentioned in the SGP and have reached a fiscal deficit of about 6.0%

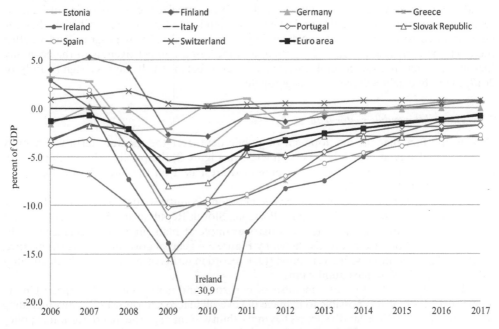

Figure 2.   Development of general government overall balance (% of GDP).
Source: Graph set out from Eurostat data (2006–2017).

of GDP and 3.8% of GDP, respectively in 2006 when the overall fiscal deficit for eurozone counties as a whole was only 1.3% of GDP.

The critical moment for the development of public finance, but mainly for overall general government balance was the year 2009. All the countries mentioned in the graph have significantly increased their fiscal deficit, namely the Greek and Irish deficits reached 15.6% and 13.9% of GDP, respectively.

In terms of fiscal sustainability, Ireland is a very specific case. In 2006, Ireland reached a fiscal surplus of around 2.9% of GDP. However, by 2010 the fiscal deficit increased to 30.9% of GDP. This huge increase of deficit was connected with additional expenses required to clear the bad loans in the banking sector. In 2011, the fiscal deficit was at a level of 11.2% of GDP and based on "Troika—EC, ECB and IMF", the adjustment program will gradually reduce it to a sustainable level by 2017 and might reach a fiscal deficit at a level of around 1.3% of GDP. In comparison with some other countries, which are also under the strong adjustment program with "Troika", Ireland's authorities are committed to the program itself and take accountability in stabilizing the national economy.

A specific case in the analysis is Switzerland. The country was hit by real external shocks; however, Switzerland is an extraordinary textbook example, because it was able to manage both economic and fiscal policies, mainly the general governance overall balance, for the whole period in an appropriate way.

This is a very particular case in terms of managing economic and fiscal policy, including the overall macroeconomic policy mix. Switzerland does not have any commitments to the Stability and Growth Pact unlike all the eurozone countries and is maintaining public finance and the overall national economy in a very good shape even during the crisis period. How is this possible? The answer might be the decent leadership of the country in combination with orientation on export performance with high productivity of growth, including accepting the rules of game, a relatively high level of transparency and a low level of corruption.

The good news is that based on this optimistic scenario, all eurozone countries will be able to reach a fiscal deficit of below 3% of GDP by 2017. However, this scenario has some unknown variables, which are influenced by uncertainties concerning the recovery of the world economy, how individual countries will be able to finalize the restructuring of banking sector, how countries will be able to deal with the high level of public debt, whether they will be committed to adopting and implementing the structural reforms program and whether there will be increased productivity growth and increased competitiveness in the economy. Despite the fact that the general government overall balance will improve significantly by 2017, the public debt will reduce only gradually.

2.2 *Outlook for public sector debt*

Despite the fact that the development of public debt over the medium-term will slightly decline, it will stay at relatively very high level. In Figure 3, the public debt development is presented from 2006 until 2017. Data presented are based on the latest World Economic Outlook (WEO, April 2014).

To better understand the overall trends, eurozone countries such as Estonia, Finland, France, Germany, Greece, Ireland, Italy, Portugal, Slovak Republic and Spain are chosen for analysis. In addition, there is an extraordinary example of managing public finance and the economy as a whole. Therefore, as shown by Figure 3—general government overall balance, Switzerland is a country with a very good track record in terms of maintaining both an internal and external economic equilibrium.

Although Switzerland is not a member of eurozone and neither of the European Union, it demonstrates how it is possible to manage the overall macroeconomic policy mix including economic policy. Switzerland's success lies in its ability to manage the macroeconomic policy mix and its implications for the real economy in supporting competitiveness. This country has for many years demonstrated a strong export performance which is based on high productivity growth and competitiveness. All these factors have significantly contributed to a

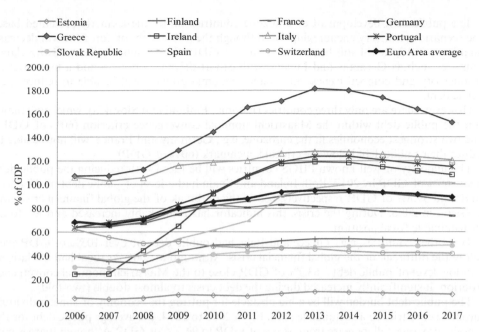

Figure 3.   Outlook for public debt in eurozone countries.
Source: Graph set out from Eurostat data (2006–2017).

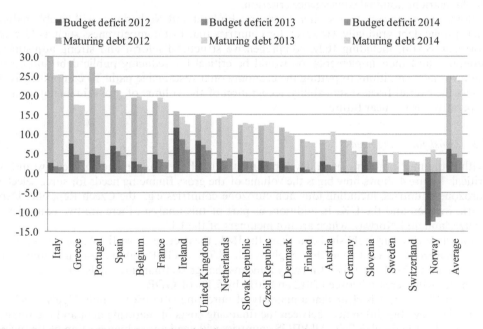

Figure 4.   Budget deficit and maturing debt in eurozone countries.
Source: Graph set out from Eurostat data (2012–2014).

relatively very high level of current account surplus. Figure 4 clearly demonstrates that historically, Switzerland's authorities were committed to maintain public finance at a manageable level of public debt even during the outbreak of the global financial crisis and the global recession. This is an unprecedentedly successful case and might be a lesson for some highly indebted countries in the eurozone.

The public debt development of eurozone countries under the formerly presented baseline scenario is not very encouraging. Even though the public debt of eurozone will decrease slightly by 2017, it will still be at almost 90% of GDP. Here again, similar to Switzerland, countries such as Germany and Finland with export driven economies and a high level of productivity and competitiveness and stable economic growth, will be able to reduce their public debt.

However, there are only three countries (Estonia, Finland and Slovakia), which will maintain the public debt within the Maastricht nominal convergence criterion (60% of GDP). The two strongest economies in the eurozone, e.g., Germany and France, will not be able to reduce their public debt below the critical benchmark (60% of GDP).

The majority of well-known PIIGS countries will be faced with reducing the public debt below 60% of GDP by 2017. Both Ireland and Spain had a level of public debt of 24.7% of GDP and 39.8% of GDP, respectively, before the outbreak of the global financial crisis and recession in 2006. During the crisis, this indicator has remarkably increased and led to an unsustainable fiscal position.

Two other countries, e.g., Greece and Italy, had public debt of over 105% of GDP even before the outbreak of the global financial crisis. Although in 2006 Portugal had a relatively very low level of public debt—63.7% of GDP, close to the Maastricht nominal convergence criterion, it significantly increased during the debt crisis to almost double (two-fold).

The public debt burden will be a big obstacle for putting the economy of some individual eurozone countries to a sustainable path. From 2006 to 2014, the average public debt for the overall eurozone will increase from 68.6% of GDP to 94.7% of GDP. Although there is currently a downward trend in reducing public debt, based on latest published data (EC, ECB, IMF), the average public debt for the overall eurozone in 2017 could be higher by 30% over the Maastricht nominal convergence criterion.

Some eurozone countries such as Germany, Finland and Netherlands will be able reduce their public debt gradually. However, for countries that lost competiveness such as Greece, Portugal, Spain, including Italy, comprehensive structural reform and strong adjustment programs and their implementation would be critical for reducing public debt and creating favorable conditions to putting the economy on a sustainable path. To comprehensively assess the overall fiscal sustainability, recognition of the volume of gross financing needs is imperative for the near future.

2.3 *Gross financing needs*

Within the medium-term fiscal development, the level of gross financing needs would be critical. Figure 4 shows how big is the volume of the gross financing needs for some selected eurozone countries, including four non-eurozone countries e.g., the Czech Republic, Denmark, Sweden and the UK. In addition, as part of this analysis, there are two countries—Switzerland and Norway, which are not members of the EU.

Based on officially published data from the IMF, the average gross financing needs for the whole eurozone as a whole in 2012 might reach 18.7% of GDP and in 2014 this volume will increase up to 23.8% of GDP. In reality it means that gross financing needs for the eurozone countries will increase between 2012 and 2014 by 5.1% of GDP.

From Figure 4, it is clear that a majority of eurozone countries, mainly highly indebted countries, have big differences between the financing costs of maturing debt and the financing costs for the fiscal deficit. All PIIGS countries will need a very high portion of financing needs in 2014.

Figure 4 also shows that a majority of eurozone countries have higher financing needs than some non-eurozone countries, such as Denmark and Sweden. It might be concluded that those countries such as Norway and Switzerland, where the authorities are committed to structural reforms programs, to maintaining fiscal discipline and to supporting export driven economy based on a high level of productivity growth, have a very low level of gross financing needs. The latter related to the potential not only to the eurozone economy, but also to the global economy.

3 CONCLUSION

The outbreak of the mortgage crisis in the USA not only led to the global recession, but exceedingly contributed to the global debt crisis. Based on the data presented, one should conclude that despite some positive sign of the latest development, there are still some risks facing the global economy.

Although global economic growth has been improving in 2014 in comparison with 2013, there are some risks related to the reduction of economic growth in some emerging market economies, but in particularly in the BRICS countries. In this regard, the reductions have been pronounced with the geopolitical risks related to the Russian Federation and Ukraine. This unfavorable development could significantly contribute to the growing risks in the world economy.

The trend in reducing the global imbalances is encouraging. Despite the positive reduction of the current account deficit in the USA, a decrease in the Chinese current account surplus is a step in the right direction. However, still there is a persistence of external imbalances. To this regard, some export driven economies such as Germany, Japan and China have a relatively high level of current account surpluses. However, mostly eurozone debtor's countries still have a relatively huge current account deficit that might complicate the situation with the financing of public debt.

To reach fiscal sustainability in eurozone countries is critical. Even though there is a positive expectation in reducing fiscal deficit and improving primary fiscal balance, still there are some risks in reducing long-term public debt. Reducing public debt to an appropriate level in order to fulfill the Maastricht convergence criterion still has a long way to go. Reducing public debt in some developed countries, including the USA, is imperative for the stabilization of the world economy.

The growing role of financial derivatives is of paramount importance. The fact is that financial derivatives (over the counter market) are growing relatively very fast. In reality, the derivatives market is growing much faster than the real economy. If the derivatives market will grow, but mostly the types of financial derivatives that are less transparent, this could increase the risks for the global economy, mainly for the real sector.

There are very many other risks that the world economy is facing. For the purpose of this article, I concentrate only on some of them. All risks of the world economy are closely connected. Therefore, to reduce the risks of the world economy, collective efforts are essential. Since the potential global risks are growing, the global community has to adopt and implement all necessary measures that are imperative for stabilizing the global economy. Postponement in adopting all necessary measures and steps in reducing the risks for the global economy could lead to unprecedented damages for the human beings on the planet.

ACKNOWLEDGEMENTS

This paper is a result of VEGA research "Fiscal and Monetary Policies and their Impact on International Business Environments".

REFERENCES

Bank for International Settlements, (International banking and financial market developments), Quarterly review (1999/April 2014).
Chencherita, Ch. and Rother, P. 2010. The Impact of High and Growing Debt on Economic Growth and Empirical Investigation for the Euro Area. European Central Bank Working Paper Series No. 1237, August.
De Grauwe, P. 2005. Economics of Monetary Union. 6th ed. New York: Oxford University Press.
Dunn, R., Jr. 2008. EMU Convergence: Reason for Encouragement. Challenge, no. 51, no. 2: 85–96.
Eurostat database (2000–2013).
Kenen, P. 1969. The Theory of Optimum Currency Areas: An Eclectic View. Monetary Problems of the International Economy, ed. R. Mundell and A. Swoboda. Chicago: University of Chicago Press.

McKinnon, R. 1963. Optimum Currency Areas. American Economic Review, no. 53, nos. 3–5: 717–25.

Mundell, R. 1961. A Theory of Optimum Currency Areas. American Economic Review, no. 51: 657–65.

Sbrancia, M.B. 2011. Debt, Inflation and the Liquidation Effect. Mimeograph, University of Maryland, College Park.

Sipko, J. 2012. The European Debt Crisis. *Research Journal of Economics, Business and ICT*. 7, No. 1, pp. 21–26.

Sipko, J. 2013. Excessive Imbalances and Debt Crisis in the Eurozone. In: *ICIBET*. [Conference Proceedings.] Paris: Atlantis Press, pp. 268/271. ISBN 97890778677567. (Advances in Intelligent System Research, 26).

World Economic Forum, Committee to Improving State of the World, Global Risks. Ninth Edition, Geneva, ISBN-13: 92-95044-60-6, ISBN-10: 978-92-95044-60-9.

World Economic Outlook, April 2014, International Monetary Fund.

History of journalism theories as a research project

S.G. Korkonosenko
St. Petersburg State University, St. Petersburg, Russia

ABSTRACT: This article is focused on systematization of national journalism theories. Modern research literature entertains notions of the demise of journalism; accordingly, there is no need for journalism theory. However, one can arrive at other decisions when turning to historical roots of the press-related knowledge. Conceptual representations of journalism have emerged naturally, as a reflection of its development and growing influence. Archeology of the history of journalism reveals the origins of its theoretical differentiation into several disciplinary directions that still exist and evolve. Theories of journalism have national-cultural specificity that is not effaced in the modern world. Finally, research into their origins creates basis for the structuring of theoretical knowledge concerning journalism and media on national and world scale. Focused study in the given direction is especially topical for Central and East European countries. This project was developed at St. Petersburg State University, Russia.

1 INTRODUCTION

The article theme may seem strange and provoke questions from those readers who are not well familiar with theoretical bases of journalism. Wide academic community has long expressed skepticism towards the combination of journalism and theory. This chronic skepticism finds support in the current condition of media development with its exponential growth of net communications and emergence of legions of amateur informants and commentators who call themselves civil journalists. Accordingly, poorly grounded ideas concerning the "death of journalism" gain greater popularity.

In this context an attempt to evaluate the dynamics of journalism theory risks meeting misunderstanding and even ostracism. Not only those who are unfamiliar with the press, but also many editorial employees are sure that this field of activity belongs exclusively to the sphere of practice and cannot be analyzed in theoretical terms. A Russian media manager states categorically:

"There is just no science under the name of *journalism*, however, there is something *from a science* in journalism. I declare this science non-existent with full responsibly—as professor of journalism at three universities. Strictly speaking, I would define this discipline as follows: *theory of modern journalism is the division of modern practical political science which has some scientific basis in psychology of masses, sociology, and in politics (as a science)*" (Tret'yakov 2004, p. 92).

It is hardly necessary to engage in serious polemics with the professor who calls theory a part of practice (practical political science being essentially this) and, moreover, labels politics a science (not metaphorically, but "strictly speaking") missing other aspects of the analysis, besides sociopolitical—namely, philological, ethical, and moral. We have to add that participants of the panel devoted to the role of journalism theory in the system of knowledge quote this statement to challenge it; incidentally, they are not journalists but representatives of economic and legal areas of knowledge (Bychkova et al. 2013).

Russia is not an exception to the universal rule. German experts, who were tracking history and evolution of concepts and theories in journalism studies, have discovered:

"When researchers in the U.S. began to conduct studies with special attention to journalistic production and the journalists' labour context, their work was rather sceptically received by practitioners who labelled these efforts 'Mickey Mouse studies' ... Even though the 'high noon' of normative and individualistic ideas in journalism studies is over, they still can be found in both journalistic practice and theoretical approaches to the field" (Löffelholz & Rothenberger 2011, p. 11).

However, authors of the above mentioned research are not inclined to agree with their opponents. This is how they formulate their position:

"Despite its apparent multidisciplinary roots, 21st century journalism studies have reached a comparatively high level of disciplinary institutionalization across the globe, as evidenced by the large number of specific schools, professorships and professional associations" (Ibid, p. 8).

The present article is also based on the historical-evolutional approach to the theory (theories) of journalism. When looking through the evolutionary prism it becomes clear that this area of theoretical reflection has evolved naturally, so no one can make subjective decisions on its "cancelation" or renaming. The primary *aim of the article* is to produce an overview of the existent approaches to building a research project devoted to the origins and early development of the journalism theory. The project has been initiated at St. Petersburg State University and has realistic prospects for prolongation. The analysis is carried out mainly on the basis of the history of Russian science, with the necessary references to the international experience.

2 ESSENCE OF THE PROJECT

It is worth noting that those supporting the idea of theoretical approach to the phenomenon of journalism can be found not only among media researchers, but also among representatives of other branches of science. One of them, Doctor of Economics and University provost, earlier a practicing scientific journalist, gives a highly competent discussion of the subject:

"Journalistic science is a multi-aspect, complex and independent branch of knowledge the object of which lacks accurate definition. In its importance it can be compared to such fields of humanities as history, sociology, psychology, and economy. The research of its method carried out at universities is obviously not enough to create a comprehensive theory. Deeper fundamental research is necessary. In my opinion, we need a research center under the aegis of the Russian Academy of Sciences ... Besides, it is necessary to establish a special division within the Russian Academy of Science focused on scientific research in the sphere of journalism, akin to similar branches in economy and history" (Astashonok 2012, p. 20–21).

There may arise question weather it is expedient to study journalism theory in *historical-evolutionary* aspect. Does retrospective analysis provide substantial benefits considering that press is subject to the powerful influence of the time, that its forms are not simply historically determined but depend on a current social order, political regime, trends of mind and fashion, everyday practices, information technologies, etc.? Perhaps theoretical view of journalism possesses no continuity as its subject progresses discretely, in a dotted line.

However, such doubts do not discourage the above cited German authors from their consistent research work. Let us turn to adjacent fields of science and humanities that pointedly and thoroughly develop historical-evolutionary approach in their spheres of knowledge. Apart from a set of axiomatic references to the history of philosophy and the history of jurisprudence a great number of other disciplines could me mentioned. It is indicative that this approach is being introduced into educational practice and is realized in textbooks on sociology, cultural studies, history, economy, etc. One could hardly argue that these disciplines are to a lesser degree dependent on the changing time and place than journalism.

Our point is that journalism theory is subject to the general law of scientific knowledge expressed in Imre Lakatos's well-known saying (though made on another occasion): "Philosophy of science without history of science is empty; history of science without philosophy of science is blind" (Lakatos 1971, p. 91). This formula touches on the general laws of epistemology: studying theories on "horizontal" time plane makes no sense if we do not attempt to see their genesis and past development, their evolution, changing and alternation, their modifications under the influence of diverse factors. That is to say that factors of development can constitute a primary research object. However, this would not be a historical study as such, in the direct sense of the word, as attention would be focused not on the events in the "biography" of theories, but on the evolution of meanings and values, on the reasons and circumstances of priorities redistribution, and also (this is, possibly, the most important) on revealing how the past influences the modern state of theory and its future prospects. One cannot deduce the present condition of a theory neither from the time interval observable for the researcher de visu, nor from the theory per se, ignoring life contexts into which it is inevitably and necessarily immersed.

Another vital issue is the study of theoretical representations rooted in a *particular country*, Russia in our case. Global domination of Anglo-American concepts and the leveling of national-cultural distinctions do not meet universal approval. The authoritative researchers emphasize that the map of media and communications theories is in fact quite diverse. According to John Downing, besides Britain and the United States "the other nations on whose experience and culture media communication theory has mostly been based have been Germany, France and Italy, although the rapid growth of media studies in the Canadian and Australian academies has recently added those nations to the list" (Downing 1996, p. x).

D. McQuail adds to the list some other original national schools: a) France and the francophone area; b) United Kingdom; c) Germany; d) Scandinavia; e) the Mediterranean region with Italy leading and Spain following (McQuail 2009, p. 288). Of course, among these countries we do not see Russia as well as, say, Asian or Central and Eastern European states. It means that only scholars from these "forgotten" regions are capable of making the map of media theories more pluralistic and realistic.

The framework of the present project does not cover as its special subject the current state of the theory (theories) of journalism. Chronologically our research is limited by the period of the 18th and 19th centuries when press in Russia had reached a high level of development as a social institute and a field of professional activity while in the public sphere there appeared distinct ideas of the press that can be related to the category of theoretical thinking. However, our research team is certainly aware of the correlation in the sphere of knowledge between the past, the present and the future. The "evolutionary" phase is part of a long-term project "Theories of Journalism in Russia" that is being developed and carried out by the Theory of Journalism and Mass Communications Department of St. Petersburg University. Consequent project stages will be directly connected with the present with the current stage serving as their theoretical basis. In many respects we share the concerns of those local experts who consider that:

"Today there is no classification of main theoretic divisions of mass communication research in Russia. On the one hand, we can see a terminological mess in the objects of theorizing … On the other hand, researchers sometimes do not know what to call theories, concepts, approaches, traditions of analysis, schools and research works, principles, scientific divisions, paradigms, methodologies, methods, etc." (Dunas 2013, p. 89).

It should be added that similar estimations can be related to the science of journalism and mass communications on a world scale.

Coming back to our main subject it is necessary to explain the choice of research directions and the project structure. Attempts aimed at the historical study of the Russian press theory have already been made (Stan'ko 1986). Collaborators of the project certainly rely on their predecessors' works of. However, it should first be noted that a large-scale theme cannot be *completed* after publication of several individual works, even the most thorough ones. A couple of decades ago review of existing publications gave basis for the following conclusion: "Modern research literature is increasingly interested in the historical becoming of the theoretical understanding of journalism" (Stan'ko 1987, p. 14). The present project can serve as an example of the tendency that had been outlined in the earlier years.

Second, as the academic community well knows, journalism theory synthesizes *a number of disciplinary approaches* and actively interacts with adjacent thematic fields. Subsequently, its understanding as a homogeneous matter lacking differentiation into thematic branches can be acceptable only at an earlier, preliminary stage of research. Any encounter with concrete problems of the press immediately demands diversification of the methodological toolkit. For example, the fundamental question of freedom of press is treated variously, from philosophical and sociological to legal and ethical viewpoints. Therefore it makes sense to apply spectral analysis of sorts to the set of theoretical views on the press. This concerns contemporary state of affairs; however, when turning to the past from the modern point of view researchers should still adhere to the uniform methodological approach and follow those lines of differentiation that draw borders within the spectrum of journalism theory today.

In the suggested project we plan to pursue the following theoretical directions within a holistic framework of theoretical reasoning on journalism:

- Historical-theoretical study of journalism;
- Normative theories of journalism;
- Social-philosophical theories of journalism;
- Political theories of journalism;
- Sociological theories of journalism;
- Psychological study of journalism;
- Cultural study of journalism;
- Artistic-aesthetic study of journalism.

Naturally this list does not cover the whole spectrum of disciplinary differentiation within understanding of journalism. However, it reflects interests and resources of the research team and in this respect meets the requirement for the instrumental interpretation of a basic concept. In the course of the project development this list will (and should) be extended and probably reconsidered.

It is obvious that this task is quite difficult. Contemporary and even more so past theoretical views on the press often present themselves as an inseparable whole; it requires a teleological analytical effort to distinguish in them elements, say, of political knowledge from the sociological. As a result different disciplinary sections of the project will repeatedly turn to the same names of the outstanding figures of Russian journalism, science and culture. It could hardly be called a genetic flaw of the plan and its realization. The research agenda does not cover chronological reconstruction of the history of journalism; our aim is to track the formation of certain *theoretical directions* on the axis of continuity and evolution. To give an example one might mention Mikhail Lomonosov, a real encyclopaedist, traditionally considered the founder of Russian science in the fields of chemistry and linguistics, physics and mineralogy, geography and history and so on.

Third, at the early stages of a science development we can see not theories as such (sociological, cultural, aesthetic, etc.), but only their *preconditions*, the germs, the initial manifestations. Therefore the titles of the project sections look "wider" than their actual subject matter; they rather mark vectors than measure the level of a science maturity.

Nevertheless, without such data it is impossible to clarify the problem of national schools dealing with the theory of journalism in its different subject spheres. The majority of classical scholarly disciplines—both in natural sciences and humanities—generally recognize the existence of theoretical schools designated by the names of authoritative national researchers. Strangely, journalism, at least in Russia, presents a different situation. If a "personalized" school or paradigm is referred to in a theoretical work (like a thesis), it almost without exception is attributed to a Western scholar while local researchers are mentioned as autonomous individuals who "contributed to the analysis of a question".

Such attitude to predecessors is, to say the least, irrational. In our opinion, the research of journalism in any theoretical aspect always relies on a 'base capital', whether we want it or not. This capital is made by that volume of information (in the broad sense of the word) that a given science has accumulated. In other words, researchers as a community and as separate individuals should not and cannot make successful progress if they ignore the "previous" knowledge. Otherwise they are doomed either to rotating in a closed intellectual circle, with only the illusion of advance, or to revealing only the situational, momentary characteristics of journalism without past or future projections.

3 CONCLUSIONS

We hope that the accumulating and structuring of the material concerning the origin and evolution of journalism theories in Russia will contribute to the solution of the described problems. Certainly, for a small research group it is difficult to cover all the disciplinary directions within the field. Our team is able to carry out a reasonably limited amount of work hoping to continue the launched process in the future, with the involvement of new participants. Moreover, we are sure that similar approaches to the construction of systems of scientific knowledge are vital for different countries, in particular Central and East European. They encounter nowadays similar problems of structuring of the national theoretical heritage in the field of journalism and mass communications. This, in turn, gives us ground for cooperation and hope for the joint projects.

ACKNOWLEDGEMENT

The article is written with financial support from St. Petersburg State University, project № 4.23.2204.2013.

REFERENCES

Astashonok, E. 2012. Aleksandr Suhodolov: Teoriya zhurnalistiki trebuet fundamental'nyh nauchnyh issledovanii (Alexander Sukhodolov: The theory of journalism demands fundamental scientific research). *Voprosy teorii i praktiki zhurnalistiki* (Questions of the theory and practice of journalism) 1: 14–21.

Bychkova, A.M., Rachkov, M.P. & Suhodolov, A.P. 2013. Pochemu Sergei Kapica ne stal akademikom RAN (Why Sergey Kapitsa did not become the academician of the Russian Academy of Science). *Voprosy teorii i praktiki zhurnalistiki* (Questions of the Theory and Practice of Journalism) 1: 215–218.

Downing, John D.H. 1996. *Internationalizing media theory: Transition, power, culture—Reflections on media in Russia, Poland and Hungary, 1980–95.* London: Sage Publications.

Dunas, D.V. 2013 Mapping mass communication theories in contemporary Russia. In Elena L. Vartanova (ed.), *World of Media. Yearbook of Russian Media and Journalism Studies*: 88–107. Moscow: Faculty of Journalism of Lomonosov Moscow State University; MediaMir.

Lakatos, Imre 1971. History of science and its rational reconstruction. In R. Cohen, R. Buck (eds), *Boston Studies in the Philosophy of Science* 8: 174–182.

Löffelholz, M. & Rothenberger, L. 2011. Eclectic continuum, distinct discipline or sub-domain of communication studies? Theoretical considerations and empirical findings on the disciplinarity, multidisciplinarity and transdisciplinarity of journalism studies. *Brazilian Journalism Research* 7(1): 7–29.

McQuail, D. 2009. Diversity and convergence in communication science: The idea of "national schools" in the European area. In N. Carpentier, P. Pruulmann-Vengerfeldt, R. Kilborn, T. Olsson, H. Nieminen, E. Sundin, K. Nordenstreng (eds). *Communicative approaches to politics and ethics in Europe: The intellectual work of the 2009 ECREA European media and communication doctoral summer school*: 281–292. Tartu: Tartu University Press.

Stan'ko, A.I. 1986. *Stanovlenie teoreticheskih znanii o periodicheskoi pechati v Rossii (XVIII v.—60-h gg. XIX v.)* [Becoming of theoretical knowledge on the periodical press in Russia (XVIII century—60 s of XIX century)]: Dr. Dissertation. Leningrad: Leningrad State University.

Stan'ko, A.I. 1987. Istoricheskaya rekonstrukciya teoreticheskih znanii o zhurnalistike: (itogi i problemy) (Historical reconstruction of theoretical knowledge of journalism: results and problems). In E.A. Kornilov (ed.). *Tipologiya zhurnalistiki: voprosy metodologii i istorii* (Typology of journalism: questions of method and history): 14–19. Rostov-on-Don: Rostov University.

Tret'yakov, V.T. 2004. *Kak stat' znamenitym zhurnalistom* (How to become a popular journalist): Lectures on the theory and practice of modern Russian journalism. Moscow: Ladomir.

Current Issues of Science and Research in the Global World – Kunova & Dolinsky (Eds)
© 2015 Taylor & Francis Group, London, ISBN 978-1-138-02739-8

Multi-cloud hosting IoT based big data service platform issues and one heuristic proposal how to possibly approach some of them

Peter Farkaš
Faculty of Informatics, Institute of Applied Informatics, Pan-European University, Bratislava, Slovakia
Faculty of Electrical Engineering and Information Technology, Institute of Telecommunications,
Slovak University of Technology in Bratislava, Bratislava, Slovakia

Eugen Ružický
Faculty of Informatics, Institute of Applied Informatics, Pan-European University, Bratislava, Slovakia

ABSTRACT: The trends in development in the Internet of Things and Big data areas has to be taken into account by proposing and shaping cloud computing systems of the future. Huge amounts of information streaming from and to distributed sensors and actuators will create difficult challenges for the cloud infrastructure which will store process and transport this information and will have to support useful services. Underlying cloud infrastructure has to be proposed allowing efficient management of voluminous blocks of globally distributed structured, semi-structured and non-structured data generated at very high rates. A global multi-cloud and multi-archives service platform interconnected by high throughput communications networks would be needed, capable to handle these new challenges. For the services these global cloud platform technologies should appear as one uniform platform. This paper will summarize current issues in this area of research and sketch a heuristic vision how to approach the research and later maybe also the design of such global multi-cloud infrastructure in future. It was inspired by water circulation and management on our planet. Sustainability and usefulness for human beings are the main motivations and constraints for this effort.

Keywords: cloud; big data; internet of things; mobile; e-heath; multi-cloud; global cloud platform technologies; efficient management; sustainable; human

1 INTRODUCTION

Multi-Cloud was identified by EC recently as a natural choice for hosting IoT based Big Data service platform [1] and at the same time as an area of research worth to invest into it at least 6 million Euro this and following two Years in Horizon 2020 Program. In this paper an overview of the state of the art, current issues in this area is first presented. Later a sketch of ideas is drawn, which could lead to a research approach and later proposal and design of a sustainable architecture for federated multi-cloud IoT and Big Data global infrastructure.

The paper is organized as follows. In Section 2 the state of the art is presented, and the challenges (issues) and open problems are reviewed, in Section 3 our proposal to use water circulation and water supply systems as an inspiration by thinking about a Multi-cloud Big Data platform is presented. In Section 4 some concluding remarks are made.

2 STATE OF THE ART IN COMBINING BIG DATA AND INTERNET OF THINGS IN A CLOUD

We will first deal with Issues of Big Data, later with IoT and than with their combination in different subchapters.

2.1 Issues and state of the art in big data technology in connection with clouds

Big Data refers to huge datasets [2] that are difficult to acquire, store, search visualize and analyze. The complexity of Big Data is increasing quickly and it becomes immense at the present time.

Behind the volume also other properties (challenges) could be used to characterize Big Data, namely:

Value: the challenge is how to mine or extract the value which is hidden in Big Data for different services and tasks;

Variety or absence of formats: structured, semi-structured and unstructured data. The challenges are how to integrate, merge, make fusion of various types of data [3]. It is estimated that unstructured data will grow one order times faster than structured [4]. It is much more difficult to handle such unstructured sets of data than structured.

Velocity: Big Data has to be usually processed, stored in real time and this is a big challenge in connection with rapid increase and unpredictability of the volumes generated.

From the point of view of usefulness for services the types of data in Clouds are partially also relevant. In [4] the following classification the data types in Big Cloud was given:

- Enterprise data
- Machine generated/sensor data
- Social data.

There are many approaches which nowadays try to solve the challenges, which sometimes seem to be complementary. For example because the capacity of the transmission networks is still an expensive resource in recent time the "function shipping rather than data shipping" become popular [5]. On the other hand, significant progress was made also in Relational Database Management Systems (RDBMS). Structured Query Language (SQL), which is usually used nowadays prevalently, has some limitations and therefore lot of new non SQL approaches usually based on the paradigm Map-Reduce were developed recently. Map—Reduce exploits massively distributed and relatively cheap hardware. Hadoop is its free implementation. In [6] it is argued that: "*It is in fact cheaper to duplicate (for reliability) and to over compute (process duplicate data) as communication is relatively more expensive than storage and computational resources (and this gap is increasing).... It should be noted that Map-Reduce has weaknesses in that it is not, by design, general-purpose, but rather was designed for something very specific: keyword processing and access.*"

2.2 Issues and state of the art in internet of things

The historical milestones on the way to Internet of Things (IoT) posted on [7] contain also statements which Nikola Tesla made in an interview with Collier magazine in 1926: "*When wireless is perfectly applied the whole earth will be converted into a huge brain, which in fact it is, all things being particles of a real and rhythmic whole ... and the instruments through which we shall be able to do this will be amazingly simple compared with our present telephone. A man will be able to carry one in his vest pocket.*"

His vision started to materialize later when mobile communications, Wireless Sensor Networks (WSNs), RFID networks and hybrid WSNs as well as actuators, MEMS circuits and the wireless communications techniques become mature enough few years ago. Thanks to this progress different small scale networks were developed for so called sensing intelligent environments, but in the first phase these networks formed isolated islands. Later introduction of micro IP stack and IPv6 over Low power Wireless Personal Access Networks (6LowPAN) [8] enabled to connect these devices and isolated networks into Internet of Things (IoT). The research community agrees that the IoT encompasses in some degree also the idea of integration of the physical world with the virtual world of Internet [9]. The distributed sensing devices for data collection and actuators for actuation tasks in different applications have to be addressed individually. In this way physical objects connected to Internet will become smart. They could report on their status or transmit batch of locally captured data to data centers, which will provide the smartness to them. IoT is also seen by many as an important class of Cyber-Physical System (CPS) and defines it as a network, which interconnect physical objects with addresses via Internet or other telecommunications networks. In [10] the author even argues that: *"From the definition, one could mathematically conclude that IoT is a subset of CPS. However, it is difficult to provide an example of a CPS that is not IoT. Consequently, the two concepts are practically equivalent."* And later: *"Machine to Machine (M2M) communications well describe most existing CPSs. M2M uses a device (such as a sensor or meter) to capture an event (such as temperature, inventory level, etc.), which is relayed through a network (wireless, wired or hybrid) to an application (software program) that translates the captured event into meaningful information, which can trigger an actuation."* Step by step the network becomes not only the tool for communication, but also the place where the computing and decisions take place. However there are many challenges and unsolved problems on the way to a sustainable global platform for IoT. Probably the main is the problem how to handle the explosive increase of amount of connected devices and consequently the Big Data generated by them. It is estimated that in last year more than 10 billion devices were connected and till 2020 the number should increase by a magnitude of five [11]. There is now a relatively lot of knowledge in different books on Wireless Sensor Networks and RFIDs, which will be prevalent components in IoT together with actuators [12, 13, 14]. The knowledge obtained during research and development of WSN is concentrated on different technical and technology issues, such as overall architecture, routing, node, localization, energy management, Medium Access Control (MAC), security, protocols in different layers, synchronization and many others [15], [16], [17]. For our architecture proposal the most related works from this are concentrated on developing different applications, databases and operating systems based on WSN technology [18], [19]. However they were not designed with the goal to share seamlessly data collected from distributed sources in a global scale. On the other hand the WSN issues will be utmost important for our future research work on this issues, because they will form a substantial part of sources and nodes in a future IoT.

2.3 Issues and state of the art in combining big data and internet of things technology

Probably [20] was the first platform for IoT, which allowed to access sensor data remotely in real time. Interoperability and integration of the data feeds from different sources in IoT was partially addressed in [21], [22] by proposing to use semantic annotation for them.

Observing the development of IoT and Big Data, it could be expected that in future the data streams delivered via Internet could become resources for business models which will make from them a services. In connection with this prognosis, in [23] there is argued that: *"… the federated sensor services paradigm demands many advanced capabilities that are lacking in the current state of the art. First and foremost, sensor service platforms will necessarily have multiple distinct stakeholders. There will be multiple sensor service providers possibly with very different capabilities and multiple service consumers with distinct sensor feed requirements. In addition, the sensor service platforms need to have auxiliary functionalities such as feed processing (e.g., filtering, sampling, etc.), feed storage and access control. These auxiliary services may again be operated by multiple service providers. The platform should be able to seamlessly support these multiple stakeholders."* They conclude that federated clouds are necessary to be exploited for such tasks and also that data quality from the sensors has to be introduced. In this paper they also show a high level architecture of Sensor Service Cloud Framework. This architecture has following three layers:

- physical resources and devices,
- virtualized services
- domain applications.

In it all sensors feeds, disks, computational elements etc are represented as Data Quality (DQ) aware virtualized services or with other words, data services. The authors proposed also that a DQ aware catalog of virtualized services have to form a basis for their Cloud architecture.

This excellent paper contains many additional smart ideas how the DQ paradigm could influence the cloud architecture [23]. They address the following issues: Models for feed contents and quality, techniques for feed discovery, composition and adaptation, Pricing models and SLAs. Especially the concept of DQ and DQ catalogs for different data sources is very important.

2.3.1 *Stream processing issues*

Stream processing issues are also very relevant and important for our aims. They were tackled in Aurora, Borealis and Stream projects [24], [25], [26]. The future research work on a global cloud for IoT and Big Data has to study carefully the ideas and concepts proposed in these projects and select from them such, which could help to achieve sustainability of the solutions.

Two kinds of cloud based technologies are important for big data analytics:

- Stream (on-line) processing (projects Aurora, Borealis, Cayuga, Stanford Data Stream manager, System S have explored many issues associated [27].
- Map-reduce-based technologies for batch processing (The optimality of the generated mapper and reducer codes for IoT Big Data remains an open problem [23].)

2.3.2 *Massive sensor feeds storage issues*

In the future global multi-cloud Big Data IoT platform it will be necessary also to have access to (historical) stored data including stored data feeds. Two kinds of technologies from the area of Big data research have to be considered as potentially useful for these goals in the area connected with this problem which are very relevant:

- NoSQL document stores (For example MongoDB and Apache CouchDB [28]).
- store and key-value stores (For example BigTable, Dynamo, Apache Cassandra, [29], [30]

It is open question which of these technologies will be optimal or at least best suited for our aims. It is possible that some combination taking the most adapted futures could be adequate solution for this problem.

Our ideas on multi-cloud research proposed in this paper will be sketched in a next section. They were used as a basis for a new project proposal [31].

3 CLOUD INSPIRED MULTI-CLOUD RESEARCH PROPOSAL

3.1 *Heuristic description of the main idea*

The generation and consumption of data from huge amount of sensors and actuators in future IoT have many parallels to water circulation on our planet. Water evaporation on our planet Earth could be seen as data generation from sources distributed on the globe. The evaporated water is collected in "local" clouds, which travel, join with other clouds and disappear by raining seemingly stochastically. The rain fall moistures the surface on Earth and the flowing water is again collected in seas, reservoirs etc. However the main difference with information and the natural sustainable process is that the amount of information collected by human beings is increasing over time and the other aspect is that the resources the human kind has available to deal with storing and processing the ever increasing volume of information seems to be constrained. Therefore instead of storing all information collected, it would be much more efficient to store only "useful" information for future. The question which information will be useful in future could be not answered today easily in connection with Big Data, it could be only estimated or guessed. On the other hand it is not sustainable to store all information in top gear (expensive) clouds from which it could be made usable in real time.

The other big challenges connected with handling Big Data today is how to treat people as human beings and not additional things in the federated Big Data—IoT—Multi-Cloud. Nowadays: "*People are treated as small elements in a bigger information machine …*" [32]

Therefore our first main idea is to propose a new architecture (infrastructure) with federated cooperating "clouds" and "reservoirs". The clouds mean evolved efficient data centers which will store the information which is needed to be "on line" all the time and used with small delay. The reservoirs would be archives with less expensive storage systems connected with broadband connections into the overall ecosystem.

The second idea is to propose a new management of Big Data, inspired by water circulation or better to say by todays and future water processing and supply systems and networks. Simply to say it is about dealing with information in similar way like with water. The goal is to get efficiency and sustainability into the overall architecture. Water, which is useful for us comes on one side in more or less stochastic manner from real clouds and on other hand from water supply systems, which are today far from ideal. But never the less, we can be inspired by the effort to clean the water right at its source, to store it and after usage to clean it again, when it goes to our big reservoirs to the sea. It is evident that our inspiration would be quite restricted if we would take into consideration only the existing water supply systems. We can easily imagine future water supply systems, which will use information technology and Big Data as well. For example good weather prognostics could improve the efficiency of distribution of water. The goal in this area will be to find similar tools as weather prognostics for estimation how much information from sensors will come from different areas of the globe, what are the main social, political and other factors.

I. Energy efficiency: storing, transmitting and processing information costs energy.
    It would be necessary to find a balance between resources necessary per bit for each of these tasks in dependency on time. Such task is a very challenging one, because today we do not have adequate measures for quantitative and qualitative estimations of these expenditures.
    It will involve some prediction methods for classification of information into different classes. Maybe A, B, C ..., classified for example as follows:
    A. Could be for the top gear information (actual often used and very valuable), which has to be stored in "best" clouds;
    B. Second class with questionable value, which however could be useful in future and therefore should be stored in reservoirs from which it could be moved into clouds if necessary, via broadband connections or a new clouds could be set into existence which will be feed from this reservoir "sucking" and later mining information from this specific reservoirs;
    C. Information which after some time could be deleted.
II. The architecture and its management has to be humanistic or human centric. The human beings should not be treated as other things in IoT. Therefore lot of effort has to be assigned to research of the concern of people and taking measures to protect their security, privacy and quality of experience.

3.2   *The way from state of the art to beyond the state of the art: Building entities of the future architecture*

In case that the project proposal [31] will be selected for financial support, the research connected with it will be based on the recent achievements described in [23], but it will go beyond it as this architecture did not take in to account important global constraints such as sustainability and human oriented approach.

Our architecture will reflect these inevitable constraints of the future. Let's start with description of the basic broker oriented service architecture. It is depicted in Figure 1.

In this architecture the main enablers are the following entities:

• Virtualized sources (sensors/actuators ...);
• Virtualized Auxiliary services;
• Data quality notion.

The application domain will sit on a layer of virtualized data sources (sensors, actuators, storage systems ...), which will hide the physical layer underneath it. We have to distinguish

43

between auxiliary services and services for end users connected with application domain. The internal virtualized services will represent for example servers and storage systems via their Quality Parameters such as: Storage capacity (used and free), volume of RAM, Processing resources and their speed, operating system, etc.). The virtualization will hide the technical details of the physical devices in IoT from higher layers. All these devices and data will be represented with *augmented DQ data*. The DQ aware catalog of all virtualized services will be one of the most important entities in the architecture. Behind the state of the art not only such attributes will be considered such as accuracy, error rate, availability, latency, validity, thrust—worthiness but also the following which will position our architecture beyond the state of the art:

Location of the origin, original source characteristics, energy demands, cost demands, security, privacy and risks and others. These parameters will be taken into account in DQ at origin (source), DQ at storage, DQ at processing entity. The other parameter will be the aging (time dependent DQ parameters) profile estimation of the data, which will allow to discriminate and better to mange (eventually also delete) the data in a global multi-cloud.

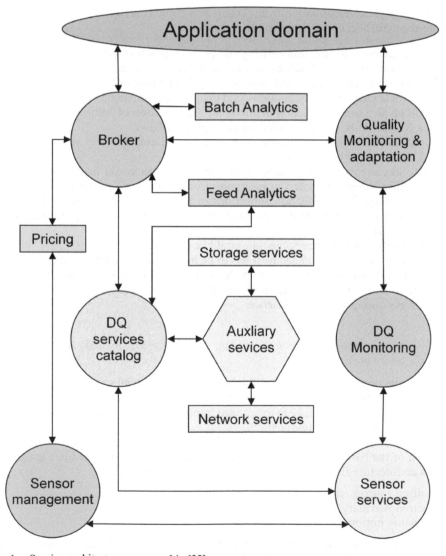

Figure 1. Service architecture proposed in [23].

The multidimensional quality parameters will be expressed as dynamically changing vector in multidimensional space. It is necessary to enhance all entities of the architecture proposed in [23] beyond the state of the art in order to take into account the new requirements and the new dimensions and the dynamic character of DQ vectors with the goal to make it sustainable.

On the other hand to make it more human friendly following aspects will be taken into account:

The concerns and requirements of the people obtained via questionnaires and research;
The existing law and standard requirements in order to protect privacy and security;
The effort will be to make the interfaces usable also by not specialized programmers and professionals from related areas. For this the PRogramming Intentbased Cloud-scale IoT Applications PatRICIA [33] framework which is currently in development in TU Vienna, will be studied and its usability estimated. In [33] the authors describe this framework main goal as follows: *"The core idea of the PatRICIA is to enable the development of value-added IoT applications, which are executed and provisioned on cloud platforms but leverage data from different sensor devices and enable timely propagation of decisions, crucial for business operation, to the edge of the infrastructure."*

4 CONCLUSION

In this paper some main issues in today's research of the complex problems connected with multi-cloud technology for Big Data coming from IoT were mentioned. A heuristic discussion of a possible approach to research and design of future federated multi-cloud architecture was also presented which has the aim to fulfill following basic constraints. It has to be sustainable and humanistic.

ACKNOWLEDGMENT

This work was supported by, Slovak Research and Development Agency under contracts SK-AT-0020-12 and SK-PT-0014-12, by Scientific Grant Agency of Ministry of Education of Slovak Republic and Slovak Academy of Sciences under contract VEGA 1/0518/13, by EU RTD Framework Programme under ICT COST Action IC 1104 and by Visegrad Fund and National Scientific Council of Taiwan under IVF–NSC, Taiwan Joint Research Projects Program no. 21280013 "The Smoke in the Chimney—An Intelligent Sensor—based TeleCare Solution for Homes".

REFERENCES

[1] EU-Japan Research and Development Cooperation in Net Futures, H 2020—EUJ-1-2014: http://ec.europa.eu/research/participants/portal/desktop/en/opportunities/h2020/topics/3051-euj-1-2014.html.
[2] Oracle Company, "Oracle Information Architecture: An Architect's Guide to Big Data", August 2012. http://www.oracle.com/technetwork/topics/entarch/articles/oea-big-data-guide-1522052.pdf
[3] D. Charles, "The Power of Habit: Why We Do What We Do in Life and Business", Random House, 2012; M. S. Viktor, N. C. Kenneth, Big Data: A Revolution That Will Transform How We Live, Work, and Think. Houghton Mifflin Harcourt Publishing Company, 2013.
[4] Zaiying Liu; Ping Yang; Lixiao Zhang, "A Sketch of Big Data Technologies," *Internet Computing for Engineering and Science (ICICSE), 2013 Seventh International Conference on*, vol., no., pp.26,29, 20–22 Sept. 2013 doi: 10.1109/ICICSE.2013.13.
[5] Tekiner F., Tsuruoka Y., Tsujii J., Ananiadou S., Keane J., "Parallel Text Mining for Large Text Processing", pages 348–353, in Proceedings of IEEE CSNDSP 2010, 21–23 July, Newcastle, UK.
[6] Tekiner, F.; Keane, J.A., "Big Data Framework," *Systems, Man, and Cybernetics (SMC), 2013 IEEE International Conference on*, vol., no., pp.1494, 13–16 Oct. 2013 doi: 10.1109/SMC.2013.258.

[7] A brief History of Internet of Things, http://postscapes.com/internet-of-things-history.

[8] N. Kushalnagar, "IPv6 over Low-Power Wireless Personal Area Networks (6LoWPANs): Overview, Assumptions, Problem Statement, and Goals", RFC 4919.

[9] Lu Yan, Yan Zhang, Laurence T. Yang, Huansheng Ning, "The Internet of Things: From RFID to the Next-Generation Pervasive Networked Systems", 2008 by Auerbach Publications, 336 p., ISBN 9781420052817.

[10] Stankovic, J.A., "Research Directions for the Internet of Things," *Internet of Things Journal, IEEE*, vol. PP, no. 99, pp. 1,1 doi: 10.1109/JIOT.2014.2312291.

[11] Horizontal Nature of Internet of Things, http://www.fiercesmartgrid.com/story/horizontal-nature-internet-things/2014-02-19.

[12] Holger K., Wilig A. *Protocols and Architectures for Wireless Sensor Networks.* Wiley, Chichester, 2005, ISBN: 978-0-470-09510-2.

[13] Ahson S., Ilyas M. RFID Handbook. CRC Press, Boca Raton, 2008, ISBN: 978-1-4200-5499-6.

[14] Janocha H. Actuators: Basics and applications. Springer 2004, ISBN 3540615644.

[15] I. Akyildiz and M. Vuran, Wireless sensor networks. Wiley, 2010, vol. 4.

[16] J. Stankovic, "Wireless sensor networks," Computer, vol. 41, no. 10, 2008.

[17] D. H. Kim et al., "Sensloc: sensing everyday places and paths using less energy," in SenSys, 2010.

[18] S. Madden et al., "Tinydb: an acquisitional query processing. system for sensor networks," ACM Trans. Database Syst., vol. 30, no. 1, 2005.

[19] S. Madden, "Database abstractions for managing sensor network data," Proceedings of the IEEE, vol. 98, no. 11, 2010.

[20] COSM—Intenet of Things Platform," https://cosm.com.

[21] A. P. Sheth, C. A. Henson, and S. S. Sahoo, "Semantic sensor web," IEEE Internet Computing, vol. 12, no. 4, 2008.

[22] L. Li and K. L. Taylor, "A framework for semantic sensor network services," in ICSOC, 2008.

[23] Ramaswamy, L.; Lawson, V.; Gogineni, S.V., "Towards a Quality-centric Big Data Architecture for Federated Sensor Services," *Big Data (BigData Congress), 2013 IEEE International Congress on*, vol., no., pp.86,93, June 27 2013–July 2 2013 doi: 10.1109/BigData.Congress. 2013.21.

[24] D. J. Abadi, "The design of the borealis stream processing engine," in Proceedings of CIDR, 2005.

[25] D. Carney et al., "Monitoring streams: A new class of data management applications," in Proceedings of VLDB, 2002.

[26] L. Brenna et al., "Cayuga: a high-performance event processing engine," in SIGMOD Conference, 2007.

[27] A. Arasu et al., "Stream: The Stanford stream data manager," IEEE Data Engineering Bulletin, 26(1), 2003.

[28] mongoDB," http://www.mongodb.org.

[29] F. Chang et al., "Bigtable: A distributed storage system for structured data," ACM Trans. Comput. Syst., vol. 26, no. 2, 2008.

[30] G. DeCandia et al., "Dynamo: Amazon's highly available key value store," in SOSP, 2007.

[31] EMBEDIT, Project for H 2020 EUJ 1 call, 2014.

[32] Lanier J. Who owns the future. Simon & Schuster, 2013 ISBN: 978-1451654967.

[33] Nastic, S., Sehic S., Vogler M., Truong H. Dustard S. PatRICIA—a Novel Programming Model for IoT Aplications on Cloud Platforms. IEEE 6th International Conference on Service-Oriented Computing and Applications, 16. 12.—18. 12. 2013, Kauai, Hawaii.

Current Issues of Science and Research in the Global World – Kunova & Dolinsky (Eds)
© 2015 Taylor & Francis Group, London, ISBN 978-1-138-02739-8

Emigration as a way to freedom of the journalist Ladislav Mňačko

Jozef Leikert
Faculty of Mass Media, Paneuropean University, Bratislava, Slovakia

ABSTRACT: The expert study is dedicated to a certain epoch in the life of a remarkable personality in the post-war journalism and literature. It looks into positions of Ladislav Mňačko in the historical breaking year of 1968, especially after Warsaw Pact army intervention in Czechoslovakia. The infamous date, which entered history as stymied hopes for society democratization, followed by difficult times of normalization, which changed the destinies of many people. One of them was the renowned journalist rebel and controversial writer Ladislav Mňačko, who in protest emigrated to Austria—a country behind the iron courtain, which became his second home for twenty years.

Were there different times in Slovakia—I mean more favourable for culture and arts—as they are not, the year 2014 would definitely be the year of Ladislav Mňačko. That is to say this year marks the 95th anniversary of his birth and 20 years of his death. Despite that Ladislav Mňačko, was not officially remembered by anybody, nor noted, as an excellent writer and journalist, except two very short articles in the newspaper. I can imagine what other countries would have organized having such a great personality in their history. Celebrations, conferences, almanacs, reeditions of books, collections and other various events. And what have we, Slovaks, done? Absolutely nothing, as if Mňačko had never existed, never written anything, done nothing good for our nation.

I do not know, if the Czechs—our closest neighbours and brothers—are smarter or whittier—but they knew how to honour their writer Bohumil Hrabal and named the year 2014 the year of Hrabal or the century of Hrabal. Not mentioning dozens of events related to his anniversary of birth... Because we live in Slovakia, where we do not respect personalities very much, I will at least briefly remind of who Ladislav Mňačko was and why I am talking about him at the conference in Vienna, in the country where he found a refuge and lived and created over twenty years.

Short notice: Mňačko was one of the greatest Slovak journalists and writers, even though no square, school or street has been named after him, as on the contrary was the case with the mentioned Bohumil Hrabal in the Czech Republic. It is difficult to compare these two artists, after all, this is not possible with any person. However, they have one thing in common: they were not good pupils and they even flunked the class in secondary school. Naturally, this is not important, what matters is that they became great writers of former Czechoslovakia, works of whom are known in Austria as well.

Mňačko got renowned first as a journalist. He was an editor of Rudé právo (Red law), Pravda (Truth) and Kultúrny život (Cultural Life). In Kultúrny život he he took up the position of the editor-in-chief twice in the time periods of significant changes in Czechoslovakia, which resulted in the Prague spring in 1968 stampeded and rolled-over by Soviet tanks with the assistance of driven allied soldiers of the Warsaw pact. All bad came from Moscow, which was afraid that in Czechoslovakia socialism with human face, also formed by the articles and books of Ladislav Mňačko, would win.

When others were scared and were servile—as many times before and after—Mňačko was the first Slovak writer, who in his works pointed at the infirmities of the society. His books Kde končia prašné cesty (Where dust roads end), Oneskorené reportáže (Belated reportages),

Ako chutí moc (What power tastes like), but partially also Smrť sa volá Engelchen (Death is called Engelchen), represent the peak of resistance against the powers of that day. He viewed the problems of the times then not only by his eyes, but especially by the eyes of common, poor people, who were crushed by the communist machinery. From the beginning he was a convinced communist journalist, who helped to fortify the power by his articles, he even agreed with the execution of the Czechoslovak minister of foreign affairs Vladimír Clementis, but overnight, when he was writing a reportage from a construed process with the general director of Mills and Bakeries Ladislav Galan, he pulled himself together. Nevertheless, it is not possible. A fiery communist became a fiery anti-communist.

He continued to believe the ideals of a good life for all people, but no longer believed the commnist function holders and their assistants, who only cared about themselves and their power. He criticized them in his articles and books. And not only that, the truth and the feelings of the majority of the Slovak and Czech nations, was expressed by his voice, loudly too. For example, he asked the Czechoslovak president and the first Secretary of the Central Committee of the Czechoslovak Communist Party Antonín Novotný in front of all diplomatic corps, so they could all hear well, to resign, because the nation felt no respect to him.[1] Or he slapped the representative of the Department of Culture of the Central Committee of the Slovak Communist Party Tomáš Slouka, who was harming artists—had their books shreded and their exhibitions closed—in front of all the Central Committee of the Party, and said outloud he deserved it for purposefully harming artists.[2]

Or: When Soviet Union in 1967 interrupted diplomatic relations with Israel, Czechoslovakia—loyal butler, did the same three minutes later. Majority of Slovaks and Czechs did not agree with it, but they remained silent. Only Mňačko stood in resistance and demonstratively left to Israel, although he was not a Jew, what many people have thought until today. He only sympathized with the Jewish nation and took their side publically as well. Mňačko emigrated on 10 August 1967 and first stopped in Vienna, where he published an extensive proclamation for Frankfurter Allgemeine Zeitung[3] and other international media, in which he stated his reasons for emigration. Then he provided a 13-minute interview to the Viennese TV. Following his arrival in Israel he organized a press conference, which was attended by many foreign journalists accredited in Israel. Before the end of the press conference the organizers handed Ladislav Mňačko a folded piece of paper. He opened it and read it outloud stating the world press agencies quoting the Czechoslovak press agency brought the news that Ladislav Mňačko was rid of state citizenship, his title merited artist was taken away from him and he also lost all state distinctions and he was excluded from the Communist Party.[4]

Mňačko himself was surprised how many nice and supportive letters he received in Israel, although many people did not even know his exact address. And still every letter found him. As I have dedicated several years to Ladislav Mňačko and I wrote a large monography about him, I have originals of the letters at home, which Mňačko received in Israel. For instance Ernest Aschner wrote on the envelope: "Mr. Ladislav Mňačko, Czechosl. writer—journalist, Tsechsl. Schiftsteller, Tel Aviv, Israel."[5] Jan Kuncíř wrote even less, only adding Tel Aviv next to his name[6] and Lubo Mikula added Radio Tel Aviv.[7]

Tibor Katrenčik thought he was in some kibbutz. He sent two letters simultaneously: one to kibbutz Masaryk and the other to kibbutz Haogen.[8] Mňačko received both letters, even

[1]Interview of the author with Ladislav Mňačko (24 February 1993).
[2]Archive of the Association of Slovak Writers organizations and Association of Slovak Writers. Minutes from the meeting of the Committee of the Association of Slovak Writers, held on 19 October 1961. More details can be found: MŇAČKO. L.: Siedma noc (The seventh night). Bratislava : Práca 1990, p. 137.
[3]Franfurter Allgemeine Zeitung, no. 184/1967.
[4]Archive of the Association of Slovak Writers organizations and Association of Slovak Writers. Information ASW no. 1, p. 14–15.
[5]Archive of Jozef Leikert. Letter of Ernest Aschner to Ladislav Mňačko of 15 August 1967.
[6]Archive of Jozef Leikert. Letter of Jan Kuncíř to Ladislav Mňačko of 20 August 1967.
[7]Archive of Jozef Leikert. Letter of Lubo Mikula to Ladislav Mňačko of 16 August 1967.
[8]Archive of Jozef Leikert. Letters of Tibor Katrenčik to Ladislav Mňačko of 17 August 1967.

though he was not in any kibbutz at that time. As a matter of interest I am going to mention that similarly people were writing to Mňačko even after 1989, when he permanently returned from Austria to Czechoslovakia. Paulína Bednáriková from Zlín for example on 12 May 1990sent a correspondence card to the address: "Writer Mr. Mňačko in Prague or Bratislava."[9] And this letter found him too.

Out of the array of letters Mňačko received in Israel, I am selecting a couple of quotations:

Harry S. Deutsch wrote: "Esteemed Mr. Mňačko, I consider the obligation to thank you for your position, your words and your actions, to be very pleasant. I do not believe this is going to change the opinions of the Czechoslovak authorities (at least an earthquake would be needed), but that does not belittle your heroic attitude. Thank you a thousand times."[10]

Irma Polaková: "Esteemed Mr. Mňačko! [...] I would like to tell you how much you encourage a person in faith that truth and good will win in the world and all people who know your words and actions must be grateful to you."[11]

Josef Volf: "... I appreciate your position and decision, which you took. I know it concerns in addition to Israel also all of us, our mother country, truth and justice, and last but not least, you personally. It is not easy, I know it very well, it is much easier to keep quiet and live on the top of successes at home, write something 'more bitter' and 'satisfy' the people, but still go hand in hand with those who are responsible for all evil done at home—for example by only keeping still! Therefore I want to thank you for your courage, your decision, your uncompromising protest. You are the first Czechoslovak writer after many years, who truly is a writer not only by hand and mind, but also conscience."[12]

Božena Čechová: "Dear, honoured and by all loved Ladislav Mňačko! Hopefully this letter of mine, written from the bottom of an honest and respectful heart, gets into your hands. Maybe it will please you to know that an insignificant Czech woman, will voice greetings and loyalty of thousands. We are very proud of you and glad to have you in the country of David as our spokesperson. [...] What high class Ladislav Mňačko is regarded in, I must not write. You are our second Havlíček Borovský, but much, much more courageous. Your books are in unimaginable demand, people borrow books from each other, and if they were sold somewhere there would be the longest and the most patient lines ever. We share your belief you will return and we look forward to seeing you. I have been saying that Czechoslovakia only has women and cowards, because real men died out, that it is not possible for them to suffer all this, if they could be at least a bit brave and suddenly—look, a Human appears."[13]

By the arrival of Prague Spring, which began n Bratislava and we can comfortably speak of Bratislava pre-spring, Ladislav Mňačko decided to return to Czechoslovakia, even though many had tried to talk him out of it.

Oto Reiss was convinced that if he returned they would imprison and torture him back at home. He adviced: "Remain abroad, where you can breathe good air—you can enjoy the freedom of speech and you will be much happier than in the police Czechoslovak state."[14]

Those against the return of Mňačko back to Czechoslovakia mostly lived abroad. Those people, who lived in Czechoslovakia, were of different opinion. They were hoping that his arrival would change and start many things. Božena Čechová wrote: "When you return, I will come to you with a bouquet of flowers and a handkerchief full tears of joy. [...] Keep healthy and come back, we need you! But beware, not into the lions' den..."[15]

---

[9]Archive of Jozef Leikert. Correspondence card of Paulína Bednáriková of 12 May 1990.

[10]Archive of Jozef Leikert. Letter of Harry S. Deutsch to Ladislav Mňačko of 17 August 1968.

[11]Archive of Jozef Leikert. Letter of Irma Polaková to Ladislav Mňačko of 21 January 1968.

[12]Archive of Jozef Leikert. Letter of Josef Volf to Ladislav Mňačko of 22 September 1967.

[13]Archive of Jozef Leikert. Letter of Božena Čechová (pseudonym) to Ladislav Mňačko of 22 September 1967.

[14]Archive of Jozef Leikert. Letter of Oto Reiss to Ladislav Mňačko of 18 August 1968.

[15]Archive of Jozef Leikert. Letter of Božena Čechová (pseudonym) to Ladislav Mňačko of 22 September 1967.

Many people warned him of the lions' den, even the famous Czech writer Pavel Tigrid, who emigrated from Czechoslovakia immediately after the putsch in February 1948. "Molden [publisher in Vienna—author's comment] told me (and your latest proclamation in the press confirms it), that you have not given up on the intention to return as early as possible to the Czechoslovak Socialist Republic, maybe even illegally and let yourself be sentenced. I am afraid if you decided today for this kind of return, it would be a vain gesture: no process, even less public, would take place, because you would perhaps without a trace disappear in the prison, or—at best—border guards would 'throw' you back over the border. Many of your friends at home agree that the State security service is the right hand of the power holders and that—temporarily –'hard line' was in place. Believe me, I have no ulterior intentions, when I appeal to you not to rush in this regard, build yourself armour of patience and wait for a more suitable moment, when whenever you do anything it will have a political impact without reasonable doubt and you will not make a useless sacrifice. I know, this is not easy, Moldem told me about your personal problems and I understand your restlessness. On the other hand you have all chances to return legally and at the zenith—it will suffice when the current Prague Israeli policy is revised."[16]

When the party and government in Czechoslovakia noticed that Mňačko wanted to return, they were raging and the press had no nice word for him. Because my contribution cannot encompass all Mňačko's life, I will "skip" one epoch and mention his emigration to Austria in 1968 at the time the Warsaw Pact armies headed by the Soviet Union invaded Czechoslovakia.

Following his return from Israeli emigration Mňačko worked on the book Agresori (Aggressors). During holiday time Bratislava depopulated and he perceived political events, which began moving faster and faster in Czechoslovakia, more intensively. He claimed that something was going to happen. He said it even in a TV talk show, where he was invited by the editor Jozef Vranka. And he said to me: "When he spoke out loud, we were saying, he was exaggerating. It was not so bad in our country... Mňačko was right, he knew the political situation better than we did, we were sitting at home all the time."[17]

The change and a radical one occurred on the night of 21 August, when out of the blue a lightning hit.

The situation was tense since the first hours, people knew no details, they could not explain what was happenning and how long this was going to last. They began to sense that soon people will be arrested. Several people were mentioned that the Soviets were interested in. Mňačko's name was among them. Since that time he became an outcast, who feared to sleep two nights in a row in the same bed.[18] As his niece Jarmila Keprtová revealed, he was convinced they were after him and they wanted to eliminate him, as several people warned him.[19] If not right away, certainly later they will imprison him, or take him to the Russian labor camps.

With coming days he was considering his options and his prospects of remaining in Czechoslovakia intensively. He did not have much of a choice. In his book Siedma noc (The seventh night), which was published in Vienna in 1970, he wrote: "I could hide in the mountains, I knew where, I did not have to fear betrayal. My friends offered me a comfortably furnished cottage on one of many islands on the Danube, where they could not find me so easily. There were more options. But to what avail? [...] And to wander in the mountains fearing I will be once caught, sooner or later, they get everybody, this is not an appealing outlook."[20]

---

[16]Archive of Jozef Leikert. Letter of Pavel Tigrid to Ladislav Mňačko of 9 September 1967.
[17]Interview of the author with Jozef Vranko (5 June 2007).
[18]Interview of the author with Ladislav Mňačko (28 January 1993).
[19]Interview of the author with Jarmila Keprtová (15 April 2004).
[20]MŇAČKO, L.: Siedma noc (The seventh night). Bratislava: Práca, 1990, p. 156.

Still more and more he was teased by the idea to leave. Something was telling him—go, do not wait even an hour, what if they close the borders and you will never get out again? His alter ego was persuading him not to go, not to leave his nation, which he loved so much. Exactly the seventh night of occupation began and he went to walk in his beloved city despite the risk he could be recognized and seized. In the mentioned book Siedma noc (The seventh night) he writes: "When somewhere I heard the roar of the engine, I hid in the nearest door. It could be only a Russian patrol on an armoured vehicle. That night no personal car drove through town. Here and there people were standing in front of houses and semi-loudly discussed what would happen. When I walked past them they distrustfully looked at me. The old psychosis is here again, everybody fears each other. Russians are here. They will no longer leave. We pointlessly hope for their departure. They are here to stay. [...] There is no safe place for me in this country, no hope. Even though I was making some illusions, the eyes of my friends, acquaintances, and above all the people who love me, speak to me clearly. They do not have to. I will go. I only wanted to wait until this ending, I wanted to see it, feel it and experience it."[21]

For a moment he sat on a bench in the park where once a bronze statue of J.V. Stalin stood. Someone once wrote on the crumbling pedestal in chalk: "Proletarians of all countries, unite, else I will shoot!" And that decided. Even though he had been back home from Israeli emigration only for several months, he knew he had to leave.

He went home and asked his girlfriend Eva Okályiová, who he lived with, to pack only the most important things, as they were leaving. She did not protest, because she shared the same feelings and was convinced it was better for them to go. Mňačko put his shaver in one bag along with several books of his favorite writers. He took five of his books too. The woman put underwear in another bag and grabbed her handbag in the other hand. Then Mňačko called a taxi and without useless words he told the taxi driver: "To the Austrian border!"

After years Ladislav Mňačko revealed: "I thought he would refuse, but without saying anything he started the car and we moved. He did not say one word during the entire journey..."[22]

Mňačko did not feel like debating either. The more he was tortured by thoughts whether he did well and whether he was not supposed to stay at home. When the taxi stopped at the border crossing Petržalka—Berg, as in a romantic movie, he hugged the woman and they crossed the border. They stayed in Vienna for several weeks, then for a shorter while they moved to Israel, to leave to Italy, and finally settle in a small Austrian village Grosshöflein, located not so far away from the Slovak borders. They captivated attention wherever they stopped, because Mňačko was one of the most famous Slovaks, who left Czechoslovakia after 21 August 1968. In 1993 he said to me: "I did not think I was so popular in the world and that people knew me behind borders as well. They invited us into their homes, offered us dinner, lunch. Some people even wanted to give us money, which we fundamentally refused. We lived off royalties paid for my books, which were published abroad, luckily, I had a bank account in Austria. In the beginning I received compensation for lectures about Czechoslovakia and my literary work too. From time to time newspapers and magazines paid me for interviews I gave. I perceived it more or less as a pleasant favour, at that time the world symphatized with emigrants from Czechoslovakia very much and helped us. This did not last too long though, naturally, and everybody had to find their living in their own way. We could survive thanks to my royalties from new books and scripts. I must admit the Austrian Chancellor Bruno Kreisky helped us then immensely by providing us good contacts and acknowledging me as a writer."[23]

---

[21] MŇAČKO, L.: Siedma noc (The seventh night). Bratislava: Práca, 1990, p. 194–195.
[22] Interview of the author with Ladislav Mňačko (24 February 1993).
[23] Interview of the author with Ladislav Mňačko (25 February 1993).

Mňačko with Eva Okáliyová, who he married in Vienna, lived satisfied over twenty years in Austria. (Later, her mother, who was the wife of the well-known poet Ivan Krasko, joined them.) They returned to Czechoslovakia after the change of he political regime following 17 November 1989, when they sold their house in Grosshöflein and came home. Even though they had built their home in Austria, they felt called to Slovakia.

Today neither Ladislav Mňačko, nor his wife are among the living. As I became his court historian, the only one, who studies his life and work, I can say with full worth: "Austria, Austrians, great thank you for helping Ladislav Mňačko and his family, and providing them with a refuge for such a long period!" And I would like to add: "I would be happy and really grateful if you could name a street after this great European writer, because in Slovakia, we are so indifferent and disgraceful! Thank you, respected and dear Austrians!

REFERENCES

Archive of Jozef Leikert. Letter of Ernest Aschner to Ladislav Mňačko of 15 August 1967.
Archive of Jozef Leikert. Letter of Jan Kuncíř to Ladislav Mňačko of 20 August 1967.
Archive of Jozef Leikert. Letter of Lubo Mikula to Ladislav Mňačko of 16 August 1967.
Archive of Jozef Leikert. Letters of Tibor Katrenčik to Ladislav Mňačko of 17 August 1967.
Archive of Jozef Leikert. Correspondence card of Paulína Bednáŕiková of 12 May 1990.
Archive of Jozef Leikert. Letter of Harry S. Deutsch to Ladislav Mňačko of 17 August 1968.
Archive of Jozef Leikert. Letter of Irma Polaková to Ladislav Mňačko of 21 January 1968.
Archive of Jozef Leikert. Letter of Josef Volf to Ladislav Mňačko of 22 September 1967.
Archive of Jozef Leikert. Letter of Božena Čechová (pseudonym) to Ladislav Mňačko of 22 September 1967.
Archive of Jozef Leikert. Letter of Oto Reiss to Ladislav Mňačko of 18 August 1968.
Archive of Jozef Leikert. Letter of Božena Čechová (pseudonym) to Ladislav Mňačko of 22 September 1967.
Archive of Jozef Leikert. Letter of Pavel Tigrid to Ladislav Mňačko of 9 September 1967.
Archive of the Association of Slovak Writers organizations and Association of Slovak Writers. Minutes from the meeting of the Committee of the Association of Slovak Writers, held on 19 October 1961. More details can be found: MŇAČKO. L.: Siedma noc (The seventh night). Bratislava: Práca 1990, p. 137.
Franfurter Allgemeine Zeitung, no. 184/1967.
Archive of the Association of Slovak Writers organizations and Association of Slovak Writers. Information ASW no. 1, p. 14–15.
Interview of the author with Jozef Vranko (5 June 2007).
Interview of the author with Ladislav Mňačko (28 January 1993).
Interview of the author with Jarmila Keprtová (15 April 2004).
Interview of the author with Ladislav Mňačko (24 February 1993).
Interview of the author with Ladislav Mňačko (25 February 1993).
Mňačko, L. 1990. *Siedma noc (The seventh night)*. Bratislava: Práca, 1990.

*Faculty of law, section current issues
of legal theory and legal practice*

Current Issues of Science and Research in the Global World – Kunova & Dolinsky (Eds)
© *2015 Taylor & Francis Group, London, ISBN 978-1-138-02739-8*

Right of peoples to self-determination within the context of international law of armed conflict

A. Ďurfina

Faculty of Law, Pan-European University, Bratislava, Slovakia

ABSTRACT: The need for enhanced effectiveness of laws regulating armed conflict is substantiated by increasingly varied warfare at a global level, as well as by progressively unpredictable positioning of parties involved in armed conflict. Currently, the practice of international law, especially law relating to armed conflict and to humanitarian law, is exposed to new pressures stemming from novel forms of warfare. Under such circumstances, it is necessary to redefine and modify the principles, and rules of international law guiding armed conflict. Such adjustment or reform of international law of armed conflict becomes highly important when considering the right of peoples to self-determination. Increasingly, this facet of international law is utilized within both, minor and major conflicts, characterized by long-term nature and asymmetry within warfare. Right of peoples to self-determination is one of principal human rights. Global application of this right to all people is currently extremely problematic.

1 INTRODUCTION

The need for enhanced effectiveness of laws regulating armed conflict is substantiated by increasingly varied warfare at a global level, as well as by progressively unpredictable positioning of parties involved in armed conflict. Currently, the practice of international law, especially law relating to armed conflict and to humanitarian law, is exposed to new pressures stemming from novel forms of warfare. Under such circumstances, it is necessary to redefine and modify the principles, and rules of international law of armed conflict. Such adjustment or reform of international law of armed conflict becomes highly important when considering the right of peoples to self-determination. Increasingly, this facet of international law is utilized within both, minor and major conflicts, characterized by long-term nature and asymmetry within warfare. Right of peoples to self-determination is one of principal human rights. Global application of this right to all people is currently extremely problematic. Throughout the development of human society, people have often lived in conflict, the most apparent example of this being the 20th century, which was marked by two consecutive World Wars. This period was further complemented by various minor conflicts, revolutions, coups, and decolonization struggles around the globe. Because of this, the need to honor international law, which was essentially created during the 20th century, was a logical and necessary solution to the pressing questions of global security. The United Nations undertook a challenge, to create a system of international law that was based on principles of collective international security, a system that functions as a basis of legal interaction within the international community up until the present day.

2 RIGHT OF PEOPLES TO SELF-DETERMINATION AND LAW IN ARMED CONFLICT

Contemporarily, there are many nations that are fighting for independence, attempting to exercise their right of peoples to self-determination (Buchanan, 2007). Historically, the concept of the right of peoples to self-determination dates back significantly, but for the purposes of this paper, we will define the beginnings of this doctrine to the year 1918. After the end of the First World War, nations began to comprehend the necessity of creating a new international system, one that would be based on principles of collective safety and respect for international law. United Nations (UN) under the UN Charter significantly modified some areas of law applied between nations, including the concept of equality and self-determination of peoples. Subsequently, these ideas had a great impact on further development of international relations, for example, the creation of the Jewish state after the Second World War, at a time when Jewish people were strongly dependent on the existence of such international law.

States are the basic subjects of international law, legally understood as members of the international community of states. International law currently assists states in conflict resolution, as well as limits states' freedom to act unilaterally, through fundamental norms that are the basis of international law. On one hand, states are the subjects of law but on the other, they are also the creators or the sources of such law. They are significantly involved in the interpretation and change of international law, which is formed by various rules, and thus, ensuring legitimacy of their actions within the international arena. Historically, within traditional international law, one could distinguish the Law of War as a special branch of international law, which contained all the rules concerning the conduct of war. This law included both, the right to go to war, as well as the laws which were in effect during the war itself. Currently, these forms of law are collectively referred to as the 'law of armed conflict', which is directly related to international humanitarian law. Its objective is to protect the lives of all endangered or vulnerable civilians during wars or armed conflicts.

The first attempt to codify international law of war was an attempt to ratify a treaty in 1874 in Brussels. Subsequently, in 1899 the first Hague Peace Conference was held. Its primary result was the adoption of the Convention for Pacific Settlement of International Disputes (first written rules that addressed mediation to resolve disputes; the establishment of the Permanent Court of International Justice, the International Commission of Inquiry, as well as, the adoption of the Convention with respect to the Laws and Customs of War on Land). Furthermore, in 1907 the second Hague Peace Conference was held, which resulted in a new version of the Convention for the Pacific Settlement of International Disputes, the new version of the Convention with respect to the Laws, as well as Customs of War on Land. However, the Hague Peace Conference was followed by two World Wars, influencing the development of international law, within the subfield of humanitarian law, which was regulated by each of the Geneva Conventions and their additions (Additional protocols I. and II. to Geneva Conventions of 1949).

This article solely focuses on the context of law regulating armed conflict, as a set of standards contained in international treaties, conventions, regulations, rules, and customary international law, all of which are used in times of hostility during such conflict. Currently however, existing international rules and norms governing armed conflict do not reflect the real state of modern conflicts, which are mainly led by other means of struggle and are mostly asymmetric.

Presently the link between the law of armed conflict and the law of peoples to self-determination can be seen with the way in which some nations seek to receive and apply these rights. Many nations are de facto trying to exercise such right by the use of arms and force, and thus, creating conflict within the international community. Several cases of the Declaration of Independence, which the Security Council of the UN did not recognize in its resolutions of the right to self-determination, are clear examples; the Resolution 216 (1965)[1], Resolution 217 (1965)[2] and Resolution 787 (1992)[3]. In response to those resolutions of the Security

---

[1]Resolution 216 (1965)—referring to the situation in Southern Rhodesia. The Security Council decides to condemn the unilateral declaration of independence made by a racist minority in Southern Rhodesia and Decides to call upon all States not to recognize this illegal racist minority regime in Southern Rhodesia and to refrain from rendering any assistance to this illegal regime.
[2]Resolution 217 (1965)—referring to the situation in Southern Rhodesia.
[3]Resolution 787 (1992)—referring to the situation in the Republic of Bosnia and Herzegovina.

Council, the Court of Justice of the European Union noted, that at any given time when the Declaration of Independence was in question, the Council's decisions in these cases depend on particularities of each situation, case-by-case approach. The illegality of the Declaration of Independence is not based solely on the unilateral statement itself, but also on the fact that the Declaration was associated with illegal use of force or other extraordinary violation of norms of general international law (in particular mandatory rules—jus cogens).

As mentioned at the beginning of this article, the concept of the right to self-determination of nations did not begin to develop fully until 1918, after the end of the First World War, when the international political arena started looking for ways to modify the rules of armed conflict between sovereign states. At the time, the initiator of the agenda of the League of Nations, W. Wilson, created fundamental principles that had been summarized into 14 clear elements, including the right of peoples to self-determination. It was supposed to allow European nations to exercise this right, by providing assistance during the creation of nation states, as well as by ensuring international peace and stability. It turned out however, that the agenda of the League of Nations had failed, since efforts to create an international system based on collective security were thwarted by the rise of B. Mussolini and A. Hitler. After the end of the Second World War, building on the foundations set up by the League of Nations, a new international institution was created—the United Nations—defined in the UN Charter as an institution needed for international peace safekeeping, through the establishment of friendly relations and cooperation between states.

An important example of use of right to self-determination was the creation of the Jewish state, after the Second World War, created inside Palestinian Territories that were occupied primarily by an Arab population with religious and cultural association to Islam. In 1948, despite serious opposition from Arabic nations, due to intense support from the international community, the establishment of the state if Israel was declared. Unfortunately however, the emergence of Israel created a large number of disputed, mainly stemming from religiously-based territorial claims. Presently, the Arab-Israeli conflict remains unresolved, despite continuous international efforts for diplomatic solutions of this challenging situation.

UN Charter provisions also gave claim to various liberation movements that tried to lift themselves from the colonization of Western powers. Widest decolonization efforts took place after the Second World War, mainly in the countries of Africa and Asia. In these cases the UN subsequently supported and confirmed their independence, and collective right of peoples to self-determination in colonized regions. The right to self-determination of peoples is also defined by the International Convention on Economic, Social and Cultural Rights, and in the International Convention on Civil and Political Rights. The first article of the pact states that "All peoples have the right to self-determination. On the basis of that right they freely determine their political status and freely pursue their economic, social and cultural development" (United Nations, 1966). Right of peoples to self-determination is a fundamental principle that is also well-defined within the Declaration on Principles of International Law, governing friendly relations and cooperation between states (United Nations, 1970).

Despite wide adaptation of rights in international legal collaboration, their practical applicability, their enforcement, and even clarification of the nature of the law are presently insufficient. Decolonization period had become history and thus, presently the application of international law should be managed within the context of globalization, not in the context of post-World War. It is extremely important to define this right within a context of rights in armed conflict. As an example, following is a description of determinants of conflicts in the Caucasus, among other nations in the geographical location.

3 CONFLICTS IN THE CAUCASUS AND THE STRUGGLE OF NATIONS FOR SELF-DETERMINATION

Conflicts in the Caucasus are determined mainly by geopolitical determinants of this region. Geographically, the Caucasus can be divided into North Caucasus, which is formed by the southern regions of the Russian Federation, and South Caucasus, consisting of Armenia,

Azerbaijan and Georgia. Caucasus largely consists of states that together create a border between Europe and Asia, located in proximity to the Caspian Sea with rich oil deposits. Struggle for influence in the region was eventually won by Russia who dominated it for decades. In terms of cultural and religious inconsistency, the region is characterized by extremely high heterogeneity of the populations, a prerequisite for the emergence of nationalist and separatist conflicts (Potuček, 2005a).

In the past, the concept of the right of peoples to self-determination was changed in the Soviet Union in their constitution, under the leadership of Stalin. It was strongly influenced by the communist ideology. Following the process of Russification, ethnic groups living under Soviet governance were given solely an imaginary possibility for national independence and state autonomy. Inside Soviet ideology of political unity of its territories, ethnic diversity was gradually disappearing. With the fall of the Soviet Block however, ideological unification vanished, giving rise to tensions between previously unified ethnic groups, providing cause for many armed conflicts. In the early 90's of the last century, Caucasia region became saturated with high concentration of weapons, controlled by various ethnic groups.

Currently, North Caucasus consists of several autonomous states that fall under the administration of the Russian Federation, and are constantly a source of potential conflict. The least stable is the region of Chechnya, suffering several armed conflicts. In 1991, Chechen President D. Dudayev led the country to independence, but due to a quick military strike, the territory was yet again dominated by the Russian Federation in 1994. The conflict ended with a peace treaty, signed in 1997, only to escalate again in 1999, based on military aggression from Chechnya toward neighbouring Dagestan. After the peace treaty of 1997, Chechnya was expected to adapt to democratic developments (Potuček, 2005b), but proceeded to be governed in accordance with Islamic Shari'a. This resulted in Russian efforts to regain its' power over the region, by force if necessary. This in turn resulted in massive death tolls, thousands of refugees, and Chechnya's further political, economic, and social isolation. This is only one example of peoples whose struggle for self-determination resulted in a bloody conflict without a resolve.

In North Caucasia, the Republic of Ingushetia and North Ossetia are other regions with conflict present. The situation in Chechnya has created a domino effect, further destabilizing the situation within these countries. Additionally, North Ossetia contributes to conflict between South Ossetia and Georgia, in South Caucasus. After the end of the Cold War, territories in South Caucasus increased their efforts for self-determination. After the breakup of the Soviet Union, there was a possible pathway for these nations to democratically develop, a similar situation was experienced by Baltic states. In contrast to Estonia, Lithuania, and Latvia; Georgia, Armenia, and Azerbaijan continued on the path of armed conflict.

South Caucasus is currently ruled by a number of ethnic conflicts fought among many national groups. Meanwhile in Georgia, there are only few ethnic groups and about 20% of its territory consists of areas without a claim to stateship. The main conflict zones in Georgia are Adjara, Abkhazia, and South Ossetia. While Adjara is a relatively stable country, which has achieved the status of an autonomous republic and recognizes the authority of Georgia, the regions of Abkhazia and South Ossetia are ruled by separatist efforts and enforced through armed conflicts. Historically, the Abkhaz nation was autonomous but during the creation of the Soviet Block, and a change to its autonomy occurred under the influence of discriminatory Soviet unification policies. With disintegration of the Soviet Union and Georgia's independence, the struggles for national self-determination in Abkhazia fully flared. In 1992, Abkhazia unilaterally declared independence from Georgia. The declaration of independence gave rise to armed conflict, resulting in intervention from United Nations. The Russian Federation became the mediator for this conflict, deploying troops and peacekeeping missions for military bases and other critical areas. Despite the 1994 ceasefire, the conflict between Abkhazia and Georgia is still current, due to the efforts of president Saakashvili (Georgia) to unify Georgia and Abkhazia again, with Abkhazia rejecting such proposal.

South Ossetia, while another conflict region of Georgia, has in comparison to Abkhazia focused its national interests in the opposite direction. While Abkhazians directed their efforts to establishing a sovereign state, South Ossetia was interested in collaborating with their neighbors, North Ossetia, a part of the Russian Federation. South Ossetia declared independence in 1991.

However, independence was followed by armed conflict with military operations in place until agreements between the Russian Federation and Georgia were signed in 1992. The Dagomyss agreements were supposed to halt combat operations and establish a regulatory body that would help cease and regulate conflicts in these areas. On part of the Russian Federation, their condition was the inclusion of Russian troops in peacekeeping operations within the conflict zone. As in the case of Abkhazia, the official cease-fire does not guarantee a permanent solution to the causes of conflict in a region.

Despite the fact that Abkhazia and South Ossetia had to fight for their independence, the vast majority of countries do not legally recognize them as independent states. Future unification of Georgia with such countries will most likely be possible only through use of armed force. Moreover, Georgia has currently no support from its allies, which ultimately proved true with the 2008 war between Georgia and the Russian Federation, nicknamed the Olympic war. Despite assistance received by Georgia from the United States, who sought at least a diplomatic victory over Abkhazia and South Ossetia, the real winners in the struggle for national self determination and independence are precisely these states. The Russian Federation, unlike Georgia, strengthened its position in the geopolitical space of the Caspian Sea and managed to expand its sphere of influence inside the buffer zone between Europe and Asia. This situation can have various negative effects on the region, including continuous growth of separatist movements and conflicts.

From the perspective of international law, the absence of an enforced, international authority is particularly problematic, if the international community wants to create a world with globally maintained peace and security. Many times it is impossible to tell which state will go against the rules, making violations against the prohibition of war or direct aggression in between states very difficult to enforce in practice. Subsequently, the very principles of international law, particularly the principles of law of armed conflict are existent only ideologically. While states are required to comply with them, it is effectively impossible to prove their breach in international courts of law, as well as enforce a punishment from a unified international community of independent nations.

4 CONCLUSION

The concept of the right of peoples to self-determination was created by a new wave of ideological principles appearing after the experience of the First World War. The concept was a necessary platform for the process of decolonization, particularly in Africa and Asia, for the creation of the Jewish state—Israel, and for the formation of new states occurring after the collapse of the Soviet Block in 1989. Contemporarily however; the right of peoples to self-determination has been mainly unenforceable internationally, and as such, in direct opposition to the right of people to statehood. Furthermore, exercising the right of peoples to self-determination closely correlates with international law in armed conflict, given that many countries can only achieve independence using armed force.

States, the basic subjects of international relations, create the rules and principles of international law. However; such law must incorporates and enforce the right of peoples to self-determination. There is currently no authority that can easily prove the guilt to either party in armed conflict, nor enforce compliance of countries with the principles of international law. It is essential that the principles and rules of international law continue to develop, and that the capacity of institutions to enforce them becomes effective not only ideologically, but also practically.

REFERENCES

1. Baar, V. 2002. Národy na prahu 21. Století. Emancipace nebo nacionalismus? 415 pp., Ostrava: Tilia.
2. Buchanan, A. 2007. Justice, legitimacy, and self-determination: moral foundations for international law. 520pp. Oxford: Oxford University Press.
3. Bukovan, Š. 2008. Konflikt v Južnom Osetsku. Johnson, <http://forum.valka.cz/viewtopic.php/p/275926#275926>
4. Potuček, J. 2005a. Kavkaz: regionální konfliktní complex—1. část. In E-polis. <http://www.e-<polis.cz/mezinarodni-vztahy/101-kavkaz-regionalni-konfliktni-komplex-1-cast.html>
5. Potuček, J. 2005b. Kavkaz: regionální konfliktní komplex—2. část. In E-polis <http://www.e-polis.cz/mezinarodni-vztahy/102-kavkaz-regionalni-konfliktni-komplex-2-cast.html>
6. United Nations. 1966. Medzinárodný pakt o hospodárskych, sociálnych a kultúrnych právach. <http://www.amnesty.sk/wp-content/uploads/2012/01/Medzinárodný-pakt-o-hospodárskych-sociálnych-a-kultúrnych-právach.pdf>
7. United Nations. 1970. Resolution adopted by the General Assembly *[Adopted on a Report from the Sixth Committee (A/8082)]* 2625 (XXV). Declaration on Principles of International Law concerning Friendly Relations and Co-operation among States in accordance with the Charter of the United Nations.
8. United Nations. 1965. Resolution adopted by the Security Council. Resolution 216 (1965). <http://www.un.org/en/ga/search/view_doc.asp?symbol=S/RES/216(1965)>
9. United Nations. 1965. Resolution adopted by the Security Council. Resolution 217 (1965), <http://www.un.org/en/ga/search/view_doc.asp?symbol=S/RES/217(1965)>
10. United Nations. 1983. Resolution adopted by the Security Council. Resolution 787 (1983) <http://www.un.org/en/ga/search/view_doc.asp?symbol=S/RES/787(1992)>

European framework of administrative decisions control mechanism

B. Pekár
Faculty of Law, Comenius University, Bratislava, Slovakia

ABSTRACT: This paper deals with control mechanisms and their functioning in relation to the institutions of the European Union, its bodies, offices and agencies; their decisions, procedure and errors in respect of maladministration. Great emphasis is placed on two institutions that activities contribute to the control of particular European institutions. First of them is the European Ombudsman, that was established as a control mechanism by the Treaty on European Union (Maastricht Treaty), and acting on the basis of complaints from citizens and other entities of the European Union which are registered or established in the territory of the European Union. The second institution of control is the European Parliament, that also performs control activities and receives petitions from citizens and other entities under the Article 227 of the Treaty on the Functioning of the European Union. Both of these control mechanisms belong to the top institutions of the European Union which are dealing with submissions from citizens of the European Union, where particular institutions act contrary to the EU law. The Ombudsman investigates appeals on the grounds of maladministration in institutions and bodies of the European Union. At the same time any EU citizen may appeal with a petition at the European Parliament. The author stresses particular institutes representing control mechanism, doing so within their historical definition as it was in case of Ombudsman, through their initiation to the institutions of the European Union, to the use of them by citizens of the European Union and by other entities that are registered or have been established in the European Union. The paper presents a simplified view on these two institutes of extrajudicial control of the EU institutions.

1 INTRODUCTION

Currently, we encounter increase of issues of good administration at the level of authorities of the Member States or at the level of institutions and bodies of the European Union, its principles and its particular providing mechanisms. Good public administration is one of essential principles of the European administrative law and its application in the context of European administrative law and within the European administrative space has its specific features which should be noted.[1]

Right to good administration is embodied as well in the Article 41 of the Charter of Fundamental Rights of the European Union, according to which: every person has the right to have his or her affairs handled impartially, fairly and within a reasonable time by the institutions and bodies of the Union.[2]

In this article the Charter particularly classifies the following rights:

a. the right of every person to be heard, before any individual measure which would affect him or her adversely is taken;

---

[1]Pekár, B.: Európske správne právo a európsky správny priestor—analýza vybraných inštitútov, Bratislava: EUROIURIS—Európske právne centrum, 2012, page 63.
[2]Charter of Fundamental Rights of the European Union, Art. 41.

b. the right of every person to have access to his or her file, while respecting the legitimate interests of confidentiality and of professional and business secrecy;

c. the obligation of the administration to give reasons for its decisions.

In cases of violation of these rights every person has a right for compensation for damages caused by institutions or employees of the European Union when exercising their functions, while the injured party has a right to address the institutions of the Union in one of the languages of the Treaties. Party at stake shall receive a reply in the same language. In these cases we encounter mainly two institutions to which eligible persons may apply in cases of maladministration in the EU institutions, which are both of a non-judicial nature. These two institutions are the European Ombudsman and the European Parliament with its Committee on Petitions.

2 THE EUROPEAN OMBUDSMAN

The institution of Ombudsman was established for the first time in Europe, Sweden by the king Charles XII., who derived this institute from the Turkish model of Chancellor of Justice, who exercised control over the government and handled complaints addressed to him. This system existed in Islamic countries since the 7th century during the reign of the second Caliph Omar. In Europe, the institution of Ombudsman was for the first time established in 1713 under the name Office of the Swedish Chancellor of Justice. The office worked as a body of royal power, and was in charge of supervising behavior of royal officials. Such royal officials were, for example, judges, tax collectors, administrators and others who carried out their activities in the name of the king.

Nearly 100 years later, by a decision of the Constitutional Committee made in 1809, this institution become codified in the Constitution of Sweden. The origin of the word ombudsman is based on the Swedish word "umbup", which belongs to the medieval Swedish language and means the power or authority and can be translated as "proxy holder." Another analogical word that has the same base is the word "ombud", which in translation means a person who is acting in the interests of others. Since its codification in the Swedish constitution, the institution of Ombudsman has been gradually spread and established also in other countries of Europe.

The European Ombudsman was created by the Treaty on European Union (Maastricht Treaty from 1992). The first European Ombudsman, elected on 12th of July 1995 was Jacob Soderman, a former Ombudsman and Minister of Justice of Finland, who took office on 27th of September 1995. He was re-elected to his office for a second term. Unlike previous adaptations of the institution of Ombudsman, which operated only in their country, currently the Ombudsman constitutes a particular exception that its function go beyond the level of one country, while he is in his position equal to any other Ombudsman who carry out his functions in the particular EU Member State. Likewise, his activities are similar to the relationship and to the activities of "standard national" ombudsman. The main objective of his activities is primarily to monitor whether the rights that flow from the particular treaties adopted by the European Union and the Member States are being adopted. He carries out these duties on the basis of his activity and on the basis of complaints received when finding satisfactory solutions to problems, reviewing administrative behavior of the institutions of the European Union and their improvement, conducting investigations, making recommendations and finally reporting to the European Parliament.

The status and function of the Ombudsman is also defined by the Charter of Fundamental Rights of the European Union, which characterizes him, as follows:

"Any citizen of the Union and any natural or legal person residing or having its registered office in a Member State has the right to refer to the Ombudsman of the Union cases of maladministration in the activities of the Community institutions or bodies, with the exception of the Court of Justice and the General Court acting in their judicial role."[3]

---

[3]Charter of Fundamental Rights of the European Union, Art. 43.

The Ombudsman's term of office has a duration of five years and is created in election made by new Members of the European Parliament, who are elected for the same five-year period. The main role of the European Ombudsman is to investigate complaints against the institutions, bodies, offices and agencies of the Union whose actions have infringed the law of the European Union in their course of action. The names of candidates for the post are proposed exclusively by the members of the European Parliament. The election is a secret ballot, based on a two-round majority system and the election shall require an absolute majority of votes cast. The newly elected European Ombudsman swears before the Court of the European Union, the term of his office is parallel to the mandate of the European Parliament and his office may be repeatedly held. An exception is made for the Court of Justice of the European Union and the General Court whose activities are to a great extent exempted or even entirely exempted from the supervising power of the Ombudsman.

With regard to the method of his election, the Ombudsman report annually on its activities to Parliament. During the term of his office, the Ombudsman shall not take any other paid or unpaid placement. At the request of the European Parliament, the Court may call off the Ombudsman if he does not meet the conditions required for the performance of his duties or if he is guilty of a serious misconduct.[4] The European Parliament on the basis of the Commission's opinion and with consent of the Council shall adopt in form of a regulation general conditions and regulations governing the performance of the Ombudsman's duties by which he is bound.

Every citizen of the Union, who feels that his or her rights have been violated by the improper activities of the EU institutions, has the right to complain to the European Ombudsman. Eligible entities include in addition to natural persons also legal entities (companies, associations, organizations and other entities) which are registered or incorporated in a Member State of the European Union. Making a complaint about the EU institutions to the European Ombudsman shall be made by post or electronically.

It is important to note that Ombudsman's investigation, in other words, his control activities, does not begin solely on the basis of the complaint, but he may take action also on his own initiative.

The Lisbon Treaty has broadened and strengthened the mandate of the European Ombudsman. Due to this extension, the European Ombudsman can deal also with complaints about procedures of all institutions of the European Union, and about all violations of the Charter of Fundamental Rights of the European Union made by particular institutions such as the European Commission, the EU Council and the European Parliament, the European Court of Auditors, the European Economic and Social Committee, EU Committee of the Regions, the European Investment Bank, European Central Bank, the European Personnel Selection Office (EPSO), the European Anti-Fraud Office (OLAF), the European Police Office (Europol) and decentralized agencies. However, while exercising their jurisdiction, the European Court of Justice, together with the General Court and the European Union Civil Service Tribunal are excluded from the scope of Ombudsman's competences. In all other cases, the European Ombudsman is authorized to investigate and to investigate the submitted complaints.

On the other hand, the European Ombudsman is not authorized to investigate complaints against national, regional and local authorities of the Member States, even in that case, if such investigation concerns matters of the European Union. Particular examples are ministries, state agencies and bodies of a local governments, courts or ombudsmen. Further, his control power does not concern businesses and private individuals.

If a complaint is filed, the European Ombudsman firstly has to examine whether the complaint is justified. After examining whether the case is well-founded, he will begin an investigation. During the investigation, the institutions of the European Union are obliged to provide information and make available relevant documents. In this case, the Member States

---

[4]Treaty on the Functioning of the European Union, Art. 228.

similarly must provide information that may assist in the investigation of cases of maladministration in the EU institutions.

The European Ombudsman in its investigation mainly deals with the cases of maladministration, poor or failure of administration, cases where an institution fails to comply with a law, does not respect the principles of good administration, or violates human rights. The most common reasons for initiation of investigation are contractual disputes, problems related to public procurement, refusal of access to documents, delay, violation of fundamental rights, administrative errors, discrimination, abuse of power by public officials, refusal to provide information, unnecessary delays in process and others.

When the investigation of a complaint or proceeding on Ombudsman's own initiative comes to the conclusion that shows maladministration, the Ombudsman is required to find an acceptable, mutually beneficial solution with the institution or body of the European Union in order to eliminate the problem and thus to resolve the complaint lodged.

In addressing a particular complaint, the Ombudsman is entitled to inform the institution of the European Union subject to the complaint and ask it to solve the problem, or Ombudsman may as well also attempt to mediate the dispute in order to resolve the complaint.

In the case when the amicable resolution of the problem is not possible, whether for the reason of inadequacy with regard to the case, or because of the refusal of such treatment by the institutions of the European Union, the European Ombudsman may close the case with a critical note or leave the case open and subsequently submit a specific proposal.

In the case when that specific proposal has not been satisfied, or another proposal which could solve maladministration has not been submitted by the institutions of the European Union within three months, during which it shall deliver an opinion on the matter, the Ombudsman may submit a special report to the European Parliament or he may criticize in his annual report the institution that erred. Likewise, the Ombudsman may close a particular case by adding the negative comments about the outcome of the case. In the case when the subject matter has been met, the Ombudsman closes the case. The European Ombudsman notifies a person who lodged the complaint about the results of the investigation.

In the case when a complaint against an authority of the Member state, or regional or local government of a Member State is lodged, the European Ombudsman forwards the complaint to a member of the European Network of Ombudsmen or submits the complainant to this network.

In practice, in some cases it may happen that the European Ombudsman does not find any maladministration during his investigation and thus closes the investigation. Nevertheless, such a decision is not necessarily a negative result, but can also be a beneficial, especially as a form of independent and impartial analysis of the case. As a particular example, we can mention the case 2062/2010/JF. In this case an Irish citizen domiciled in the Netherlands became ill and his doctor prescribed him a medicine containing the substance of cannabis. After some time, the complainant wanted to get permission to visit and go back to Ireland. However, the Irish authorities reject to issue permission to visit because after entering the country he would be arrested for the possession of illegal drugs.

He addressed this problem to the Commission. The Commission, having examined the issue, did not find any breach of the European Union law on the side of Ireland. Therefore, he turned to the European Ombudsman with a complaint against maladministration done by the Commission. In this case, the European Ombudsman did not find any maladministration either. The reason was primarily the fact that the entire case concerned provisions of the Schengen acquis which were not generally binding for Ireland at that time and this state could apply its own laws to govern the unlawful possession of drugs. Despite the fact that the Ombudsman has not found any violation of right administration and the whole case has been closed, he turned to his colleague, an Irish Ombudsman and also to the President of the Irish Human Rights Commission and informed them about the case and invite them to take steps which they consider to be effective.

In the proceedings before the European Ombudsman it is not required to be represented by an advocate as it is necessary before the Court of Justice of the European Union. It is sufficient to describe in the complaint facts alone, without any legal classification of the case.

Referring to the previous, complainants who are represented by a lawyer in proceedings before the European Ombudsman carry all the expenses by themselves and thus bear the costs of such legal representation. Clear similarity can be seen in decisions of the General Court confirmed by the Court of Justice, for example in the decision from 28th of June 2007 in Case C-331/05 P Interanationaler Hilfsfonds v Commission of the European Communities. In this case the Court ruled that costs associated with the Ombudsman, should be distinguished from those incurred in the proceedings. On this basis, it is not possible to claim a compensation for the costs of legal representation in proceedings before the Ombudsman.

European Network of Ombudsmen was founded in 1996 and consists of national or regional ombudsmen. This platform serves as an effective mechanism for cooperation in dealing with individual complaints from citizens and bodies of the European Union.

Importantly, the decisions of the Ombudsman are not legally binding and does not create enforceable rights for the complainant. Therefore, it is precisely up to the institution of the European Union that has committed maladministration, in its own initiative, to eliminate the error incurred and to avoid it in the future. From the above-stated facts, one might conclude that the European Ombudsman does not have sufficient powers and could be considered as unnecessary. But the truth is that the decisions and comments of the Ombudsman are greatly accepted by the various institutions of the European Union which are guilty of such misconduct.

These decisions of the Ombudsman in the overall scale does not only deal with specific cases, but provide a means of influencing the activities of other institutions of the European Union in the exercise of its activities through internal processes and methods of proceedings in other cases.

Arising from the assumption that the Ombudsman gives decisions on cases, they are of no binding nature, whatsoever. To state the obvious, it was confirmed in judgment of the Court of Justice of the European Union in its finding from 25th of June 2009 in Case C-580/08 P Devrajan Srinivasan against the European Ombudsman, which follows the finding of the General Court from 3rd of November 2008 on the same subject matter under the case number T-196/08. In this finding, the Court expressed the opinion that the Ombudsman has no power to adopt binding measures, and on that basis his decisions do not have any legal effects on third parties. In the next part the Court concludes that the special reports which Ombudsman drafts and addresses to the European Parliament represent only the fact that he had found a maladministration in the activities of the institutions of the European Union, but the nature of the report is not binding for Parliament. The European Parliament on the basis of this report may decide freely how to proceed further and which steps it will make. Above stated fact also applies for the Annual Reports which are presented to the Parliament.

Since it's establishment, the European Ombudsman has undergone an extensive changes. As noted, through his activities the Ombudsman has built a strong position among the institutions and bodies of the European Union, whereas he represents a non-judicial way of dealing with maladministration and violations of individual rights guaranteed in Treaties adopted by the European Union and the Member States.

3 PETITION TO THE EUROPEAN PARLIAMENT

Another means by which the citizens and other entities of the European Union can seek protection of their rights violated by institutions and bodies of the European Union or Member State's authorities and regional authorities. A petition can be submitted no matter whether it is maladministration, or even violation of fundamental rights guaranteed by the Treaties, right to petition is guaranteed. Right of petition to the European Parliament is one of the fundamental rights of European citizens and other entities of the EU. The petition can be filed in writing or electronically. Each petition must contain the name, nationality, permanent address of the petitioner and, if it is signed by more natural or legal persons, identification data of their common representative or substitute are also required.

Right to file a petition to the European Parliament is based on Article 227 of the Treaty on the Functioning of the European Union and it is for every citizen of the European Union or a person residing in one of the countries of the European Union. The petition can be filled not only by a citizen of the European Union himself, but it can also be filled together with other persons and thereby at any time exercise their right enshrined in the law of the European Union. Among the entities that are entitled to file a petition to the European Parliament under Article 227 of the Treaty on the Functioning of the European Union, except of the citizens of the European Union, also companies, organizations or associations with its headquarters in the European Union. The right of petition is also governed by the Charter of Fundamental Rights of the European Union, in particular in the Article 44, as follows:

> Any citizen of the Union and any natural or legal person residing or having its regis-
> tered office in a Member State has the right to petition the European Parliament.[5]

In the course of its duties, the European Parliament may, at the request of a quarter of its component Members, set up a temporary Committee of Inquiry to investigate, without prejudice to the powers conferred by the Treaties on other institutions or bodies, alleged contraventions or maladministration in the implementation of Union law, except where the alleged facts are being examined before a court and while the case is still subject to legal proceedings.[6] Temporary Committee of Inquiry completes its work by submission of a report. The enforcement of the law to investigate is strictly regulated by regulations that shall be subject to approval by the Council and the Commission.

Subject matter of the petition shall always comprise of a substance that falls within the activities of the European Union, which directly relates to the citizen and the persons who filed this petition. The areas of EU activities in which it is possible to file a petition include especially: rights of citizens of the European Union set out in the Treaties; environmental issues; consumer protection; free movement of persons, goods and services and the internal market; issues related to employment and social policy; recognition of professional qualifications; and other matters related to the exercise of the rights guaranteed by the European Union in the Treaties.

Treaty on the Functioning of the European Union recognizes two forms of petitions, namely the form of complaint and request, and may relate to issues of public or private interest. In view of all the above, the petition may constitute an individual application, complaint or comment related to the application of European Union law, or it may contain an invitation to Parliament to give its opinion on the specific case that was covered in the petition filed.

The possibility to petition the European Parliament represents a particular control mechanism through which the Parliament draw attention to any breach of the guaranteed rights of the European citizen or entity from the European Union, which has been made by institutions, Member States and local authorities.

After the petition is referred to the Committee on Petitions, this Committee will assess whether its subject matter relates to the field of activities of the European Union. If the Committee on Petitions gives a positive decision and declares the petition as admissible, then the Committee, in accordance with the Order, decides on the further steps to be taken.

Committee on Petitions may, when dealing with the petition, follow the bellow stated procedure:

Ask the European Commission to carry out a preliminary survey, request information concerning compliance with the relevant Community legislation or contact SOLVIT;

Forward the petition to other committees of the European Parliament for the purpose of informing or taking further steps;

In a special cases may prepare and submit a report to Parliament, which is determined for a voting in plenary, or organize a fact-finding visit to the relevant Member State, region and on this base elaborate a report with its findings and recommendations for the Committee;

---

[5]Charter of Fundamental Rights of the European Union, Art. 44.
[6]Treaty on the Functioning of the European Union, Art. 226.

Similarly, it may also adopt any further action if deemed appropriate, if the solution of the case requires so or provide an appropriate response to the petitioner.

If a petition does not fall within the field of activities of the European Union, the Committee on Petitions will assess it as inadmissible. A reason may be mainly the fact that the subject matter of the petition falls within the competence of the Member States. In this case, the petition is only registered, and the Committee on Petitions does not take any further action. With regard to the subject matter of petition the Committee on Petitions may recommend to the entitled person to apply to another authority, which is not an institution of the European Union or to a national authority.

As it is in the case of the Ombudsman, also the Committee on Petitions cannot change or annul the final decisions of the bodies and institutions of the European Union or the Member States, respectively to interfere in their decision-making process.

In practice, it is possible to see a case where a petition is filed by a person who is not a citizen of the European Union, is not domiciled in a Member State nor established there. In this case, the petition is registered and archived in a special way.

4 CONCLUSIONS

Finally, it can be said that both the European Ombudsman and the Committee on Petitions of the European Parliament represent the means of control mechanism of institutions and bodies of the European Union. They start their activity on the basis of complaints and petitions from citizens of the European Union and other entities established or registered in the European Union. Both institutions are considered as the means of out-of-court dispute resolution which arise from the nature of the activities of the institutions of the European Union in cases where there has been maladministration or breach of individual rights enshrined in the Treaties of the European Union.

Their activity as a control mechanism contributes to improving the administrative operation of the institutions, the elimination of cases of maladministration and respect for the rights set out in the Treaties and regulations adopted by the European Union.

ACKNOWLEDGEMENTS

This article was written with the support of grant awarded by the Agency for Support of Research and Development, No. APW-0448-10 and is part of the research project.

REFERENCES

Charter of Fundamental Rights of the European Union.
Craig, P. 2006. *EU Administrative Law*. Oxford: University press.
http://www.ombudsman.europa.eu/
http://www.euroinfo.gov.sk/
http://www.europarl.europa.eu/
Pekár, B. 2012. *Európske správne právo a európsky správny priestor [Analýza vybraných inštitútov]*. Bratislava: EUROIURIS—Európske právne centrum.
Treaty on the Funcioning of the European Union.

Current Issues of Science and Research in the Global World – Kunova & Dolinsky (Eds)
© *2015 Taylor & Francis Group, London, ISBN 978-1-138-02739-8*

Co-existence of various instruments protecting fundamental rights in European arrest warrant proceedings: Court of justice of the European Union case of *Melloni* (C-399/11)

Libor Klimek
Faculty of Law, Criminology Research Centre, Pan-European University, Bratislava, Slovakia

ABSTRACT: The Framework Decision 2002/584/JHA on the European arrest warrant and the surrender procedures between Member States respects fundamental rights and observes the principles recognised by the Treaty on European Union and reflected in the Charter of Fundamental Rights of the European Union. However, it does not deal with the co-existence of various instruments protecting fundamental rights. The conference contribution deals with the Court of Justice of the European Union Case of *Melloni* (C-399/11), which gives answer. At the outset the contribution emphasises the subject matter of the analysed case. As a starting point for further analysis, it presents the dispute in the proceedings. As a consequence, there are introduced questions referred to the Court of Justice. Further, the contribution examines the opinion of the Court of Justice and the view of Advocate General. At the end, the contribution is concluded by the Court's rulings.

1 INTRODUCTION

In the case of the Court of Justice of the European Union of *Melloni* (C-399/11); the request for a preliminary ruling concerned the interpretation and the validity of Article 4a(1) of the Framework Decision 2002/584/JHA on the European arrest warrant and the surrender procedures between Member States (hereinafter 'Framework Decision on the EAW; Official Journal of the European Communities, L 190/1 of 18.7.2002) as amended by the Framework Decision 2009/299/JHA [...] on grounds of infringement of the fundamental rights of the person concerned guaranteed by the national constitution.

Moreover, the Court of Justice of the European Union was asked to define, for the first time, the scope of Article 53 of the Charter of Fundamental Rights of the European Union (hereinafter 'EU Charter').

The request was made in proceedings between Mr. Melloni and the Italian Ministry of Finance (*Ministerio Fiscal*) concerning the execution of an EAW issued by the Italian authorities for the execution of a prison sentence handed down by judgment *in absentia* against Mr. Melloni.

2 DISPUTE IN THE MAIN PROCEEDINGS

The Spanish High Court (*Sala de lo Penal of the Audiencia Nacional*) authorised the extradition to Italy of Mr. Melloni, in order for him to be tried there in relation to the facts set out in arrest warrants issued in May and June 1993. After being released on bail Mr. Melloni fled, so that he could not be surrendered to the Italian authorities.

The Italian court (*Tribunale di Ferrara*) declared that Mr. Melloni had failed to make appearance in court and directed that notice should in future be given to the lawyers who had been chosen and appointed by him. By judgment Mr. Melloni was sentenced *in absentia* to 10 years' imprisonment for bankruptcy fraud. By judgment of June 2004, the Italian Supreme

Court (*Corte suprema di cassazione*) dismissed the appeal lodged by Mr. Melloni's lawyers. In June 2004, the Italian Public Prosecutor's Office (*Procura Generale della Repubblica*) issued an EAW for execution of the sentence imposed by the Italian court.

Mr. Melloni was arrested by the Spanish police in August 2008. He opposed surrender to the Italian authorities, contending, first, that at the appeal stage he had appointed another lawyer, revoking the appointment of the two previous lawyers, despite which notice was still being given to them. Second, he contended that under Italian procedural law it is impossible to appeal against sentences imposed *in absentia*, for which reason the execution of the EAW should, where appropriate, be made conditional upon Italy's guaranteeing the possibility of appealing against that judgment.

In September 2008, the Spanish High Court authorised surrender of Mr. Melloni to the Italian authorities in order to serve the sentence imposed upon him by the Italian court as perpetrator of a bankruptcy fraud. It considered that it was not proved that the lawyers appointed by Mr. Melloni had ceased to represent him as from 2001, and that his rights of defence had been respected, since he had been aware from the outset of the forthcoming trial, deliberately absented himself and appointed two lawyers to represent and defend him, who had acted in that capacity at first instance and in the appeal and cassation proceedings, thus exhausting all remedies.

Mr. Melloni filed a petition for constitutional protection before the Spanish Constitutional Court (*Tribunal Constitucional*). In his submission, the very essence of a fair trial had been vitiated in such a way as to undermine human dignity, as a result of allowing surrender to countries which, in the event of very serious offences, validate findings of guilt made *in absentia*, without making surrender subject to the condition that the convicted party is able to challenge them in order to safeguard his rights of defence.

According to the Spanish Constitutional Court the difficulty arose from the fact that the Framework Decision 2009/299/JHA [...] repealed Article 5(1) of the Framework Decision on the EAW and introduced therein a new Article 4a. Article 4a precludes a refusal to execute the EAW issued for the purpose of executing a custodial sentence or a detention order if the person did not appear in person at the trial resulting in the decision where the person concerned, being aware of the scheduled trial, had given a mandate to a legal counsellor, who was either appointed by the person concerned or by the State, to defend him or her at the trial, and was indeed defended by that counsellor at the trial. The national court pointed out that it was established that Mr. Melloni had appointed two trusted lawyers, whom the Italian Curt notified of the forthcoming trial, so that he was aware of it. It was also established that Mr. Melloni was actually defended by those two lawyers at the ensuing trial at first instance and also in the subsequent appeal and cassation proceedings.

3 QUESTIONS REFERRED FOR A PRELIMINARY RULING

For the Spanish Constitutional Court, the question therefore arose whether the Framework Decision on the EAW precluded the Spanish courts from making surrender of Mr. Melloni conditional on the right to have the conviction in question reviewed. In the light of those considerations, it decided to stay the proceedings and to refer the following questions to the Court of Justice for a preliminary ruling:

> "*Must Article 4a(1) of the Framework Decision on the 2002/584/JHA* [i.e. the Framework Decision on the EAW], *as inserted by the Framework Decision 2009/299/JHA, be interpreted as precluding national judicial authorities, in the circumstances specified in that provision, from making the execution of an EAW conditional upon the conviction in question being open to review, in order to guarantee the rights of defence of the person requested under the warrant?*
>
> *In the event of the first question being answered in the affirmative, is Article 4a(1) of the Framework Decision 2002/584/JHA [i.e. the Framework Decision on the EAW] compatible with the requirements deriving from the right to an effective judicial remedy*

and to a fair trial, provided for in Article 47 of the Charter of Fundamental Rights of the European Union, and from the rights of defence guaranteed under Article 48(2) of the Charter?

In the event of the second question being answered in the affirmative, does Article 53, interpreted systematically in conjunction with the rights recognised under Articles 47 and 48 of the Charter, allow a Member State to make the surrender of a person convicted in absentia conditional upon the conviction being open to review in the requesting State, thus affording those rights a greater level of protection than that deriving from EU law, in order to avoid an interpretation which restricts or adversely affects a fundamental right recognised by the Constitution of the first-mentioned Member State?" (emphasis added).

By its first question, the Spanish Constitutional Court wished to know, in essence, whether Article 4a(1)(a) and (b) of the Framework Decision on the EAW as amended by the Framework Decision 2009/299/JHA [...] is to be interpreted as precluding the executing judicial authority, in the circumstances specified in that provision, from making the execution of an EAW conditional upon the person who is the subject of the warrant being able to apply for a retrial in the issuing Member State.

By its second question, the Spanish Constitutional Court asked the Court of Justice to rule whether Article 4a(1) of the Framework Decision on the EAW as amended by the Framework Decision 2009/299/JHA [...] is compatible with the requirements deriving from the second paragraph of Article 47 and Article 48(2) of the EU Charter.

By its third question, the Spanish Constitutional Court asked the Court of Justice, in essence, to rule whether Article 53 of the EU Charter allows an executing judicial authority, in accordance with its national constitutional law, to make the execution of an EAW subject to the condition that the person who is the subject of the warrant is entitled to a retrial in the issuing Member State, even though the application of such a condition is not authorised by Article 4a(1) of the Framework Decision on the EAW as amended by the Framework Decision 2009/299/JHA [...]. That question therefore invited the Court of Justice to define the legal substance and scope to be given to Article 53 of the EU Charter.

4   OPINIONS OF THE COURT OF JUSTICE OF THE EUROPEAN UNION

As far as the first question is concerned, the Court of Justice argued that the literal interpretation of Article 4a(1) of the Framework Decision on the EAW as amended by the Framework Decision 2009/299/JHA [...] is confirmed by an analysis of the purpose of the provision. The object of the Framework Decision 2009/299/JHA is, firstly, to repeal Article 5(1) of the Framework Decision on the EAW (original version), which, subject to certain conditions, allowed for the execution of an EAW issued for the purposes of executing a sentence rendered *in absentia* to be made conditional on there being a guarantee of a retrial of the case in the presence of the person concerned in the issuing Member State and, secondly, to replace that provision by Article 4a. That provision henceforth restricts the opportunities for refusing to execute such a warrant by setting out, as indicated in recital 6 of the Framework Decision 2009/299/JHA [...], 'conditions under which the recognition and execution of a decision rendered following a trial at which the person concerned did not appear in person should not be refused'. It follows that Article 4a(1) of the Framework Decision on the EAW as amended by the Framework Decision 2009/299/JHA [...] must be interpreted as precluding the executing judicial authorities, in the circumstances specified in that provision, from making the execution of an EAW issued for the purposes of executing a sentence conditional upon the conviction rendered *in absentia* being open to review in the issuing Member State.

As regards the second question, the Court of Justice argued that the national court asked whether Article 4a(1) of the Framework Decision on the EAW as amended by the Framework Decision 2009/299/JHA [...] is compatible with the requirements deriving from the *right to an effective judicial remedy and to a fair trial*, provided for in Article 47 of the EU Charter and from the rights of the defence guaranteed under Article 48(2) of the EU Charter. Regarding the scope of the right to an effective judicial remedy and to a fair trial provided for in Article 47 of the EU Charter, and the rights of the defence guaranteed by Article 48(2) thereof, it should be observed that that right is not absolute. The accused may waive that right of his own free will, either expressly or tacitly, provided that the waiver is established in an unequivocal manner, is attended by minimum safeguards commensurate to its importance and does not run counter to any important public interest. In particular, violation of the right to a fair trial has not been established, even where the accused did not appear in person, if he was informed of the date and place of the trial or was defended by a legal counsellor to whom he had given a mandate to do so. This interpretation of Articles 47 and 48(2) of the EU Charter is in keeping with the scope that has been recognised for the rights guaranteed by Article 6(1) and (3) of the Convention for the Protection of Human Rights and Fundamental Freedoms by the case-law of the European Court of Human Rights.

Furthermore, as indicated by Article 1 of the Framework Decision 2009/299/JHA, the objective of the harmonisation of the conditions of execution of EAWs issued for the purposes of executing decisions rendered at the end of trials at which the person concerned has not appeared in person, effected by that framework decision, is to enhance the procedural rights of persons subject to criminal proceedings whilst improving mutual recognition of judicial decisions between Member States. Accordingly, Article 4a(1)(a) and (b) of the Framework Decision on the EAW as amended by the Framework Decision 2009/299/JHA [...] lays down the circumstances in which the person concerned must be deemed to have waived, voluntarily and unambiguously, his right to be present at his trial, with the result that the execution of an EAW issued for the purposes of executing the sentence of a person convicted *in absentia* cannot be made subject to the condition that that person may claim the benefit of a retrial at which he is present in the issuing Member State. This is so either where the person did not appear in person at the trial despite having been summoned in person or officially informed of the scheduled date and place of the trial or, as referred to in Article 4a(1)(b), the person, being aware of the scheduled trial, deliberately chose to be represented by a legal counsellor instead of appearing in person. Article 4a(1)(c) and (d) refers to circumstances where the executing judicial authority is required to execute the EAW, even though the person concerned is entitled to a retrial, because the arrest warrant states that the person concerned either did not ask for a retrial or that he will be expressly informed of his right to a retrial. Article 4a(1) of the Framework Decision on the EAW as amended by the Framework Decision 2009/299/JHA [...] does not disregard either the right to an effective judicial remedy and to a fair trial or the rights of the defence guaranteed by Articles 47 and 48(2) of the EU Charter respectively.

In case of the third question, the Court of Justice argued that it is true that Article 53 of the EU Charter confirms that, where an EU legal act calls for national implementing measures, national authorities and courts remain free to apply national standards of protection of fundamental rights, provided that the level of protection provided for by the EU Charter, as interpreted by the Court, and the primacy, unity and effectiveness of EU law are not thereby compromised.

It should also be borne in mind that the adoption of the Framework Decision 2009/299/JHA, which inserted that provision into the Framework Decision on the EAW, was intended to remedy the difficulties associated with the mutual recognition of decisions rendered in the absence of the person concerned at his trial arising from the differences as among the Member States in the protection of fundamental rights. Consequently, allowing a Member State to avail itself of Article 53 of the EU Charter to make the surrender of a person convicted *in absentia* conditional upon the conviction being open to review in the issuing Member State, a possibility not provided for under the Framework Decision 2009/299/JHA, in order to avoid an adverse effect on the right to a fair trial and the rights of the defence guaranteed by the constitution of the executing Member State, would undermine the principles of mutual trust and recognition which that decision purports to uphold and would, therefore, compromise the efficacy of that framework decision.

Advocate General *Bot* in his opinion argued, as regards the first question, that a reading of Article 4a(1)(a) and (b) of the Framework Decision on the EAW as amended by the Framework Decision 2009/299/JHA [...] reveals that the wording of those two points makes no mention of the requirement that the person concerned must, in those circumstances, be able to apply for a retrial in the issuing Member State. An examination of all the provisions of Article 4a(1) of the Framework Decision shows that the situations referred to in points (c) and (d) of that provision, which constitute the second category, are in fact the only cases where the person concerned may be entitled to a retrial.

Article 4a(1)(c) and (d) of the Framework Decision removes the discretion of the executing judicial authority, which must rely on the information contained in the EAW. The executing judicial authority is therefore required to execute it where this states, in essence, either that the person concerned, after being served with the decision and being expressly informed about the right to a retrial, expressly stated that he or she did not contest the decision or did not request a retrial within the applicable time frame, or that the person concerned was not personally served with the decision but will be personally served with it without delay after the surrender and will be expressly informed of the right to a retrial and of the time frame within which he or she has to request such a retrial. In points (a) and (b) of Article 4a(1) of the Framework Decision, the EU legislature confirmed that, if the person concerned was aware of the scheduled trial and was informed that a decision might be handed down if he or she did not appear for the trial, or if, being aware of the scheduled trial, he or she had given a mandate to a legal counsellor to defend him or her, that person must be regarded as having waived his or her right to appear at the trial, so that he or she could not invoke a right to a retrial.

By adopting the Framework Decision 2009/299/JHA [...], the EU legislature intended to remedy the defects in the scheme laid down in Article 5(1) of the Framework Decision on the EAW (original version) and to perfect it, so as to achieve a better balance between the objective of enhancing the procedural rights of persons subject to criminal proceedings and the objective of facilitating judicial co-operation in criminal matters, in particular by improving mutual recognition of judicial decisions between Member States.

Faced with those uncertainties which might reduce the effectiveness of the mechanism for mutual recognition of judicial decisions rendered *in absentia*, in the opinion of *Bot* the EU legislature considered that it was necessary 'to provide clear and common grounds for non-recognition of decisions rendered following a trial at which the person concerned did not appear in person'. The Framework Decision 2009/299/JHA [...] was aimed, therefore, 'at refining the definition of such common grounds, allowing the executing authority to execute the decision despite the absence of the person at the trial, while fully respecting the person's right of defence'. By removing the possibility of conditional surrender provided for in Article 5(1) of the Framework Decision on the EAW (original version), the EU legislature wished to improve mutual recognition of judicial decisions rendered *in absentia* while enhancing a person's procedural rights. The solution which it found, consisting in providing an exhaustive list of the circumstances in which the execution of an EAW issued in order to enforce a decision rendered *in absentia* must be regarded as not infringing the rights of the defence, is incompatible with any retention of the possibility for the executing judicial authority to make that execution conditional on the conviction in question being open to review in order to guarantee the rights of defence of the person concerned.

As far as the second question is concerned, Advocate General *Bot* argued that the second paragraph of Article 47 of the EU Charter corresponds to Article 6(1) of the European Convention, and Article 48(2) of the EU Charter corresponds more particularly to Article 6(3) of the European Convention. Under Article 52(3) of the EU Charter, in so far as the EU Charter contains rights which correspond to rights guaranteed by the European Convention, the meaning and scope of those rights are to be the same as those laid down by the said convention. However, that provision does not preclude EU law from providing more extensive protection. Advocate General *Bot* considered that the level of protection provided by the EU legislature is adequate and appropriate for achieving the aforementioned objectives and that observance of the second paragraph of Article 47 and Article 48(2) of the EU Charter did not require it to opt for a more extensive protection of the right to a fair trial and the rights of the defence, for example by making the right to a retrial an absolute requirement irrespective of the conduct of the person concerned. Apart from the fact that he did not discern reasons for going further than the balanced attitude taken by the European Court of Human Rights, the Court of Justice could not rely on the constitutional traditions common to the Member States in order to apply a higher level of protection. In his view, the validity of Article 4a(1) of the Framework Decision on the EAW as amended by the Framework Decision 2009/299/JHA [...] was not called into question by the second paragraph of Article 47 or by Article 48(2) of the EU Charter.

In case of the third question, in view of *Bot*, Article 53 of the EU Charter is not to be regarded as a clause designed to regulate a conflict between, on the one hand, a provision of secondary law which, interpreted in the light of the EU Charter, sets a given level of protection for a fundamental right and, on the other hand, a provision drawn from a national constitution which provides a higher level of protection for the same fundamental right. In such a situation, that article has neither the objective nor the effect of giving priority to the more protective rule deriving from a national constitution. It is by no means apparent from the wording of Article 53 of the EU Charter that it is to be considered as establishing an exception to the principle of the primacy of EU law. On the contrary, it may be argued that the words 'in their respective fields of application' were chosen by the drafters of the EU Charter in order not to infringe that principle.

In order to reconcile those objectives, the EU legislature set the level of protection for the fundamental rights in question so as not to compromise the effectiveness of the mechanism of the EAW. An interpretation of Article 53 of the EU Charter which would allow an executing judicial authority, in accordance with a national constitutional rule, generally to make the execution of an EAW issued for the purposes of executing a judgment rendered *in absentia* subject to the condition that the person subject to the warrant be entitled to a retrial in the issuing Member State would upset the balance thus achieved by Article 4a of the Framework Decision on the EAW as amended by the Framework Decision 2009/299/JHA [...] and cannot therefore, be allowed.

The EU Charter is not an isolated instrument, unconnected with the other sources of protection of fundamental rights. It itself provides that its provisions must be interpreted taking due account of other legal sources, whether national or international. Accordingly, Article 52(3) of the EU Charter makes the European Convention a minimum standard below which EU law cannot fall and Article 52(4) of the EU Charter provides that, in so far as the EU Charter recognises fundamental rights as they result from the constitutional traditions common to the Member States, those rights must be interpreted in harmony with those traditions. The EU Charter thus cannot have the effect of requiring Member States to lower the level of protection of fundamental rights guaranteed by their national constitution in cases which fall outside the scope of EU law. Article 53 of the EU Charter also expresses the idea that the adoption of the EU Charter should not serve as a pretext for a Member State to reduce the protection of fundamental rights in the field of application of national law.

6 CONCLUSION (RULINGS)

As far as analysed questions are concerned, in conclusion of the case of *Melloni* the Court of Justice ruled:

> "*Article 4a(1) of the Framework Decision* [on the EAW] *as amended by the Framework Decision 2009/299/JHA of 26 February 2009, must be interpreted as precluding the executing judicial authorities, in the circumstances specified in that provision, from making the execution of an EAW issued for the purposes of executing a sentence conditional upon the conviction rendered in absentia being open to review in the issuing Member State.*

Article 4a(1) of the Framework Decision [on the EAW] as amended by Framework Decision 2009/299/JHA is compatible with the requirements under Articles 47 and 48(2) of the Charter of Fundamental Rights of the European Union.

Article 53 of the Charter of Fundamental Rights of the European Union must be interpreted as not allowing a Member State to make the surrender of a person convicted in absentia conditional upon the conviction being open to review in the issuing Member State, in order to avoid an adverse effect on the right to a fair trial and the rights of the defence guaranteed by its constitution." (emphasis added).

REFERENCES

Framework Decision 2002/584/JHA of 13th June 2002 on the European arrest warrant and the surrender procedures between Member States as amended by the Framework Decision 2009/299/JHA. Official Journal of the European Communities, L 190/1 of 18.7.2002.

Judgment of the Court of Justice of the European Union of 26th February 2013—Case C-399/11—*Stefano Melloni v Ministerio Fiscal*.

Klimek, L. 2014. *European Arrest Warrant*. Cham—Heidelberg—New York—Dordrecht—London: Springer, 426 pp, ISBN 978-3-319-07337-8.

Opinion of Advocate General Bot—Case C-399/11—Criminal proceedings against Stefano Melloni.

Reference for a preliminary ruling from the Tribunal Constitucional, Madrid (Spain) lodged on 28th July 2011—Criminal proceedings against Stefano Melloni—other party: Ministerio Fiscal (Case C-399/11).

Current Issues of Science and Research in the Global World – Kunova & Dolinsky (Eds)
© 2015 Taylor & Francis Group, London, ISBN 978-1-138-02739-8

Extraordinary appeal and principle of legal certainty

M. Siman
Faculty of Law, Pan-European University, Bratislava, Slovakia

ABSTRACT: The extraordinary appeal that can be lodged under the provisions of Civil Procedure Code by the Prosecutor General of the Slovak Republic before the Supreme Court of the Slovak Republic raises questions of compatibility with the principle of legal certainty. In accordance with the settled case-law of the European Court of Human Rights, the only reason for quashing the final and enforceable judicial decision can be a substantial procedural irregularity. The present article concerns the assessment of compatibility of the extraordinary appeal regulated by the Civil Procedure Code with the principle of legal certainty covered by the Article 6 of the Convention for the Protection of Human Rights and Fundamental Freedoms.

1 INTRODUCTION

The civil proceeding is based on several principles that are traditionally considered part of civil procedural law. The principle of equality of arms is also one of the main prerequisites of a fair trial under Article 6 of the Convention for the Protection of Human Rights and Fundamental Freedoms (hereafter referred to as "Convention"). However, is the extraordinary appeal itself in accordance with the right to a fair trial?

The Convention and the case-law of the European Court of Human Rights (hereafter referred to as "ECHR") represent binding legal sources for national law-applying authorities which must be observed by them when interpreting and applying national law. In the Slovak legal order, the binding force of such sources can be derived from both the Constitution of the Slovak Republic itself (Article 154c, paragraph 1, in conjunction with Article 144) and from the procedural codes, i.e. Civil Procedure Code (hereafter referred to as "CPC") and Criminal Procedure Code.

The right to a fair trial under Article 6, paragraph 1, of the Convention implies, inter alia, equality of the parties and effective contradictory hearing of the matter so that both parties should have the opportunity to present their case to an independent and impartial tribunal under conditions which do not give preferential treatment to one of the parties. In order to preserve a "fair balance" between the parties, it is required in this respect that each party have a reasonable opportunity to present his/her matter under conditions which are apparently not disadvantageous comparing to the counterparty, in particular as regards the procedural position of the parties, the manner and extent of evidence and other procedural conditions.

In accordance with the principle of disposition, the party shall have the opportunity to influence the subject and course of proceedings through procedural acts and means. Within this principle, the rule applied is that any proposals to the court shall be submitted in particular by the parties and not by other persons, which shall be applied all the more in the case of private disputes. Precisely the equal status of both parties in the civil proceedings makes this proceeding different from other types of proceedings (especially administrative and criminal one).

In this context, it is important to determine whether the extraordinary appeal proceedings in the light of the right to a fair trial under Article 6, paragraph 1, of the Convention respects the rule of law, including the principle of legal certainty, which implies the respect of the principle of res judicata.[1]

2 RELEVANT PROVISIONS

2.1 *Relevant Slovak legislation*

Under the Article 1 CPC, the Civil Procedure Code lays down the procedures to be applied by the courts and to be followed by the parties in civil proceedings, with a view to ensuring fair and just protection of the rights and legitimate interests of the parties, and promoting compliance with the statutory law, fulfilment of duties and respect for the rights of fellow citizens.

In accordance with Article 101 § 1 CPC, the parties to the proceedings are liable to contribute to the purpose of the proceedings, in particular by giving truthful and complete description of all relevant facts, adducing evidence, and abiding by the instructions of the court.

Under Article 159 § 3 CPC, as soon as a matter has been resolved by force of a final and binding decision or a judgment, it may not give rise to new proceedings.

However, the Prosecutor General has the power to challenge a decision of a court by means of an extraordinary appeal in points of law. He may do so upon a petition of a party to the proceedings or another person concerned or injured by the decision, provided that he concludes that the final and binding decision violated the law; provided that the protection of the rights and legitimate interests of individuals, legal entities, or the State so requires; provided that protection cannot be achieved by other means; and provided that the matter at hand is not excluded from review (Articles 243e § 1 and 243f § 2 CPC). Unless provided otherwise, the Prosecutor General is bound by the scope of the petition for an extraordinary appeal (Article 243e §§ 3 and 4 CPC). Further conditions of admissibility of an extraordinary appeal are listed in Article 243f § 1 CPC: they comprise major procedural flaws within the meaning of Article 237 CPC (in that respect see, for example, Ringier Axel Springer v. Slovakia[2]), other errors of procedure resulting in an erroneous decision on the merits, and wrongful assessment of points of law.

An extraordinary appeal is to be lodged with the Supreme Court within one year of the contested judicial decision's becoming final and binding (Article 243 g CPC).

If the Prosecutor General concludes, upon a petition of a party to the proceedings or another person concerned or injured by the impugned decision, that there is the risk of considerable economic damage or other serious irreparable consequence, the extraordinary appeal may be filed even without reasons. The reasons then must be supplied within 60 days of the lodging of the extraordinary appeal with the Supreme Court, failing which the proceedings are to be discontinued (Article 243h §§ 3 and 4 CPC).

If the extraordinary appeal on points of law is accompanied by a request that the enforceability of the contested decision be suspended, its enforceability is to be suspended at the moment when the extraordinary appeal is lodged with the Supreme Court (Article 243ha § 1 CPC). The duration of the effect of such a suspension is then regulated by Article 243ha § 2 CPC, pursuant to which that effect ceases when the request is dismissed or with the decision on the extraordinary appeal, but unless extended by the Supreme Court no later than one year from the lodging of the extraordinary appeal with the Supreme Court.

2.2 *The Venice Commission's report on the independence of the judicial system*

The report adopted by the European Commission for Democracy through Law (Venice Commission) at its 82nd Plenary Session on 12–13 March 2010 refers, in its section III (9) entitled "Final character of judicial decisions", to Principle I(2)(a)(i) of Recommendation No. R (94) 12 of the Committee of Ministers of the Council of Europe on the Independence, Efficiency and Role of Judges, which provides that "decisions of judges should not be the subject of any revision outside the appeals procedures as provided for by law".

The relevant part of the report continues that: "It should be understood that this principle does not preclude the re-opening of procedures in exceptional cases on the basis of new facts or on other grounds as provided for by law. While the [Consultative Council of European Judges] concludes in its Opinion No. 1 (at 65), on the basis of the replies to its questionnaire, that this principle seems to be generally observed, the experience of the Venice Commission and the case law of the [ECHR] indicate that the supervisory powers of the Prokuratura in post-Soviet states often extend to being able to protest judicial decisions no longer subject to an appeal. The Venice Commission underlines the principle that judicial decisions should not be subject to any revision outside the appeals process, in particular not through a protest of the prosecutor or any other state body outside the time limit for an appeal."

3   LEGAL ASSESSMENT

The ECHR has already found a violation of the principle of legal certainty and, consequently, of the right to a fair trial under Article 6, paragraph 1, of the Convention in cases where the examined extraordinary remedy was not accessible for parties to proceedings, but was only available to the Attorney-General.[3]

In *Tripon v. Romania*[4], the Court observed that, according to settled case-law, the right to a fair trial guaranteed by the Article 6, paragraph 1, of the Convention shall be interpreted in the light of the preamble of the Convention which, inter alia, declares the rule of law to be part of the common heritage of the Contracting States. One of the fundamental aspects of the rule of law is the principle of legal certainty, which requires, among other things, that where the courts have finally determined an issue, their ruling should not be called into question.[5]

Time limit for annulment of a final court decision is not, in the ECHR opinion, sufficient to justify itself the existence of the extraordinary remedy. In *SC Maşinexportimport Industrial Group SA v. Romania*[6], the ECHR stated that regardless of the one-year time limit for filing an extraordinary appeal this legal institution interfere with the principle of legal certainty.

In *Asito v. Moldova*[7], the ECHR was dealing with a similar legal institution as extraordinary appeals, which also enabled to obtain the annulment of final and enforceable court decisions, finding that such an action constituted a breach of the principle of legal certainty. A one-year time limit for filing an extraordinary appeal under the Civil Procedure Code, which starts to run from the validity of the contested court decision, is thus not relevant to the assessment of the compatibility of an extraordinary appeal with the principle of legal certainty.

In *Sitkov v. Russia*[8], the ECHR referred to paragraph 77 of the judgment in *Sovtransavto Holding v. Ukraine* according to which *"(…) judicial systems characterised by the objection (protest) procedure and, therefore, by the risk of final judgments being set aside repeatedly (…) are, as such, incompatible with the principle of legal certainty that is one of the fundamental aspects of the rule of law for the purposes of Article 6 § 1 of the Convention".*

In this regard, the use of extraordinary remedies is in general permissible only in exceptional cases. This view is shared also by the Constitutional Court of the Slovak Republic which stated that *"appellate review as an extraordinary remedy challenging the final court decision is likely to result in a revocation or amendment of such a decision which is a legally admitted exemption from the immutability of final and binding decisions". Appellate review is therefore not generally applicable and can be successfully applied only in exceptional cases. The essential idea of extraordinary remedies is based on the fact that, in the rule of law concept, the legal certainty and stability set up by a final decision (Article 1, paragraph 1, of the Constitution of the Slovak Republic[9]) can be disrupted only in extraordinary and exceptional cases."*

The ECHR also called attention to the legal uncertainty as a result of a revocation of the final court decision. For instance, in *Stere and Others v. Romania*[10], the ECHR noted that in case where a party is seeking *"a review of a final and binding judgment merely for the purpose of obtaining a rehearing and a fresh determination of the case (…) the reversal of final decisions would result in a general climate of legal uncertainty, reducing public confidence in the judicial system and consequently in the rule of law".*[11]

In *Abdullayev v. Russia*[12], the ECHR stressed that *"for the sake of legal certainty implicitly required by Article 6, final judgments should generally be left intact"* and that, consequently, *"they may be disturbed only to correct fundamental errors".* According to the opinion delivered by the ECHR, such a defect cannot be in any case constituted by the fact that there are two different views on the subject matter of the case, and the ECHR concluded that such a plea cannot be considered reasonable for the re-examination of the case.

The similar legal opinion was expressed by the ECHR in *Cornif v. Romania*[13] in which the ECHR concluded that this was *"a typical case of two conflicting views on the same matter which in no case justifies the quashing of a final and binding decision"*. Moreover, nothing in this case, which was a contract law dispute between private parties, constituted circumstances of a substantial or compelling interest that would justify the quashing of the final decision under the Ryabykh[14] principles.

In this context, other cases of the ECHR on compatibility of the extraordinary appeal with the Article 6, paragraph 1, of the Convention should be referred to.

In *Ukraine-Tyumen v. Ukraine,* the ECHR held that, *"in the absence of any special factors which could justify the use of a supervisory review in the applicant's case, the fact that the latter procedure was used to set aside the resolution (…) is sufficient to enable the Court to rule that its "right to a court" under Article 6 § 1 of the Convention was infringed"*.

In *Sutyazhnik v. Russia*[15], the ECHR however stressed that the Ryabykh judgment contained an important reservation, which, at least implicitly, admitted that supervisory review could be justified in particular circumstances. The Court said: *"(…) the review should not be treated as an appeal in disguise, and the mere possibility of there being two views on the subject is not a ground for re-examination. A departure from that principle is justified only when made necessary by circumstances of a substantial and compelling character"*.

In *Eugenia a Doina Duca v. Moldava*[16], the applicants complained that the principle of legal certainty had been breached and relied on Article 6, paragraph 1, of the Convention. They argued that the revision proceedings were an appeal in disguise because the counter-party had merely tried to obtain a rehearing and a fresh determination of the case.[17] They had relied on arguments which had been raised before, such as the issue concerning the false documents, and on the fact that they had lodged an application with the Court. The ECHR ruled that *"the Supreme Court of Justice had failed to give reasons for accepting their revision request"*.

In *Bulgakova v. Russia*[18], the ECHR held that the *"Convention in principle tolerates the reopening of final judgments if new circumstances are discovered. For example, Article 4 of Protocol no. 7 expressly permits the State to correct miscarriages of criminal justice. A verdict ignoring key evidence may well be such a miscarriage. However, the power of review should be exercised for correction of gross judicial mistakes and miscarriages of justice, and not just as an "appeal in disguise"*[19].

The reasons justifying an extraordinary remedy were formulated by the ECHR in *Sizintseva and Others v. Russia*[20]. In this case, the ECHR made it clear that the opposition of one of the parties to the assessment of the case at first instance and the appellate court is not such an exceptional circumstance that would justify the interference with the guarantees provided through the principle of legal certainty. The ECHR held that *"a lack of jurisdiction, a serious breach of procedure or abuse of power may be generally regarded as a fundamental defect, and therefore justify annulment of the decision"*.[21] In this case, the challenge of the later revoked findings of the lower courts was not justified by a *"lack or misuse of power or any other procedural defect"*. On the contrary, the only reason to cancel the final decisions in question was *"the misinterpretation and incorrect application of the provisions of the federal law courts"*. In paragraph 32 of this judgment, the ECHR further stated that *"in the absence of major malformations (…), one of the opposition parties to the assessment of the case at first instance and the appellate court is not such a circumstance substantial and serious concern that would justify revocation of a binding and enforceable decisions and "reopening" of the case"*. Therefore, the ECHR in this case found a violation of Article 6, paragraph 1, of the Convention.

Thus, when the Prosecutor General is lodging an extraordinary appeal because *"the judgment consists in an incorrect legal assessment of the case"* (Article §243c CPC) or because *"the proceedings are vitiated by a different irregularity that resulted in an incorrect decision in the case"* (Article §243c CPC) the conformity of extraordinary appeal with the Article 6, paragraph 1, of the Convention can be easily questioned. However, when the Prosecutor General is lodging an extraordinary appeal for procedural irregularities indicated in Section 237 CPC (Article 243a CPC)[22], which can be the only ones to constitute a substantial procedural error, extraordinary appeal cannot be automatically considered as being in accordance with the Article 6, paragraph 1, of the Convention. For reasons of a substantial procedural error indicated in Section 237 CPC, an extraordinary remedy may be lodged even by the party itself, namely an "appellate review".

In this respect, under the document of 26 March 2013 submitted by the Ministry of Justice of the Slovak Republic[23] in the consultation procedure concerning the recodification of the Civil Procedure Code there should be no extraordinary appeal in the new Civil Procedure Code. This document indicates, for instance: *"In this respect, it must be observed that duplicity of appeals and extraordinary appeals has been criticised for a long time, the main negative point being the structure of extraordinary appeals, which favours one of the parties (proceedings without representation and court fees, excessive time-limit, admissibility defined in unclear and vague terms). Admissibility of extraordinary appeals is defined in much broader terms than admissibility of appeals, while in certain questions (e.g. reasons) they fully overlap..."*. It follows from the document as a whole, in substance, that the Ministry of Justice of the Slovak Republic itself questions the compatibility of extraordinary appeals with the Convention, adducing similar arguments and suggesting that extraordinary appeal should be abolished in the recodification of the Civil Procedure Code mentioned above.

Finally, we can conclude that the Slovak Republic appears to violate the right of the party to a fair trial stipulated in Article 6 § 1 of the Convention, in particular the principle of legal certainty, in which the Supreme Court, in the proceedings concerning the extraordinary appeal lodged by the Prosecutor General, can quash the final and binding judgment and, before that, suspend the enforceability of this judgment. However, it seems rather necessary to await the first ECHR judgment concerning the compliance of the Slovak extraordinary appeal with the Article 6, paragraph 1, of the Convention.

ENDNOTES

1. Rosca v. Moldova, Judgement of 22 March 2005, Application No. 6267/02, § 25.
2. Judgement of 26 July 2011, Application No. 41262/05, § 62.
3. See *Tripon v. Romania*, Judgement of 23 September 2008, Application No. 36942/03, § 22; *Brumărescu v. Romania*, Judgement of 28 October 1999, Application No. 28342/95, § 62; *SC Maşinexportimport Industrial Group SA v. Romania*, Judgement of 1 December 2005, Application No. 22687/03, § 36, and *Cornif v. Romania*, Judgement of 11 January 2007, Application No. 42872/02, §§ 29–30.
4. Judgement of 23 September 2008, Application No. 36942/03, § 21.
5. See also *Ryabykh v. Russia*, Judgement of 24 July 2003, Application No. 52854/99, § 51, and *Brumărescu v. Romania*, Judgement of 28 October 1999, Application No. 28342/95, § 61.
6. Judgement of 1 December 2005, Application No. 22687/03, § 36.
7. Judgement of 8 November 2005, Application No. 40663/98, §§ 47–49.
8. Judgement of 18 January 2007, Application No. 5531/00, § 31.
9. The Article 1, paragraph 1, of the Constitution of the Slovak Republic reads as follows: "The Slovak Republic is a sovereign, democratic state governed by the rule of law. It is not bound to any ideology or religion."
10. Judgement of 23 February 2006, Application No. 25632/02, § 53.
11. *"In this context it should be noted that the rule of law as one of the fundamental principles of a democratic society, is contained in all articles of the Convention. The rule of law presupposes respect for the principle of legal certainty, particularly in relation to judicial decisions representing res judicata."*
12. Judgement of 11 February 2010, Application No. 11227/05, § 19.
13. Judgement of 11 January 2007, Application No. 42872/02, § 29.
14. Judgement of 24 July 2003, Application No. 52854/99.

15. Judgement of 23 July 2009, Application No. 8269/02, § 33.
16. Judgment of 3 March 2009, Application No. 75/07, § 28.
17. See also *Cornif v. Romania*, cited above, §§ 28–29.
18. Judgement of 18 January 2007, Application No. 69524/01, § 34.
19. See *Ryabykh*, cited above, § 52.
20. Judgement of 8 April 2010, Application No. 38585/04, 2795/05, 18590/05, 24012/07, 55283/07, § 31.
21. See also Luchkina v. Russia, Judgement of 10 April 2008, Application No. 3548/04, § 21.
22. Section 237 of the Civil Procedure Code provides:

    "An appeal may be lodged against any decision of the appellate court if (a) the decision was adopted in a case outside the powers of courts, (b) the person acting in the proceedings as a party did not have legal capacity to act as such, (c) a party to the proceedings did not have procedural capacity and was not duly represented, (d) a final decision had already been adopted in the same case or proceedings in the same case had started earlier, (e) no complaint had been filed, even though it had been necessary according to legislation, (f) a party to the proceedings was denied the possibility of acting before the court by the court's action, (g) the decision was adopted by an excluded judge or the court had a wrong composition, unless the decision was adopted by a panel of judges instead of a judge sitting alone."

23. This document is available on the website of the Ministry of Justice: https://lt.justice. gov.sk/Document/DocumentDetailsReviewEvaluationDetail.aspx?instEID=-1&matE ID=5922&drCommentDocEID=296011&langEID=1&tStamp=20130531102738900 (01/07/2014).

REFERENCES

Cohen-Jonathan, G. 1989. *La Convention européenne des droits de l'homme,* Paris: PUAM.
Grotrian, A. 1994. *Article 6 of the European Convention on Human Rights—The right to a fair trial,* Strasbourg: Council of Europe.
Mole, N. & Harby, C. 2002. *Le droit à un procès équitable*, Strasbourg: Council of Europe.
Van Dijk, P. & Van Hoof, F. & Van Rijn, A. & Zwak, L. 2006. Theory and Practice of the European Convention on Human Rights, Antwerpen-Oxford: Intersentia.
Velu, J. & Ergec, R. 1990. La Convention européenne des droits de l'homme, Bruxelles: Bruylant.

Current Issues of Science and Research in the Global World – Kunova & Dolinsky (Eds)
© 2015 Taylor & Francis Group, London, ISBN 978-1-138-02739-8

Law vs. morality—permanent phenomenon

Denisa Soukeníková
Faculty of Law, Pan-European University, Bratislava, Slovakia

ABSTRACT: The relationship between law and morality is a central issue in philosophy for centuries. By many is morality associated with religion and as morally norms are considered also religious and common norms. The dividing line between these social norms is not straightforward, because each perceives individual social norms differently. Aim of this article is to point out the ongoing conflict between these central concepts, mainly to the historical and philosophical excursus.

1   INTRODUCTION

The issue of relationship between law and morality has been the central problem of law philosophy for centuries. Many people merge morality with religion and consider religious norms and custom norms to be moral norms. The borderline between both moral and legal norms is not clear, because everyone understands these norms individually.

Before I will discuss the concept of morality and its relationship to law, I would like to analyse "ethics" as a term, to which this topic is closely related. The term "ethics" is derived from the Greek word "ethos" that primarily concerns animals and indicates location of their grazing.[1]

In relation to men, the term "éthos" refers to a "place of habitual residence". Therefore, the term "éthos" refers to habit, custom, morality and tradition. Ethics as a science is linked especially with human evolution. History dates ethics to the antique period, when it was used as mean of research and stream of thinking on how men should be and how they should act. The central thought of ethics is observation of human behaviour and decision making in relation to what is good, evil, honest or dishonest. The task of ethics is to lead an individual to good. The first philosopher, who dealt with ethics, was Aristotle, who called ethics "politics". Only later Cicero derived an adjective "morally" from the word "mos" and later then the term "ethics" was established. From this moment on it comes to a dispute if ethics and morality are of the same meaning. Ethics is an independent discipline with morality as the main theme. Ethics examines morality as a society-wide phenomenon. Today's interest in ethics has its roots in the late 17th and early 18th century as the result of the weakening religion influence. Overall, ethics can be seen as a philosophical discipline dealing with moral nature of human experience, moral reasoning on good and evil, as well as on right and wrong behaviour. Many philosophers and legal theorists have asked themselves a question: "why ethics".[2] Is ethics so important that it could establish anything binding? Since the ancient past moral judgments have been presented as in personal so in political sphere. There are many areas that we need to focus on as "pure" moral areas. It comes to the sphere of asylum, the rights of foreigners, prohibited or allowed abortions, euthanasia, and rights of disabled as well as moral commitments to animals and others … Ethics became more interesting for people in the early 18th century, when there was a religious decline.

2   DISTINCTION BETWEEN LAW AND MORALITY

The fundamental distinction between law and morality must be mainly seen in the **diversity of the nature of law and moral norms:** many authors consider the sanction as the fundamental distinction; in the **diversity of the method of creation of law and moral norms:** this difference is apparent from the etymological meaning of the term "morality" as habit or custom, where moral norms are the result of repeated compliance with certain rules that have been proven by society and have been respected by the members of the society as well. On the other hand, legal norms arise from legislative activity of state authorities that are constantly changing; in the **diversity of application of legal and moral norms:** diversity is a result of external and internal regulation of human behaviour. While legal norms represent external regulation which is the regulation of mutual relationships between people, moral standards represent internal regulation through subjective motives of man.[3]

Conflict between law and morality can be seen in the literary work called Antigone, in the scene when a dispute arises between Creon and Antigone as Antigone buried her brother Polynices (enemy of the country) despite Creon's ban (anyone who disobeyed Creon's orders was immured). Antigone broke the Creon's ban because according to the religion of ancient Greek soul cannot find peace in the underworld until the dead body is buried. At the same time Antigone tried to make her sister Ismene follows her in this crime. After burying Polynices, Antigone came to Creon and said: "Zeus did not announce those laws to me. And Dike living with the gods below sent no such laws for men. I did not think anything which you proclaimed was strong enough to let a mortal override the gods and their unwritten and unchanging laws. They're not just for today or yesterday, but exist forever, and no one knows where they first appeared. So I did not mean to let a fear of any human to lead to my punishment among the gods. I know all too well I'm going to die how could I not? It makes no difference what you decree. And if I have to die before my time, well, I count that a gain. When someone has to live the way I do, surrounded by so many evil things, how can she fail to find a benefit in death? And so for me meeting this fate won't bring any pain. But if I'd allowed my own mother's dead son to just lie there, an unburied corpse, then I'd feel distress. What's going on here does not hurt me at all. If you think what I'm doing now is stupid, perhaps I'm being charged with foolishness by someone who's a fool."[4]

There are many situations when law and morality comes into a mutual conflict and our reaction depends on our individual consideration. These situations relate in particular to issues as euthanasia, abortion, positive discrimination, segregation, "murderous heirs" and others. The difference is in the fact that morality assesses certain act as good and bad, while law distinguishes between legality and illegality of such act.

As the central problem in relationship between morality and law is the conflict between natural and positive law, conflict can be seen as well in the structure of the legal and moral norms. As the example, I will mention only the absence of sanctions enforceable by state authority. Since moral norms may be characterized as habits and customs providing a system of social control, which is executed unconsciously and spontaneously. Nevertheless, moral norms in their nature should not interfere with norms of objective law.[5]

Even though current science understands law and morality as independent separated institutes, there are three possible situations regarding the relation between legal and moral norms.

1. Legal norms so as moral norms are identical in evaluation of certain behaviour. For example cases of bodily harm, murder, theft, robbery and other.
2. Evaluation of law and morality differs. This particularly relates to abortion, euthanasia, divorce … (we can see a great connection with religion).
3. Legislation is morally indifferent; respectively morality does not evaluate certain behaviour or act. For example technical norms.

The number of conflicts related to law and morality is high. Of course, anyone can have a different view on a particular issue as our opinions arise from our religion and moral beliefs, as every one of us is affected by society and family.

3   HART'S PERCEPTION OF RELATIONSHIP BETWEEN MORALITY AND LAW

According to Hart, a claim that between law and morality necessarily exists a relationship has many forms. The most extreme one is associated with the Thomist tradition of natural law which was based on two arguments. Firstly, there are certain principles of morality and justice that are of divine origin and which can be discovered by human mind even without the aid of revelation. Secondly, there are human laws which are in conflict with previous principles. Some of the supporters of the second argument did not understand morality as unchanging principles of behaviour, but as something that can be discovered through mind, as expression of human attitudes, which may vary in different social environments and different individuals. At the same time they differ in explanation of so called "necessary" relationship between law and morality. In their opinion there must exist recognition of the moral commitment to obey the law as the premise for existence of law system.[6]

Hart is not a supporter of the idea that moral rules are the other social rules staying aside the legal system. According to him, any act that is in accordance with moral rules is considered being a matter of course (it is similar with legal rules) and is not exaggeratedly praised. According to Hart's theory, compliance of moral and legal rules lies in their conception so that moral and legal rules shall bind individuals regardless of their consent and shall be supported by serious social pressure, which shall lead to observance of these rules. Their respect both the moral and the legal rules is considered to be granted.

Under the moral rules, someone who unintentionally, makes something that moral rules ban shall be acquitted. In contrary, in the legal system where objective responsibility may apply, someone who "unintentionally" makes something in contrary with law shall be subjected to punishment. Hart summarized the diversity of legal and moral rules in four points:

1. **Importance:** This feature of moral rule appears in many cases. Moral rules are constantly maintained even if they require sacrifice of personal interests and they appear also in the strength of social pressure, which shall ensure that moral norms are taught from generation to generation.
2. **Immunity to intentional change:** It is characteristic for the legal system that legal rules or norms are changing. On the other hand, moral rules cannot be so easily deleted or removed. Due to this principle morality cannot be entirely distinguished from other social norms, since this principle is not exclusively typical only for morality.
3. **The voluntary nature of moral offences:** If the person who violated a moral rule or principle proves that he or she acted unintentionally, that person is deprived of moral responsibility.
4. **The form of moral pressure:** This rule is based on the opinion that manners concern "the conscience". The legal rules have the threat of physical punishment, while the moral rules comes to respect for the rules to something that in itself is important and what is shared with all. Pressure refers to the conscience and relies on sense of guilt and of remorse,[7] which are often stronger and operate more efficiently than physical punishment.

4 CONCLUSIONS

The truth is that everyone's morality is formed with our personal development; we are influenced by society we live in, our family, work environment, friends who affect us, and belief. Therefore, every person has also his/ her own morality according which they live their lives. Therefore, it should be noted that it is necessary to return to traditional Christian values on which our (European) society is built. What happens in cases when there is a dispute between law and morality? Cases concerning in particular divorces, abortions, euthanasia … How to solve them? To which side should we incline? There is no universal rule, each case is different and unique. Any form of extremism, whether natural law or legal positivism (or morality and the law) is not correct and legal positivism must have been in its nature affected by natural rights, which preceded it. Act strictly with moral and natural law is not desired as well, since it loses legal certainty and enforcement of law … This relationship between law and morality affects our everyday lives, only recently Belgium passed a law on children euthanasia. However it is complicated to find the perfect imaginary line between these two concepts I think that law without morality and morality without law could not exist. Society must respect the law, but no one can turn the law against morality!

ENDNOTES

1. Anzenbacher, A. *Úvod do etiky.* 1. vyd. Praha: ZVON. 1994. S. 17. ISBN 80-7113-111-3.
2. Anzenbacher, A. *Úvod do etiky.* 1. vyd. Praha: ZVON. 1994. S. 17. ISBN 80-7113-111-3.
3. Harvánek, J. a kolektív *Teorie práva.* 1. vyd. Plzeň: Vydavatelství a nakladatelství Aleš Čeněk, 2008. S. 59–60. ISBN 978-80-7380-104-5.
4. Sofokles *Antigona.* Martin: Matica slovenská, 1940. S. 33–34.
5. Wieslaw L. *Prawo i moralność:* Warszawa. Państwowe Wydawnictwo Naukowe. 1989. S. 19. ISBN 83-01-08825-7.
6. Hart, H.L.A. *Pojem práva.* Praha: Prostor. 2004. S.158. ISBN 80-7260-103-2.
7. Hart, H.L.A. *Pojem práva.* Praha: Prostor, 2004. S. 174–180. ISBN 80-7260-103-2.

REFERENCES

Anzenbacher, A. 1994. Úvod do etiky. 1. vyd. Praha: ZVON. ISBN 80-7113-111-3.

Hart, H.L.A. 2004. Pojem práva. Praha: Prostor. ISBN 80-7260-103-2.

Harvánek, J. a kolektív. 2008. Teorie práva. 1. vyd. Plzeň: Vydavatelství a nakladatelství Aleš Čeněk, ISBN 978-80-7380-104-5.

Sofokles. 1940. Antigona. Martin: Matica slovenská, 1940.

Wieslaw, L. 1989. Prawo i moralność: Warszawa. Państwowe Wydawnictwo Naukowe. ISBN 83-01-08825-7.

Faculty of economics and business, section current topics of economic theory and practice in international business

Comparison of Czech and foreign managers from the perspective of their leadership and management styles

J. Dědina
Pan-European University, Bratislava, Slovakia

K. Dědinová
University of Economics in Prague, Czech Republic

ABSTRACT: The aim of the research was to conduct a survey and to compare leadership and management styles of Czech and foreign managers. We compared the differentness of the leadership and management styles of managers from several European countries. A partial goal was to compare the impact of different cultures on behaviour, leadership, management and decision making of managers in a given country. The research was carried out in the Czech Republic, Slovakia, Austria, Italy and Germany. After a research analysis of particular literature sources of national cultures (Vesely, 2014; Lang & Szabo & Catana, 2013; Sternad, 2011; Hermann & Hüneke & Rohrberg, 2012; Müller, 2005), we defined the hypothesis of the expected prospective prevailing leadership and management style in individual countries. We verified these hypotheses by an experiment, which was based on the collection of information on management and leadership style that we obtained through an empirical research about foreign managers in each country. The survey of leadership and decision styles was conducted according to the model of Vroom and Yetton (Vroom & Yetton, 1973) and the managerial grid by Blake and Mouton (Blake & McCanse, 1995).

1 INTRODUCTION

1.1 *Aims and methodology of an empirical research*

One of the goals of this research was to conduct a survey and to compare leadership and management styles of Czech and foreign managers. We compared the diversity of leadership and management styles in several European countries. Another goal was to compare the impact of different cultures on behaviour, leadership, management and decision making of managers in a given country. We tried to deduce the dependence of change of leadership and management styles on economic changes in company environment in particular state of research.

In the research were included following countries: the Czech Republic, Slovakia, Austria, Italy and Germany. We separated these countries into groups according to the geographical layout. We acquired information about national cultures by an exploration of Czech and foreign literature. After an analysis of literature sources of particular national cultures (Dedina & Dedinova, 2013; Ötzbek-Potthoff, 2013; Homma & Bauschke, 2010) we stated hypotheses of expected prevailing leadership and management styles in each country. These hypotheses were later verified by an experiment, which was based on a collection of information on management and leadership styles, which we obtained on the basis of empirical research of foreign managers using questionnaires and case studies in particular countries.

Questionnaires were the same for each country with the following structure. In the first part of the questionnaire was general information about the company and the manager. The second part of the questionnaire was devoted to the leadership and management styles of the manager. The respondents of the questionnaire were subordinates of the assessed managers.

In each country, 150 questionnaires were sent out. Response rate was about 70–80%, however it slightly changed in particular countries. The time of response was up to one to two months. The questionnaires were statistically evaluated and there was graphical output and conclusions.

The leadership and management survey was conducted according to the model of Vroom and Yetton (Vroom & Yetton, 1973) and the managerial grid by Blake and Mouton (Blake & McCanse, 1995).

The outcomes of the up-to-date research in the literature are as follows. Managers in the Czech Republic tend to lead according to internal rules and regulations, especially in public administration and government. However employees are reluctant to accept formal proceedings and rather try to improvise (Vesely, 2014). In the business sector prevails informal leadership and leads to a stronger corporate culture.

Managers in Slovakia communicate more with employees both in the public and private sector and thus leads to greater acceptance of their decisions by the employees (Lang & Szabo & Catana, 2013).

Managers in Austria use both consultative and authoritative style of leadership according to specific type of decision problem. Employees know exactly what is expected of them. Consultative leadership style is prevailing in Austria these days (Sternad, 2011).

Managers in Italy are communicative, open and use a rather authoritarian leadership style. Subordinates try to discuss before accepting an order (Müller, 2005, p. 41).

Managers in Germany use authoritarian management style, but before making a decision they consult the problem with subordinates. The subordinates obey the orders (Hermann & Hüneke & Rohrberg, 2012).

2 THEORETICAL BASIS OF RESEARCH AND DETERMINATION OF HYPOTHESES

American authors Vroom and Yetton (Vroom & Jago, 1988, p. 32) continued the work of Likert (Likert, 1967) and found two new factors that significantly affect the effectiveness of management control. These factors are: quality of the decision-making and willingness of subordinate managers to implement the given decision. Furthermore, they added some new managerial styles, which were originally specified by Likert into five groups:

- AI—strongly autocratic style—manager decides alone and relies solely on his own information,
- AII—autocratic style—manager decides alone again, but he uses selected information from his subordinates to prove or complete what he knows. He is not interested in opinions or advice of his subordinates,
- KI—consultative style—manager decides by himself, but he asks opinions of his subordinates, consults the problem individually and considers all ideas and suggestions for the final decision,
- KII—highly consultative style—manager decides by himself, but he discusses the problem with his subordinates at a meeting and accumulates necessary information,
- SII—participative style, group decision-making—the decision of manager is based on discussion with subordinates, where they are all considering possible solutions together. Manager stays more for moderator, coordinates discussion and tries to bring it to a solution, which is accepted by all.

It has to be noted, that the general trend according to the model of the authors Vroom and Yetton is that the manager, if possible, should try to use more participatory decision-making styles KI, KII to SII for specific situation. Empirical research by these authors shows that the efficiency of decision and leadership of managers is rising towards the choice of styles KII and SII, or this represents the move from an authoritarian to a group leadership and decision-making. We considered the aforementioned statement as one of the research hypotheses, which we tested.

Other authors Robert Blake and Jane Mouton (Blake & McCanse, 1995, p. 80) developed the managerial grid to help managers to identify and improve their interpersonal management style. According to them, the manager behaviour is a function of two variables:

– Social aspects (interest in people)—here we can include maintaining the confidence of colleagues, creating good working conditions, maintaining good interpersonal relations etc.
– Production aspect (concern for production)—includes opinions of manager on a number of issues such as level of decision-making efficiency, work efficiency, production techniques and processes.

These variables can be expressed in a graph where each axis has a scale from 1 to 9 and indicates in an ascending order the growing weight of each factor. Managerial grid will be formed as an intersection of these values.

Managerial grid "nine to nine" is used to locate the manager's approach to interpersonal relationships in one of 81 possible variants. Blake and Mouton for reasons of simplicity were concentrated on four positions at the corners and the style in the middle of the grid:

– Style 1, 1—depleted management—manager is not oriented towards people or production, he is focused mainly on himself,
– Style 1, 9—country club management—manager is strongly focused on people, but has little interest in production, trying to create interpersonal relationships, a friendly atmosphere and an appropriate pace of work, but he cares few about production targets, production is a secondary concern, the important thing is to avoid conflict and maintain harmony among employees, he believes that happy employees will do what is expected from them.
– Style 9,1—manager is predominantly a technocrat, he is authoritative, strongly oriented towards production and little interested in people, he focuses on management of production operations, the creation of interpersonal relations is not taken into great consideration, but an efficient job performance is still achieved, he relies on a centralized system and employees are considered as a means of production,
– Style 9, 9—team manager—he has a significant interest in people as well as production, he can combine the concern for the production tasks with the concern for good interpersonal relations and satisfaction of employees, managers discuss issues with employees and seek their ideas and give them freedom of action,
– Style 5, 5—manager in the middle of the road—he is interested in people and production on average, managers tend to avoid problems.

In 1991, there were two additional styles (Blake & McCanse, 1995, p. 203–267):

– Opportunistic management—the organizational performance is based on an exchange system, the effort is based on an exchange system, the effort is only exerted as an exchange for the same value, opportunist uses any style in the grid for his own interests and self-promotion, he adapts to the situation, so that he can benefit the maximum,
– Style 9 + 9—paternalistic management—rewards and approval are provided in return for loyalty and obedience, a punishment follows for maladjustment, paternalist strives for excellent results and uses rewards and punishments to achieve harmony, shows great interest in employee, rewards obedience and punishes disobedience.

Authors of this model speak about the optimal manager focusing on the areas 9.9 and 5.5. We claim this statement as one of the hypotheses that we tested.

3 TYPOLOGY OF CZECH CULTURE AND RESULTS OF THE SURVEY ON A LEADERSHIP AND MANAGEMENT OF CZECH MANAGERS

The Czech Republic is the most western Slavic country, which is also reflected in culture. Values, attitudes and beliefs are most affected by the German culture that has long been in contact. In the Czech culture formal relations and social status play an important role. Overall, the status is assigned to an individual according to his origin, education, contacts with influential persons and the family financial situation. It is typical to use the academic titles, to show respect to superiors and socially elder persons. The position of manager is seen as a social and material privilege. Power is concentrated in the hands of managers and company owners, employees are not included in decision-making. Inequality in a company is mainly seen by the large difference between the salaries of managers and other employees. In the Czech companies managers often underestimate the participation of employees in decision-making and fundamental changes in the organization are based on the decision of managers. Managers, first of all, give tasks and commands and monitor their performance. Managers do not use the potential of their subordinates, employees just expect instructions about what to do.

In earlier empirical and theoretical research (Meierewert, 2001) was found out that with higher management position in the Czech companies the perception of the cultural characteristics of the company is changing. According to representatives of top management, earlier, the enterprise environment was friendly and dynamic. They felt that the criterion of success are happy employees, that management is based on emphasizing the value of workers. Leadership was perceived by top management as leadership based on information, familiarizing staff with the objectives of the company, providing freedom. Top management expected by subordinates initiative, independence, creativity and result orientation.

Czechs are very sceptical and they rather improvise against hard structures. They consider themselves as inventive and flexible. They feel limited by plan, so they try to go around it or ignore it. The one, who sticks to the rules, is considered as a man who cannot think freely. Czechs have the ability to do more things at once, they do not want to miss even one chance and that is why they like to do a lot of things at the same time and change them according to their priority. For these and other reasons it is optimal to strengthen the autocratic management style (Novy & Schroll-Machl, 2003).

Czech culture is rather individualistic, an individual is more important than a group interest. The employer—employee relationship is based on mutual benefits that can give one another. When it comes to career advancement, one must demonstrate appropriate skills and abilities. The primary objective of individualistic cultures is to meet the targets even if it could be at the expense of good relations.

Generally, Czechs previously were more focused on relationships than on results. During negotiations, relationships were more important than the content, well built relationships were highly appreciated and sought to maintain. In practice, according to our conclusion from the empirical evaluation of the results in 2013 (Dědina & Dědinová, 2013) we found out, that this trend is changing with the change of business environment and change of paradigm. Therefore people-oriented relationships are changing and focusing more on results.

From the research it was found out that the position of the Czech manager is seen as a social and material privilege. Power is concentrated in the hands of managers and company owners, their employees are not included in decision-making processes. Inequality is also reflected in the large difference between the salaries of mangers and other employees. Managers often underestimate the participation of employees in decision-making and fundamental changes in the organization are made based on the decision of managers. Managers, first of all, give tasks and commands and monitor their performance. They cannot use the potential of their subordinates. Employees just expect instructions about what to do, or are afraid of what they might have done wrong again.

From the above mentioned analysis results in a hypothesis that Czech managers will be from the view of management and decision-making closer to autocratic management styles of AI and AII, they do not take into account opinions of subordinates and will rely solely on their decision (it especially applies in the civil service).

Empirical research which was made in 2013 (Dědina & Dědinová, 2013) shows that the most commonly used style among managers in the *Czech Republic is slightly autocratic style*, then the second most commonly used style is heavily autocratic style followed by both consultative styles. From these and even previous results we can conclude that the manager's decision preferably goes from himself and he is not interested in the opinions of subordinates. We also believe that this situation is influenced by historical development. Nevertheless the current situation of Czech companies plays a role, they are still dynamically affected by the global environment and, therefore their managers are increasingly pressured to make

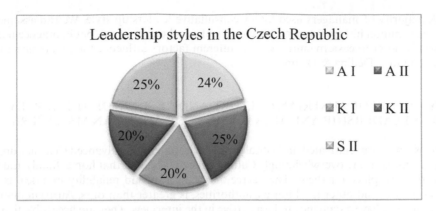

Figure 1.   Leadership styles in the Czech Republic (Dĕdina & Dĕdinová, 2013).

decisions faster and manage without consulting with colleagues. Little time remains for a classic leadership and decision-making on the basis of group meetings. Employees of companies in the Czech Republic can only act in a limited range of competence, which must not be exceeded, they lack independence, initiative, active search for solutions.

Style AII gained 25% of all votes, followed by the AI, which has 24%, both consultative styles consistently have 20% of the votes. On the last place a participative style remains. There we can see that in the Czech Republic slightly autocratic styles are the most applied, which together received nearly half of all votes.

Women in the role of a manager in the Czech Republic prefer the consultative management style. However, after consulting with a co-worker they decide by themselves and they either take into account information received or not. Generally, women in practice are expected to be more loyal, willing and conscientious.

4   TYPOLOGY OF SLOVAK CULTURE AND RESULTS OF THE SURVEY ON A LEADERSHIP AND MANAGEMENT OF SLOVAK MANAGERS

Slovakia is a landlocked country located in Central Europe. It borders with the Czech Republic and Austria to the west, Hungary to the south, Ukraine to the east and the north of Poland. Slovakia is a member state of the European Union.

Slovak society is more open, but among managers and workers can sometimes be formed a healthy distance. Organizations are more formalized and thus exploit certain rules to control managers in their work. Like in the Czech culture social status plays an important role in society. Manager's position and his level of power is based on the social background, most managers are men.

Based on our research, it was found out that managers do not underestimate the participation of employees in decision making and fundamental changes in the organization do not take place only at the discretion of managers. Sometimes they do not use so much potential of their subordinates, but also give tasks and commands and monitor their performance. Nevertheless they are still interested in the views of subordinates. Control is based on emphasizing the value of workers. Managers focus more on results. Same like the Czechs, they have the ability to multitask. It is required to come to meetings on time (Fabiánová, 2002).

The result of an empirical survey of Slovak managers does not confirm the hypothesis that the Slovak managers tend to authoritarian management style, though they are more open than the Czechs. From the results clearly outweighs strong consultative style KII. This would mean that the managers in Slovakia prefer to talk about the matter at a joint meeting, after which the manager finally decides alone. His management style is therefore strongly consultative.

This result can be explained by reliance on the national mentality. We can see a big difference between the leadership in the Czech and Slovak Republic. While in the Czech Republic it was found that the most managers have used a slightly autocratic management style, the

Slovak majority of managers used highly consultative leadership style. We can assume that given the common history, same management styles will be preferred, but in practice it is not. Slovakia is closer to eastern cultures. Also different factors influence the economic situation in each country (Dědina & Dědinová, 2013).

5 TYPOLOGY OF AUSTRIAN CULTURE AND RESULTS OF THE SURVEY ON A LEADERSHIP AND MANAGEMENT OF AUSTRIAN MANAGERS

Austria is a modern, cultivated, industrially and environmentally advanced Central European country. Austrians are overwhelmingly Catholic, it can be stated that home, family and work are on the first place for them. The degree of seriousness and reliability in a deal is high. Respect to laws and other legal norms, authorities, is greater than ours. Austrian managers love to talk and have a tendency to be assertive in the interview. They are accessible to a wide conversation circuit and hard to find a subject that they cannot speak about. However, they are strong patriots. In private companies in Austria, it is possible to record a very professional attitude to work. The level of wages and salaries is high for both private companies and the public sector. Career advancement in the hierarchical structure of the public sector is similar to ours and depends on the length of experience and education. In management positions are mostly older people. For companies, the situation is different. In small and medium-sized enterprises we can see young people in positions with full responsibility. In large concerns or banks a conservative attitude is preserved. In Austria, there is a large number of modern and highly advanced companies with advanced teamwork and corporate culture.

The Austrian partner is willing to make compromises, but he knows his limits and usually even your limits within which you can move. This often creates a high pressure on the price. Relationships among people have many characteristics as elsewhere in Central Europe, although in Austria a conservative level is maintained. The young have more respect for elders and they recognize life experience. Education is valued and we can say that there is an increasing number of people with higher education. Social status is valued and people perceive it. Austrians are proud of their intellectuals, artists and athletes. They are interested in partner's origin and live a rich social and cultural life.

In management positions are mostly men, but it is not uncommon that a team is controlled by a woman, even in the highest state and managerial positions. The relationship between supervisor and subordinate corresponds to their position and authority, but there is an open and cooperative spirit between them. Austrian business partners are friendly, open, but also factual and they know what they want to achieve. They get information about the company with which they choose to act in advance. They have great understanding of business and business negotiation tactics, they tend to be consistent. It is therefore important to prepare properly when negotiating with them. They can create a good environment for meetings. They have a considerable degree of assertiveness, they like to show off and are willing to talk about their private lives. Austrians are good hosts and conversations not connected to the subject of the meeting is a continuing the process of rapprochement and mutual confidence. It is not usually difficult to establish friendly relations. It also happens that negotiations take place right in the restaurant. They use most likely German as negotiation language, which among other things gives them an advantage. The habit of using the title has its roots in the times of the monarchy. In addition to the titles they use functional designation such as president, governor, CEO, etc. (Bretschneider, 1997; Kasper & Mayerhofer, 2005).

Czechs and Austrians have always considered their culture to be close and relative to each other, they feel close and relatively comparable, it causes a tendency to mutual competition and overall activity.

Starting from the fact that Austria is a very long time liberal, democratic country than, for example Czech Republic, we can hypothesize that the majority of Austrian managers will use more participatory leadership styles.

In Austria, a strong consultative style clearly outweighs, just as it is in Slovakia. The responsible person in Austria has far-reaching competences, he/she is thus able to make decisions on the spot of negotiations.

In most cases, managers used highly consultative leadership style KII. In conclusion it can be stated that the above mentioned hypothesis was fully verified (Dědina & Dědinová, 2013).

6 TYPOLOGY OF ITALIAN CULTURE AND RESULTS OF THE SURVEY ON A LEADERSHIP AND MANAGEMENT OF ITALIAN MANAGERS

Most enterprises are larger or smaller family company. The head is the chief (father), who is often also the owner. He is expected to be creative, inventive and to take the responsibility for the whole functioning. The ability to manage crises and improvisational talent are just as important as charisma and authority. Members of senior management (older sons) have the trust of the boss and can influence his decisions. Subordinates on the lower levels in this system can be seen as dedicated workers in the patriarchal family, who may count on a reward for a faithful service and protection and care. In small enterprises we often encounter harsh dictatorial control method, which requires a respectful subordination. The type of manager who goes from one business to another to be promoted and gain more money does not fit to the Italian formula. Being ambitious is not considered as a priority. Education, talent and good relations are considered for success factors. The most important attribute for a manager is to be flexible, which means to ignore the usual routine.

Personal ties play the key role in business. Italian business relations are based on mutual dependence. In private matters, the Italians are characterized by honesty, loyalty and devotion. Less strict moral principles are applied in contact with the authorities. Italians treat them with a deeply rooted distrust. We should therefore seek to build close and friendly relations to Italians so that they stop to look at us only as representatives of the company.

Italians are open, tolerant to others and interested in the world around them. They forgive minor mistakes if they feel that they are committed in partner's good faith. They appreciate the art of conversation, so it is important to count with a small talk about everyday matters before the main business meeting proceeds. They work with full dedication and they work for living but they do not live for working. It is said that between all nations, Italian have the biggest ability to communicate by gestures. Besides a handshake, we can expect a welcome embrace, holding the shoulders, or kisses at arrival. We must be prepared for temperament, militancy and the need to negotiate about everything. They never accept the first price offered, they are very happy to haggle. They can be very stubborn and have a strong business sense. Young people employed in modern enterprises usually call themselves with their first name. For the older generation and in relations between superiors and subordinates signor and signora are used for addressing. Other titles, functional and prestigious, have no meaning. A title is rather recognition of social status than confirmation of actual scientific achievements. Italians strictly separate work and private life. That is why we can expect an invitation to the restaurant rather than home. They emphasize outfit, business casual is a dark suit (Deresky, 2014; Gesteland, 2002).

From the above analysis it is possible to hypothesize that Italians tend to use more consultative and group decision-making style. Our experimental investigations (Dědina & Dědinová, 2013) show that the most widely used style by Italian managers is style AI, followed by the style KI, AII and KII. In this case, the hypothesis has not been verified and we believe that Italians can be in certain cases full of surprises.

7 TYPOLOGY OF GERMAN CULTURE AND RESULTS OF THE SURVEY ON A LEADERSHIP AND MANAGEMENT OF GERMAN MANAGERS

In a business, everyone respects the personal jurisdiction and responsibility to the area of work, everybody is an expert in his section. The primary business is to achieve the goal. In corporate rules, attention is focused on avoiding a situation for which no solution schemes are ready. They like situations in which there is only one correct answer. Germans realize their plans and in any situation, they seek to achieve the optimal state as much as possible. They put deliberately very high goals. They focus on top quality, minimum costs, carefully

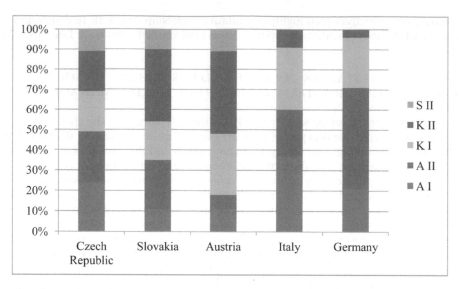

Figure 2.   Comparison of leadership styles (Dědina & Dědinová, 2013).

calculated logistic chain and time accuracy. The method to achieve optimum consists of solid organizational structures, systems and standards. So it is important to avoid disturbance of any kind. Germans are convinced that structures and standards are the key for professionalism. Standard system is a symbol of quality German work. Every thought must be brought to the smallest detail. According to them, detail is the actual core of the problem. The man in German corporate culture becomes more a robot than fallible being.

In enterprises managers decide in the frame of their competences and their subordinates and co-workers respect the division of powers. The subordinates do not interfere or overtake the power. That means that nobody does work for anyone else. Managers try to de-emphasize their higher hierarchical position. Emphasis is put on open discussion, exchange of ideas, meeting the job requirements and participation on the final decision. Privacy, tolerance or acceptance of responsibility for the failure is not an option.

Trying to say a hypothesis, we must realize that the Germans are very authoritative and bureaucracy plays an important role for them, their economy is better than the Austrian, so they will most probably vote for an authoritarian management style.

In enterprises, managers want their staff to be adequately informed. When they communicate, they express themselves clearly and distinctly, they are not afraid to oppose, point out errors and correct colleagues. Time is for the Germans one of the key issues. It is considered as a valuable asset, which should not be wasted. They believe that every task must be objectively structured, planned and precisely divided. That is why they require meeting the deadlines. A delay of a partner is perceived as disrespect, which is also transferred to a personal relationship. Time is only devoted to the important things and important people, which is confirmed by the fact that in the work sphere, one does never meet with anyone without a reason (Schroll-Machl & Nový, 2003).

From the above analysis, we assumed that in most cases the German managers use authoritarian styles of management and leadership. The experimental survey we pointed out the following order of preference: AII, KI, AI and KII. German managers were mostly using an autocratic management style and therefore our hypothesis about authoritarianism was verified (Dědina & Dědinová, 2013).

8   CONCLUSIONS

The overall comparison of empirical research shows that in terms of management styles, the Czech Republic and Germany are the closest to each other, both of these states prefer managerial style AII. Furthermore, the Slovak Republic and Austria both tagged style KII, Italy AI.

Theoretical analysis tells us that Austrian managers are the most participative and Czech managers are the most autocratic. Austrian managers do not avoid conflict discussions with subordinates and solve conflicts, Czechs are the exact opposite.

The analysis confirmed the hypothesis for the Czech Republic. It turned out that the most preferred styles are really both autocratic styles, which gained almost the same number of votes. The tendency to an autocratic management is apparently based on the historical development of the management system of our companies.

We did not totally confirm the claim that Austria uses SII style, but we have to say that from all the countries being compared in Austria participatory style is the most widely used, but despite this fact in Austria dominate both consultative styles. An explanation can be the theory, according to which Austrian managers welcome the participation of subordinates at complex problems of a technical nature, whereas organizationally simple problems they decide autocratically.

Theory is partly proved by practice in the case of Germany. The most commonly used style is AII, but the second place has a consultative style. We may seek the explanation in German perfectionism, need to have a ready solution schemes, reliable organizational structure and standards.

The result of the analysis of the Slovak Republic does not really confirm the hypothesis that Slovak managers tend to authoritarian management style, though they are more communicative than the Czechs. There clearly prevails a strong consultative style KII, which may be due to the nature of the Slovaks, who are, compared to the Czechs, more temperamental, emotional, dynamic and above all more communicative.

The country in which we absolutely did not confirm the hypothesis is Italy.

In Italy we would rather assume looser control methods. This expectation is based on the Italian lifestyle, which is not so stiff and rigid. In Italy, the ambition is not considered as a priority and the art of improvising is as important as charisma, so KII, SII better match for Italy, instead of elected strongly autocratic AI.

The aim of this research was to approach, disassembly and compare the styles of management and leadership of Czech and foreign managers.

It is clear from the entire analysis that managers in any country do not apply only participative leadership style, but according to theory by Vroom and Yetton there should be a tendency in the use of this particular style of leadership.

Manager's decision is not based on joint discussions with subordinates and they do not seek to make a common conclusion. It is also possible that the style SII is missing because decisions in this way take a long time and in turbulent times, managers are constantly exposed to greater time pressures. Another reason may be the fact that the developed economies are in some stage of the life cycle, to which they have to adapt a style of leadership and management.

Managers in Italy and Germany did not mark style SII, even though we would expect that the trend in Italy will move from the style of KI over KII to the SII.

From all the countries being compared is Austria the closest one to the general trend (using a participative management style). If we compare the German and Austrian managers, we must realize that despite geographical and linguistic proximity the management culture is in some ways different. Germans are more assertive managers, they control on the basis of facts and work exactly as prescribed. Austrians, same like Czechs, like to improvise. Austrians, unlike Czechs, have the responsible person who is able to make a decision at the point of negotiation and more women appear in managerial positions. The results of practical observation in the Czech Republic have confirmed the theory that managers prefer autocratic management. We focus on the hierarchical configuration of the relationship, the individual power and authority. Workers are seen as executors of orders and management focuses only short term. The influence of Czech managers may also be affected by the fact that the Czech culture has a strong masculine character.

REFERENCES

Blake, Robert R.; McCanse, Anne Adams. *GRID-Führungsmodell*. 3. Auflage. Wien: ECON Verlag, 1995. ISBN 3-430-11394-6. Str. 80.

Bretschneider, R.,1997. *Wirtschaftsmentalität und selbständiges Handeln*, in: Faulhaber, Theo/Hanisch, Ernst, (Hg.): *Mentalitäten und wirtschaftliches Handeln in Österreich*, Wien, 1997, p.. 220.

Dědina, J.; Dědinová, K. *Volba vhodného stylu vedení a řízení za účelem zvýšení výkonu zaměstnanců*. In Rojík, S. (ed.). In: Konkurence 2013. Jihlava: Vysoká škola polytechnická Jihlava, 2013, p. 52–65. ISBN 978-80-87035-73-3.

Dědina, J.; Dědinová, K. *Volba vhodné firemní kultury podniku za účelem zvýšení jeho výkonu*. In Majtán, Š. (ed.). In: Aktuálne problémy podnikovej sféry 2013. Bratislava: Ekonóm, 2013, p. 65–71. ISBN 978-80-225-3636-3.

Dědina, J.; Dědina ml., J. *Volba vhodného stylu vedení a řízení za účelem zvýšení výkonů zaměstnanců*. In: Karlovarská právní revue. 2011. Issue 4, p. 61–70. ISSN 1801–2191.

Dědina, J.; Dědinová, K. *The Development Of The Corporate Social Responsibility In Companies In The Czech Republic And Austria During The Last Decade*. In Management, Knowledge and Learning International Conference 2013. Zadar: ToKnowPress, 2013, s. 107–115. ISBN 978-961-6914-02-4.

Deresky, Helen. *International Management: Managing Across Borders and Cultures*. 8th ed. Harlow (UK): Pearson Education Limited, 2014. ISBN-13: 978-0-273-78705-1. S. 365–366.

Fabiánová, E. 2002. *Identifikácia sa zamestnancov s podnikom v kontexte podnikovej kultúry*: Prešov: FF PU, 2002.

Gesteland, R. *Cross-Cultural Business Behavior*. 3. reprint, Copenhagen Business School Press, 2002. ISBN 9788763000932.

Herrmann, D; Hüneke, K.; Rohrberg, A. *Führung auf Distanz*. 2. Edition. Wiesbaden: Springer Gabler, 2012. ISBN 978-3-8349-3005-7.

Homma N., Bauschke R. *Unternehmenskultur und Führung*. Berlin: Gabler Verlag., 2010. ISBN 978-3-8349-1546-7.

Kasper, H.; Mayerhofer, W. 2005. *Personální management, řízení, organizace*. Praha: Linde, 2005. p. 592. ISBN 80-86131-57-2.

Lang, R; Szabo, E.; Catana, G.A.; Konečná, Z.; Skálová, P. Beyond participation?—Leadership ideals of future managers from Central and East European Countries. *Journal of East European Management Studies* [online]. 2013, vol. 18, issue 4, p. 482–511. [vid. 2014–04–24]. ISSN 0949–6181. Access from EBSCO: http://web.a.ebscohost.com.zdroje.vse.cz/ehost/detail?vid = 3&sid = c851441b-fdad-4e5f-9686-13efba89cfbc%40 sessionmgr4005&hid = 4204&bdata = Jmxhbmc9Y3Mmc2l0ZT1-laG9zdC1 saXZl#db = bth&AN = 92574348.

Likert, R. *The Human Organization: Its Management and Value*. New York: McGran-Hill, 1967. ISBN 978-0070378513.

Meierewert, S., 2001. *Tschechische Kulturstandards aus der Sicht österreichischer Manager*; in: Fink, G.; Meierewert, S. (Hrsg.): *Interkulturelles Management—Österreichische Perspektiven*; Springer Verlag; Wien, New York; 2001.

Müller, Susanne. *Management in Europa: Interkulturelle Kommunikation und Kooperation in den Ländern der EU*. Frankfurt/ Main: Campus Verlag GmbH, 2005. ISBN 3-593-37721-7.

Nový, I.; Schroll-Machl. *Interkulturní komunikace v řízení a podnikání*, 3. Auflage, Management Press, Praha 2003. ISBN 978-80-7261-158-4.

Özbek-Potthoff, Gülden. *Implizite Führung im interkulturellen Kontext*: Stand der Forschung, Erweiterung der Theorie und empirische Analyse. Wiesbaden: Springer Gabler, 2013. ISBN 978-3-658-02233-4.

Reber G.(1992). *Modell zur besseren Entscheidungsfindung*. Aggregate of Lectures. Prague, 1992.

Schroll-Machl, S.; Nový, I. 2003. *Beruflich in Tschechien: Trainingsprogramm für Manager, Fach- und Führungskräfte*; Vandenhoeck & Ruprecht; Göttingen; 2003 herausgegeben von THOMAS, A. in der Reihe Handlungskompetenz im Ausland.

Sternad, D. *Strategic Adaptation: Cross-Cultural Differences in Company Responses to an Economic Crisis*. Wien: Springer, 2011. ISBN 978-3-7091-0454-5.

Vesely, Markus. *Gegenwärtige und zukünftige Kompetenzen tschechischer Führungskräfte*. Südwestdeutscher Verlag für Hochschulschriften. 2014. ISBN 978-3838137766.

Vroom, Victor H.; Jago, Arthur G. *The new leadership: managing participation in organizations*. Englewood Cliffs (USA): Prentice-Hall, 1988. ISBN 0-13-615030-6. S. 32.

Vroom, V. H.; Yetton, P. W. *Leadership and decision-making*. Pittsburg (USA): University of Pittsburgh Press, 1973. ISBN 0-8229-3266-0.

The impact of the global economic crisis of 2008/2009 on the national competitiveness of Central and Eastern European countries

T. Dudáš
Faculty of Economics and Business, Pan-European University, Bratislava, Slovakia

ABSTRACT: The main goal of this paper is to assess the impact of the global economic crisis of 2008/2009 on the national competitiveness of Central and Eastern European countries. The paper will use the indices of national competitiveness published in the Global Competitiveness Report and World Competitiveness Yearbook to assess the state of competitiveness in the individual countries. The paper will primarily focus on the new EU member states from Central and Eastern Europe, as national competitiveness is a key factor of economic development for them. The last part of the paper will try to provide policy implications for the governments of the countries analyzed.

1 INTRODUCTION

In order to be successful in the global economy, countries have to constantly improve their economies. To measure the competitiveness of countries in the global economy, modern indices of national competitiveness were developed. Although not all economists agree with the current definitions of national competitiveness (e.g. P. Krugman, 1994), the global reports and indices of national competitiveness play a central role in assessing the position of different countries in the global economy, and the results of these reports are often used as a basis for economic policymaking on the national level.

The tools analyzing national competitiveness are especially important for the countries of Central and Eastern Europe, as they still lag behind the economically developed countries of Western Europe. The global reports on national competitiveness can serve as a guideline for these countries in economic policymaking, as they can compare themselves to benchmark countries such as Finland, Switzerland or Austria. The position of these countries in the global competitiveness rankings is a measure of the successes and failures of the governments of Central and Eastern Europe.

The aim of this paper is to describe the changes in national competitiveness caused by the global economic crisis of 2008/2009 in Central and Eastern Europe. This crisis shook the global economy and introduced a new economic reality of low economic growth and high unemployment in many countries. Central Europe was no exception, as most of the countries still feel the effects of the crisis in the form of higher unemployment and other economic problems.

For the purpose of this paper, the state of national competitiveness will be represented by the two most widely used indices of national competitiveness—the Global Competitiveness Index published in the Global Competitiveness Report and the World Competitiveness Index published in the World Competitiveness Yearbook. The paper will focus on the group of new EU member states from Central and Eastern Europe that were accepted into the EU before the global economic crisis of 2008/09—Hungary, Czech Republic, Slovakia, Poland, Slovenia, Romania, Bulgaria, Estonia, Latvia and Lithuania. In our analysis, we will compare changes in the indices of national competitiveness of these countries before and after the global crisis, and look for possible trends and common patterns.

2 THE EVOLUTION OF THE CONCEPT OF NATIONAL COMPETITIVENESS—A LITERATURE REVIEW

The concept of national competitiveness is a relatively new concept in economics. Competitiveness was originally studied on the corporate level in the 1970s and the 1980s and it was raised to national level only in 1990 by Michael Porter in his renowned book The Competitive Advantage of Nations. Porter belonged to those researchers who have devoted a lot of effort to examine the concept of competitiveness on the corporate level, so it was a logical step for him to raise the examination of competitiveness to the national level, since states and their institutions have a significant impact on the competitiveness of domestic corporations.

According to Michael Porter, national competitiveness is based on four key areas that can boost or reduce the competitiveness of domestic firms. These are as follows (Porter, 1990):

1. Factor conditions: the nation's position in factors of production such as skilled labor, infrastructure, physical resources and technologies, necessary to successfully compete in a given industry;
2. Firm strategy, structure and rivalry: the conditions in the nation governing how companies are created, organized and managed as well as the nature of domestic rivalry;
3. Related and supporting industries: the presence or absence in the nation of supplier industries and related industries and institutions (research, education) that are internationally competitive; and
4. Demand conditions: the nature (from a qualitative and/or quantitative point of view) of home demand for the industry's products or services.

This theoretical model of Michael Porter is also known as Porter's diamond of competitive advantage and, in principle, it explains the impact of domestic business environment on national competitiveness. The main factors of the model have a very strong mutual influence and are exposed not only government policies but also to external factors.

While exploring the concept of national competitiveness, Michael Porter states that it is a dynamic model and he identified the basic stages of its development. Porter distinguishes three main phases of the development of national competitiveness: a factor-driven, investment-driven and innovation driven (Porter, 1990, p. 545). In economies that are factor-driven, competitive advantage comes solely from the factors of production (available natural resources, land suitable for agriculture and a large number of less educated but cheap labor). Companies in these economies build their competitiveness on low prices and operate mainly in technologically undemanding sectors (Porter, 1990, p. 546–547). In investment driven economies, national competitiveness is based on the state's willingness to invest aggressively, with the use of new technologies acquired from abroad (through the purchase of licensing or joint ventures). The competitiveness of local firms is based not only on factors of production, but also on more advanced business strategies. The highest level of development of national competitiveness is the innovation-driven phase. At this stage of economic development competitiveness is based on innovation, unique business strategies of domestic companies and on globally recognized products and brands. Significant outward foreign direct investments emerge at this stage, as domestic companies seek to exploit their competitive advantages abroad (Porter, 1990, p. 552, 554).

Obviously, the reality of economic development does not always copy the assumptions of economic models, what is true also for the development phases of national competitiveness. This fact is ultimately recognized also by Porter himself, who admits that most countries did not fit exactly into the model (Porter, 1990, p. 545). However, it is a very interesting and useful tool for the analysis of national competitiveness. The importance of Porter's model is ultimately reinforced by the fact that the Global Competitiveness Report has largely taken over Michel Porter's methodology and also ranks countries in three stages of economic development, which is very similar to Porter's phases—factor-driven, efficiency-driven and innovation-driven (World Economic Forum, 2010).

Of course, the examination of national competitiveness cannot be limited to the works of Michael Porter. Indeed, in recent years interesting alternatives to the model were created by various authors. However, currently the field of study is largely fragmented and there is not a universally accepted definition of national competitiveness today. The problem is that the concept of national competitiveness can be defined in different ways and can be influenced by a number of different factors. For example, Michael Porter defines national competitiveness strictly on the basis of productivity. According to him, this is the only relevant perspective to national competitiveness (Porter, 1998, p. 160).

Other authors have different approaches to the concept of national competitiveness. However, in the variety of definitions, we are able to find some unifying ideas and themes. For example, Scott and Lodge define national competitiveness as "a country's ability to create, produce, distribute, and/or service products in international trade while earning rising returns on its resources" (Scott, Lodge and Bower, 1985). Blaine uses a similar approach to national competitiveness when he describes it "a nation's competitiveness refers to its ability to produce and distribute goods and services that can compete in international markets, and which simultaneously increase the real incomes and living standards of its citizens" (Blaine, 1993). Lastly, a very similar definition of national competitiveness can be found in the World Competitiveness Yearbook. This defines it as the country's ability to create added value and thereby increase the wealth of the nation (IMD, 2010).

It is visible that the above definitions describe national competitiveness in the context of the ability to succeed in the global economy, with the ultimate goal to increase the nation's wealth and the standard of living of its inhabitants. It is also a dynamic category that is changing over time. The state is taking a major role in changes of national competitiveness, since economic policy can affect it both positively and negatively. That is why governments, economists and business entities became very interested in tracking the changes in the development of national competitiveness. The result of these efforts to define and compare national competitiveness on the global level was the emergence of global competitiveness indexes, which are trying to create a comparable marker on the international level. Currently there are two widely accepted yearly publications dealing with the topic of global comparison of national competitiveness—Global Competitiveness Report and World Competitiveness Yearbook.

Nevertheless, it should be noted that the publications offering global comparison of national competitiveness also have their opponents, who accuse them that they offer only a limited perspective of national competitiveness—a view through the prism of economic and political institutions. The Global Competitiveness Index is the most heavily criticized, which defines national competitiveness as "a set of institutions, policies and factors that affect productivity in the economy" (World Economic Forum, 2010). The reality, however, is that currently there are no better tools for the analysis of national competitiveness. These are the only global indexes of national competitiveness which include dozens of countries, thus they allow serious international comparison. That is the main reason for the utilization of these two publications in this paper.

3 THE METHODOLOGICAL BACKGROUND OF THE GLOBAL INDICES OF NATIONAL COMPETITIVENESS

The gradual development of a global economy created a demand for tools for global comparison of national competitiveness, which would be widely accepted. Since the 1980s, several approaches to measuring competitiveness at the global level began to develop, and over time two indices emerged, which are currently considered the most important and most widely accepted ones. The first is the Global Competitiveness Index and the second is the World Competitiveness Index. Both indices are published annually and include all countries analyzed in this paper. However, before we begin to evaluate the changes in national competitiveness in Central Europe based on these two indices, we will try to provide some methodological insight into their composition.

3.1 Global Competitiveness Index

The Global Competitiveness Index (GCI) is a part of the Global Competitiveness Report, published annually by the Swiss think-tank World Economic Forum. The latest edition of the report, dated 2013–2014, includes a current global ranking of national competitiveness based on the GCI. The current edition of the publication contains data on 182 countries. All the countries from Central Europe were included in the global ranking among the first states of the former Soviet bloc countries.

The GCI is a complex index, which in itself contains a large amount of numerical data, and also "soft" data collected through an international survey carried out on the level of middle and top managers of large corporations. The actual index is based on twelve pillars, which are divided into three main groups—basic requirements, efficiency enhancers and innovation and sophistication factors. It is important to note that while the weight of each group in the resulting index is not static, the changes depend on the nature of the individual countries. For poor developing countries, which are in the early stages of economic development, basic requirements are the most important group with a total weight of 60% in the GCI. On the contrary, for economically developed countries the same group has a weight only of 20%. On the other hand, the position of the innovation and sophistication factors is much more important. This approach allows a more realistic assessment of the competitiveness of less developed countries, since their economic conditions are quite different from the economically developed countries. The overall GCI index is finally obtained by aggregating the partial values in each category using the predefined weights.

3.2 *World Competitiveness Index*

The World Competitiveness Yearbook, the other highly respected publication examining national competitiveness, has been published since 1989 by the Swiss non-profit organization and business school International Institute for Management Development. This publication has a narrower range; the latest edition from 2013 examines only 60 highly developed economies. This publication also contains an international ranking of national competitiveness based on World Competitiveness Index (WCI hereafter). After the collapse of the socialist bloc, countries from Central and Eastern Europe were added fairly quickly to this yearbook.

Like the previously mentioned GCI index, the WCI is also a composite index, which includes 246 primary and 81 supporting indicators in its current form. We have to mention that WCI uses not only quantitative but qualitative indicators as well. The statistical data used in the calculations is obtained from various international and national institutions and this so-called "hard" data is used in 131 major and 81 supporting indicators. On the other hand, 115 indicators use "soft" data, which is gathered from a survey conducted among managers of major national and international corporations. These data are used to analyze areas where the availability of "hard" data is limited, such as management practices, corruption, bureaucracy or quality of life. The questionnaire survey is conducted in each state under review and the questionnaires are distributed among managers at middle and senior management level. The sample size for each state corresponds with the size of the economy. In the recent edition of the publication, 4,460 questionnaires from 59 countries were processed. Respondents evaluate each area on a scale from 1 to 6, which is subsequently converted to a standardized scale from 1 to 10.

Unlike the GCI index, WCI works with fixed weights of the sub-indices. Each area has an equal weight of 5% in the final index, so the 20 main areas of analysis together provide 100%. The report provides 327 sub-indices and the 246 most important ones create the final WCI index. This index is then used to create an international ranking of national competitiveness of the countries included in the report.

4 CHANGES IN THE NATIONAL COMPETITIVENESS OF THE COUNTRIES ANALYZED AFTER THE GLOBAL CRISIS OF 2008/09

The financial crisis that started in 2007 in the USA quickly developed into a full blown global financial crisis with deep consequences for the real economy. In 2009, the world economy was already in recession, for the first time since the recession after the oil crisis in the 1970s. All the key economic powers were in recession (except China) and global trade and investment flows contracted. The impact on Central and Eastern European economies was no different, as the whole region suffered from recession and increasing unemployment.

Table 1 shows the development of GDP in the chosen sample before and after the global economic crisis. It is clear that these countries were growing above the EU average before 2009, and even double digit economic growth was not unheard of (Slovakia, Estonia or Latvia). The first signs of economic trouble can be observed already in 2008, when economic growth started to slow in almost all countries and Estonia and Latvia slipped into recession. The situation in 2009 became dire, as all countries in Central Europe experienced a deep contraction of GDP. Poland became the only country in the region (and in the whole EU) that was able to retain economic growth even in this difficult year.

The global economy rebounded from the recession fairly quickly after 2009 and the economies in Central and Eastern Europe followed a similar pattern. Most of the countries in the region recovered quickly from the recession and returned to economic growth. The bad news is that this growth is lower than the economic growth before the crisis and it is more volatile. Most countries needed a strong fiscal consolidation which (together with weak domestic consumption) contributed to decreasing growth rates in the region. As a result, some countries even slipped into recession in 2012 and 2013 (Hungary, Czech Republic and Slovenia).

So, the available economic data confirms a deterioration of macroeconomic results in Central and eastern European countries. "The new normal" is a lower level of economic growth, a higher level of unemployment, and a higher level of public debt. The question is whether these changes translate into an unfavorable change in the national competitiveness of these countries. In order to answer this question we will use the indices of national competitiveness published in the Global Competitiveness Report and the World Competitiveness Yearbook.

Table 2 displays the position of the new EU member states from Central Europe in the global ranking of national competitiveness created using the GCI between 2005 and 2013. To observe the changes caused by the global economic crisis, we have to note the developments after 2010—as this was the first year the post-crisis data were incorporated into the GCI index. Overall, there is no common trend visible in our group of 10 states. If we compare the positions in the global ranking, only 5 countries fell in the rankings (Czech Republic, Hungary, Romania, Slovakia and Slovenia). In contrast, the other 5 countries were able to improve their national competitiveness, some of them considerably (e.g. Bulgaria climbed 19 places between 2009 and 2013). The average position of the analyzed countries fell slightly after the crisis, but not dramatically (from 51.5 to 55.6).

If we try to look for common factors among countries with declining national competitiveness, 4 out of 5 countries in this group belong to former frontrunners of economic transition in Central and Eastern Europe. Czech Republic, Hungary, Slovakia and Slovenia have suffered from below average economic growth since the crisis, and the governments of these countries have often made controversial economic policy decisions. Slovakia shows the worst

Table 1. Real GDP growth in selected Central European Countries 2005–2013 (%).

	2005	2006	2007	2008	2009	2010	2011	2012	2013
Bulgaria	6.4	6.5	6.4	6.2	−5.5	0.4	1.8	0.6	0.9
Czech R.	6.8	7.0	5.7	3.1	−4.5	2.5	1.8	−1.0	−0.9
Estonia	8.9	10.1	7.5	−4.2	−14.1	2.6	9.6	3.9	0.8
Hungary	4.0	3.9	0.1	0.9	−6.8	1.1	1.6	−1.7	1.1
Latvia	10.1	11.0	10.0	−2.8	−17.7	−1.3	5.3	5.2	4.1
Lithuania	7.8	7.8	9.8	2.9	−14.8	1.6	6.0	3.7	3.3
Poland	3.6	6.2	6.8	5.1	1.6	3.9	4.5	2.0	1.6
Romania	4.2	7.9	6.3	7.3	−6.6	−1.1	2.3	0.6	3.5
Slovakia	6.7	8.3	10.5	5.8	−4.9	4.4	3.0	1.8	0.9
Slovenia	4.0	5.8	7.0	3.4	−7.9	1.3	0.7	−2.5	−1.1

Table 2. The position of Central European countries in the global rankings of national competitiveness according to GCI 2005–2013.

	2005	2006	2007	2008	2009	2010	2011	2012	2013
Bulgaria	58	72	79	76	76	71	74	62	57
Czech R.	29	31	33	33	31	36	38	39	46
Estonia	20	26	27	32	35	33	33	34	32
Hungary	35	41	47	62	58	52	48	60	63
Latvia	44	44	45	54	68	70	64	55	52
Lithuania	43	39	38	44	53	47	44	45	48
Poland	43	48	51	53	46	39	41	41	42
Romania	67	68	74	68	64	67	77	78	76
Slovakia	36	37	41	46	47	60	69	71	78
Slovenia	32	40	39	42	37	45	57	56	62
Average ranking	40.7	44.6	47.4	51	51.5	52	54.5	54.1	55.6

performance in the post-crisis years, as it lost 31 places in the global rankings between 2009 and 2013. This effectively means that in 2013 Slovakia was the least competitive country in the selected group of Central European countries.

On the other hand, the progressive Baltic states were able to improve their national competitiveness even after the global economic crisis hit these countries especially hard. But the governments of these countries did not choose populist economic policies, and continued to foster their business environment and national competitiveness. Poland is a similar case, as the Polish government continued to improve the business environment despite the global crisis.

The World Competitiveness Scoreboard assembled by the authors of the World Competitiveness Yearbook shows similar results to the Global Competitiveness Report. The movements of the Central European countries in the rankings are less dramatic, but we have to bear mind that the World Competitiveness scoreboard analyzes a more limited sample (ex. 60 countries in 2013). Latvia had to be excluded from the analysis according to WCI, as it was included in the WCI rankings only in 2013.

The remaining 9 countries in our sample show signs of deteriorating national competitiveness after 2009. The biggest fall in the average position of the sample countries in the WCI rankings happened in 2010, when the data of the disastrous year 2009 were incorporated into the index. In 2010, the average ranking of the analyzed countries fell to 43.1 from 37.9 in 2009. Interestingly, in the following years the position of the sample countries stabilized and the average ranking oscillated around 43 from 2011 to 2013. This could mean that the global economic crisis of 2008/09 was a unique external shock that had a lasting negative impact on the national competitiveness of the Central European economies.

Once again, there are differences between the sample countries. While most of the countries declined in the rankings, Poland was able to improve its national competitiveness even in the times of complicated conditions in the world economy. In 2009 Poland occupied rank 44 in the WCI scoreboard and was improve its position to 33 in 2013. This is the biggest improvement in our sample, no other country was able to climb significantly in the rankings. We already mentioned that the Polish government was dedicated to economic policies leading to modernization and a better business environment. The positive development of Poland is visible in all global indices such as Doing Business created by the World Bank or Index of Economic Freedom created by the Heritage Foundation.

On the other hand, the WCI rankings also confirm that the underperformers of the group showed a steep decline in national competitiveness. Slovakia and Slovenia were the worst performing countries in the group after 2009, which is consistent with the results from the GCI rankings. The national competitiveness of these countries rapidly deteriorated after 2009, which shows economic policies that were unable to cope with the problems created by the crisis. The situation of the Czech Republic and Hungary was somewhat better, but both

Table 3. The position of Central European countries in the global rankings of national competitiveness according to WCI 2005–2013.

	2005	2006	2007	2008	2009	2010	2011	2012	2013
Bulgaria	–	46	41	39	38	53	54	54	57
Czech R.	36	31	32	28	29	29	30	33	35
Estonia	26	20	22	23	35	34	33	31	36
Hungary	37	41	35	38	45	42	47	45	50
Latvia	–	–	–	–	–	–	–	–	41
Lithuania	–	–	31	36	31	43	45	36	31
Poland	57	58	52	44	44	32	34	34	33
Romania	55	57	44	45	54	54	50	53	55
Slovakia	40	39	34	30	33	49	48	47	47
Slovenia	52	45	40	32	32	52	51	51	52
Average ranking	43.3	42.1	36.8	35.0	37.9	43.1	43.6	42.7	43.7

Table 4. Development of the 12 main pillars of the GCI for Slovakia 2007–2013 (rankings).

	2007	2008	2009	2010	2011	2012	2013
Institutions	50	60	73	78	89	101	104
Infrastructure	53	58	64	63	57	57	56
Macroeconomic environment	37	62	49	40	32	56	54
Health and primary educat.	65	39	44	48	45	43	42
Higher educat. and training	39	41	45	47	53	53	54
Goods market efficiency	38	35	35	32	51	51	54
Labor market efficiency	24	25	36	29	40	59	86
Financial market dev.	27	33	31	28	37	47	48
Technological readiness	33	36	36	33	34	37	45
Market size	53	57	56	57	58	58	59
Business sophistication	47	52	53	51	57	63	61
Innovation	42	51	58	68	85	96	89

countries fell several places in the WCI rankings. The Baltic states and Romania remained relatively constant after the global crisis. Lithuania is especially noteworthy in that it was able to recover its national competitiveness in 2012 and 2013 after the losses in 2010 and 2011.

Bulgaria is a very interesting case, as the GCI ranking and the WCI ranking show a different trajectory of national competitiveness for this country. In the rankings published in the World Competitiveness Report, Bulgaria improved its national competitiveness considerably in 2012 and 2013 (from 74th place to 57th place). On the other hand, in the WCI ranking shows that Bulgaria lost 19 positions after 2009. What is the cause of these different paths? One reason is the number of countries in each ranking. As the WCI ranking analyzes only 60 countries the movements of countries are not as dramatic as in the GCI ranking that contains 148 countries in its latest edition. The other reason is the different methodology used to calculate GCI and WCI that can lead to somewhat different results. Nevertheless, the conflicting results for Bulgaria call for a deeper analysis of the sub-indices that exceeds the scope of this paper.

Clearly, not all changes in national competitiveness after 2009 can be traced back to the global economic crisis of 2008/09. The methodology of the indices on national competitiveness is complex and many factors are dependent on the economic policies of the national government. In the last part of the paper we demonstrate this reality in the case of Slovakia using the Global Competitiveness Index. As we already stated, the national competitiveness of Slovakia measured by the GCI decreased considerably after the global economic crisis. The question is, is it a coincidence, or was the crisis a key factor in this development?

To answer this question we have to look at the 12-pillar structure of the GCI (Table 4). In the case of Slovakia, the ranking of the country deteriorated in 10 of these 12 pillars after 2009. Slovakia achieved the worst results in the pillars Institutions, Labor market efficiency and Innovation. The composition of these pillars clearly shows that these sub-indices respond more to economic policies than to changes in the external economic environment. Labor market efficiency is a good example, as Slovakia started to lose its position after the government of Robert Fico introduced a new less flexible labor code. Similarly, the quality of institutions is also dependent on the policies of the national government. On the other hand, the pillar Macroeconomic environment that is supposed to mirror the negative changes in the world economy, did not suffer significant losses (rank 49 in 2009 and rank 54 in 2013). So, it is safe to say that the policies of the Slovak government played a more significant role in the decreasing national competitiveness than the global economic crisis.

5 CONCLUSION

The main goal of this paper was to assess the impact of the global economic crisis on the national competitiveness of Central and Eastern European countries represented by the 10 new EU member states from this region. To assess national competitiveness we used the global rankings created using the Global Competitiveness Index and the World Competitiveness Index.

The available data from both rankings clearly indicates that there was an overall deterioration of national competitiveness across the region after 2009. Nevertheless, the decline was not universal, as there were several countries with improving national competitiveness even after the global crisis (e.g. Poland and the Baltic states). As these countries introduced progressive reforms within the time period analyzed, we can conclude that national competitiveness is more dependent on the policies of the national governments than on the state of the external economic environment. A deeper analysis of the Global Competitiveness Index of Slovakia confirms this premise, as the worst performing sub-indices after 2009 were those that were the most policy-dependent (institutions and labor market).

Ultimately, it is too early draw complex conclusions from the changes in the global indices of national competitiveness since the global economic crisis. The time series are short yet, so it is hard to distill the real impact of the crisis. It will be useful the revisit this question in several years when the lasting impact of the global economic crisis will be better measurable.

REFERENCES

Blaine, M. (October 01, 1993). *Profitability and Competitiveness: Lessons from Japanese and American Firms in the 1980s.* California Management Review, 36, 1, 48–74.
Institute for Management Development. (2010). *IMD world competitiveness yearbook 2010.* Lausanne: IMD.
Institute for Management Development. (2013). *IMD world competitiveness yearbook 2013.* Lausanne: IMD.
International Institute for Management Development. (2006). *IMD world competitiveness yearbook 2006.* Lausanne: International Institute for Management Development.
Krugman, P.R. (1994). *Competitiveness: A dangerous obsession.* Foreign Affairs, 73, 2, 28–44.
Porter, M.E. (1990). *The competitive advantage of nations.* New York: Free Press.
Porter, M.E. (1998). *On competition.* Boston, MA: Harvard Businesss School Pub.
Porter, M.E., Schwab, K., López-Claros, A., & World Economic Forum. (2006). *The global competitiveness report 2006–2007.* Basingstoke: Palgrave Macmillan.
Scott, B.R., Lodge, G.C., & Bower, J.L. (1985). *U.S. competitiveness in the world economy.* Boston, Mass: Harvard Business School Press.
World Economic Forum. (2010). *The global competitiveness report 2010–2011.* Geneva: World Economic Forum.
World Economic Forum. (2013). *The global competitiveness report 2013–2014.* Geneva: World Economic Forum.

Current Issues of Science and Research in the Global World – Kunova & Dolinsky (Eds)
© 2015 Taylor & Francis Group, London, ISBN 978-1-138-02739-8

Open innovation strategy—a complex approach in environmental innovation processes design

P. Molnár
Faculty of Economy and Business, Pan-European University in Bratislava, Slovakia

ABSTRACT: Author of a paper tries to present brand new managerial approaches in innovation management activities, in which author is a specialist. Therefore, some specific approaches (e.g. "complex innovation strategy", "spreading—chaining of impulse innovation" as far as interconnection of these with "the open innovation strategy" from the authors Cheobough, Joel and Vanhaverbeke (2006) [1] are results of author's educational and research activities, thus those are presented as author's original ones.

Prior orientation was highlighted towards environmental innovation space. Complex innovation strategy is a classical approach before introducing "open innovation" paradigm practical realization. Complexity of environmental innovation management covers particular "impacts" in innovative actions with an entire cooperation within internally and/or externally specialized subjects. Moreover, the same approach is going to be realized in additional, several other spheres in innovation management activities. This is what the ordinary and/specialized reader could use for his/her innovative actions—inspirations and realizations!

1 INTRODUCTION

1.1 *Management of innovation processes in relation with environmental requirements (Environmental Relationship Management—ERM)*

In connection with one of the major imperatives of recent period, an "up-to-date" phenomenon is creation of innovative concepts in products and services respecting environment issues. This means that innovative proposals must implement environmental actions and initiatives from the very beginning to the very end of a product life cycle. Designers are accepting existing European Union laws, national regulations and public pressure and provide "environmentally friendly proposed" innovations. Within the variety of approaches and techniques to prevent waste, the newest one is aggregated into the six main areas—the "6th Re Hierarchy", as it was former stated in Molnar and Jachymovic (2011) [5].

Innovations accepting this hierarchy have firstly to satisfy requirements in a product development stage:

1. Reduce—reduce particular amount of material, energy, water consumption, etc.,
2. Reuse—reuse and/or use whole, or a part of the product for the same purpose after its basic life cycle termination,
3. Refurbish—renovation (re-designing) of existing product and/or its parts
4. Remanufacture—reworking of existing product and/or its parts
5. Recover—use for energy purposes, composted, or the other way again to use the whole product or its parts,
6. Recycle—to use the product for the other purposes, for creating completely different product,
7. Dispose—remove, destroy, demolish and clean up the original product or its parts.

Following this hierarchy, companies have started to limit their waste treatment approach in order to set preventing waste hierarchy approach. Experience shows that the use of waste prevention and environmentally conscious purchasing lead firms to reduce costs and increase profits.

Cost savings may take different forms:

- lowering disposal costs,
- lowering costs for waste treatment,
- lowering energy costs,
- savings in materials and supplies,
- reducing cost of regulatory (mandatory) permits,
- lowering storage costs,
- returning costs through the sale of recyclable materials,
- returning costs by selling 6th Re's technology.

Small and Medium-sized Enterprises (SMEs) are decisive fields of success or failure of sustainable development. Scientists are warning now that we move toward a "point of no return" in climate change, polluting the natural compartments with too many emissions, as it was documented in Molnár and Dolinský (2012) [3]. Therefore, the aim of every responsible entrepreneurial subject has to be to force all similar or cooperative co-subjects in similar initiatives and/or actions.

In this connection, the "open innovation paradigm—strategy" has to be the new way of spreading ideas and innovations in sustainable—environmentally oriented—development.

According to authors Cheobough, Joel and Vanhaverbeke (2006) [1] "open innovation" is meant as "useful knowledge widely distributed using external knowledge sources as a core process in innovation".

Concentrating on Small and Medium sized Enterprises (SMEs), industrial innovations are the proper space for implementing the above mentioned approach. In this connection, "complex innovation" designed in Molnar (2007) [2] methodology has to be placed as analogy in achieving expected results within sustainable development initiative.

1.2 Basic issues and principles of complex innovation approach

Complex innovation methodology deals with the assumption that every particular innovation does not act in isolation, vice versa, it influences the other company activities as well as activities of the relevant neighbourhood.

The result of this approach is the notion (theory) of "**impulse (stimulating) innovation**".

The most important domain (department, venue) of impulse innovation is the design area (department) of a company.

It is a common experience, that from 90% to 95% of product innovations are created (invented) in design departments. The utility of every product innovation is based on new ideas—inventions, industrial designs, utility designs and know–how's, which are primarily invented in design departments. Innovations in design generally evoke the need to make innovations (in the same quality—the same innovation degree) in other parts of the analysed system, and the same in the wider surroundings, and all these in a relatively very short period (changes—innovations in the design of a product evoke the necessity to make innovations in technology and in the organizational aspects of production, in energy consumption area, etc.). This process is a kind of **horizontal spreading (chaining) of impulse innovation.**

After a time, the need to provide additional, mostly lower degree innovations, will come automatically, again in the design department the most, and followed in next relevant areas in company processes. This process, which was "evocated" (invented) by a lower degree designed innovation, will make the chain within following lower degree innovations in the same area of company activity, and in wider surroundings as well. This activity is a **vertical spreading (chaining) of impulse innovation.**

Horizontal and vertical spreading (chaining) of impulse innovation leads to wide area from the original (impulse innovation) in innovation management, and explicitly to mutual cooperation among tangible subjects, which corresponds to the "open innovation" strategy.

It has been considered: "the higher the degree of impulse innovation, the wider the spreading (chaining) of developed innovations could be achieved".

The wide spreading of impulse innovation in the system (company) and in its surroundings is "designing" the **operational radius of impulse innovation**.

Utilization of operational radius of impulse innovation rule could—with complexity and with maximum effective realization—provide the sort of organization system (company), which is able to enforce and ensure all developed innovations in the area of the operational radius of a complex innovation. The general process (basic chart) within a company is presented on the graph—Figure 1.

In Figure 1:

a – impulse innovation,
b – evocated innovations from impulse innovation in the same innovation spectrum degree, or the lower ones
c, d – evocated innovations from the "b" ones in the same innovation spectrum degree, or the lower ones
e, f, g – evocated innovations from the "c" and "d" in the same innovation spectrum degree, or the lower ones.

Particular arrows in the Figure 1 represent the influencing activities (initiatives) of one department (subject) to the neighbouring ones in order to achieve chaining (synergy) effects.

Operational radius of the impulse innovation means the "power" (ability) of the impulse innovation to influence the most relevant departments (subjects) in horizontally and/or vertically directions.

Analogy with the open innovation "philosophy", as it was mentioned in Cheobough, Joel and Vanhaverbeke (2006) [1], realisation of "useful knowledge", as it was stated in in

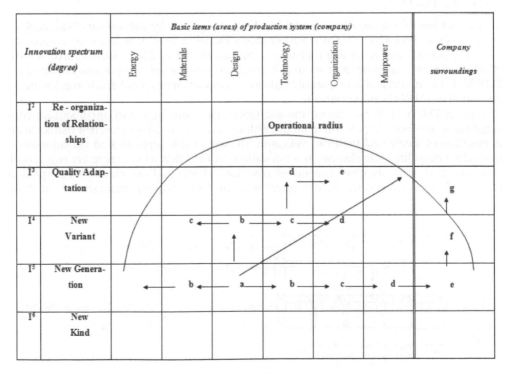

Figure 1. The operational radius of impulse innovation.

environmental friendly innovations is the utilization of abovementioned approach in all of cooperating subjects of a company—suppliers, customers, spin-offs, relevant research and academic institutions, etc.

Particularly, results in environmentally friendly achievements in core entrepreneurial subject have to be provided in entirely activities of cooperating bodies. Positive methodology results in e.g. innovations in "total electricity consumption" in the main subject has to be provided to all tangible subjects in order to synergize environmentally friendly effects towards sustainable development. Regarding the former research achievements at our university research, positive decreasing of (e)IMPACT, using this approach, will continuously influence cooperation with all the surrounding subjects, thus more environmental friendly impacting space will be achieved.

The way of application is presented in developed "universal assessment model" within the article of Molnár and Dolinský (2012) [3].

1.3 Universal assessment model

In order to present practically applied model, electricity consumption impact measuring is an example to design the way of application. Every organization (company), department, office, etc., is consuming electricity. Electricity consumption (e)IMPACT) is one of the best indicators for expressions of changes in behavior, managerial decisions (e.g.: organizational and process innovations resulting into lower electricity consumption).

In the Figure 2, projection of entire (e)IMPACT assessment of electricity consumption is presented:

– calculation of (e)IMPACT of identical electric appliances (consumptions) (1)—the First Tier,
– calculation of (e)IMPACT of all electric appliances within one department (organization subjects) (2)—the Second Tier,
– calculation of (e)IMPACT of entire organization (complete activities of an organization) (3)—the Third tier.

Categorization of the suggested model has been designated by the authors Narodoslawsky and Krotscheck (2007) [6] as a Sustainability Decision Tool. It is based on comparing two time periods and recording a change occurred. Model is able to identify extend into which organizational, process or product innovation is responsible for positive improved (e) IMPACT. The model is able to quantify influence a concrete managerial decision is having on an overall (e)IMPACT improvement.

Term (e)IMPACT improvement, the <u>size (particular value in m²) of earth surface consumed</u> is understood. Expression in m² of earth surface consumed was reached thanks to an application of Ecological footprint indicator. The use of this sophisticated indicator automatically brings life cycle element into evaluation. In the calculations, there are two sets of elements used: multiplicative elements of equation (1) and additive elements of equation (2, 3). Information delivered by the model is devoted to company management, to those

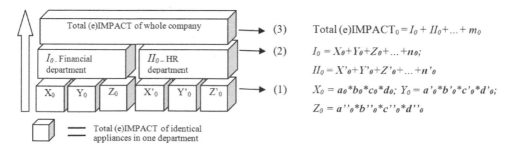

$$\text{Total (e)IMPACT}_0 = I_0 + II_0 + \ldots + m_0 \quad (3)$$

$$I_0 = X_0 + Y_0 + Z_0 + \ldots + n_0; \quad (2)$$

$$II_0 = X'_0 + Y'_0 + Z'_0 + \ldots + n'_0$$

$$X_0 = a_0 * b_0 * c_0 * d_0; \quad Y_0 = a'_0 * b'_0 * c'_0 * d'_0; \quad (1)$$

$$Z_0 = a''_0 * b''_0 * c''_0 * d''_0$$

Figure 2.   Universal three-tier model.

having decision making power to implement Corporate Social Responsibility concept into company's business strategy. When CEO spots improvement in (e)IMPACT, he/she is logically interested in knowing the reason of this improvement in order to identify inventors, i.e. author of the environ innovation with the following premium. CEO wants to know, which factor of particular environmental innovation, caused the improvement. This information will be delivered via this model. The model is able to state, which department had the biggest% contribution into overall improvement, what was calculated within the work of Molnár and Dolinský (2012) [3].

Consecutively, additional positive innovation impacts will be utilize in the future periods:

- (m)IMPACT—consumption of materials—raw materials in technology, in production, in administration, etc.
- (h)IMPACT—heat energy consumption, etc.
- (w)IMPACT—water consumption, etc.
- (ws)IMPACT—waste production, etc.
- (em)IMPACT—emission production, etc.
- …etc.

are subjects of analysis and innovation activities within all cooperating bodies as continuing processes in similar way as it was presented on the (e)IMPACT example—electrical energy consumption savings, as it is designed in the chart, Figure 3.

Particularly, the basic chart (Figure 1) will be after utilization of abovementioned innovative actions designed as shown in Figure 3.

All these initiatives could be based on a "friendly approach"—mostly for cooperating subjects (employees, suppliers, customers, etc.), and/or for competing subjects, too. In this "second party" case, providing new innovations in an "unfriendly (competitive) environment", those actions will keep the basic conditions and "philosophy" of "open innovation model", which advises that "useful knowledge is no longer concentrated in a few large organizations, but companies use innovations from outside via **licensing**, etc., and vice versa look outside their boundaries for ideas (innovations) and intellectual properties they can bring—again in accordance with the authors Cheobough, Joel and Vanhaverbeke (2006) [1]

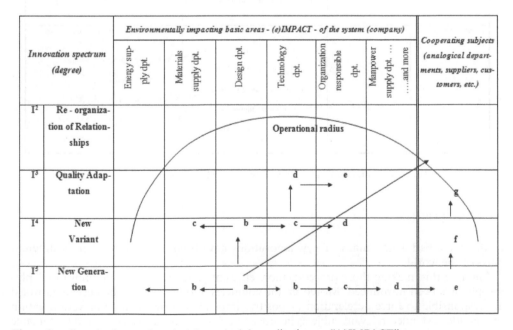

Figure 3.   Complex innovation chaining principle application on "(e)IMPACT".

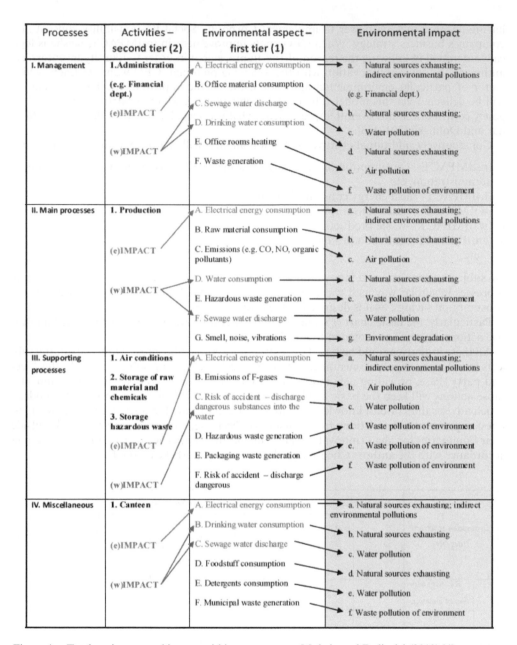

Figure 4. Total environmental impact within a company—Molnár and Dolinský (2012) [4].

2 CONCLUSION

In the developed world, many of new innovations have been invented parallel on different parts of the world.

Limiting this, making those processes more efficient generally, an "open innovation" philosophy has been introduced. Moreover, the countries and citizens of the developed world are responsible to force acceleration of positive implementation of any innovation within the broader extended relevant subjects in order to achieve more prosperous and sustainable results in human development. Presented approach is possible to apply to any innovations, and recently in environmental oriented innovation actions the most!

REFERENCES

[1] Cheobough, H., Joel, W., Vanhaverbeke,W.: *Open innovation*. Oxford University Press, 2006.

[2] Molnár, P.: *Innovation Management*. Bratislava, University of Economics in Bratislava, 2007.

[3] Molnár, P., Dolinský, M.: *Environmental Performance Assessment Model as a Sustainability Decision Tool in SMEs*. In: World academy of science engineering and technology, Issue 65. May 2012.

[4] Molnár, P., Dolinský, M.: *Total Environmental Assessment Framework in an Organization*, In: Creative and Knowledge Society. Volume 3, Issue 2. 2014.

[5] Molnár, P., Jachymovic, V.: *Manažement inovácií podniku*. Bratislava, University of Economics in Bratislava, 2011.

[6] Narodoslawsky, M., Krotscheck, CH.: SPIonExcel—fast and easy calculation of the Sustainable Process Index via computer Resources. In: Conservation and Recycling. 2007, vol. 50.

Current Issues of Science and Research in the Global World – Kunova & Dolinsky (Eds)
© 2015 Taylor & Francis Group, London, ISBN 978-1-138-02739-8

A note on the social security systems with built-in pro-fertility measure

V. Palko
Faculty of Informatics, Pan-European University, Bratislava, Slovakia

ABSTRACT: In the paper social security systems with built-in pro-fertility measures are defined. These systems do not only take into account the contributions of the pensioner, but also the number of his children and the amount of their contributions. Reflections on the need of these systems are based on the published negative effects of social security on fertility and demographic developments. This article provides a brief history of proposals and attempts to implement such systems in the policies of several countries.

1   DEMOGRAPHIC FRAMEWORK

1.1   Demographic decline

In the second half of the twentieth century we saw an unprecedented decline of total fertility rates (i.e. the expected number of children per woman) in most countries worldwide. During the recent decades, in Europe, in the Russian Federation and many other countries, the Total Fertility Rate (TFR) achieved the form of a sub-replacement fertility. Nowadays, the average TFR in the European Union countries is slightly below 1.6 (Eurostat 2014), in the Russian Federation it is 1.6. Slovakia is below the EU average, with TFR 1.4 (CIA 2014).

1.2   Unsustainability of social security and healthcare systems

Fertility decline and subsequent future decline of economically active population incorporates a discussion on the sustainability of state pension and health care systems. The following scheme is a brief result of the discussion:

Fertility decline under the sub-replacement level

⇒

Unsustainability of generous social security and healthcare systems

This unsustainability is unjust. The current contributors to the system, the future retirees, will not be reimbursed with the sum they contributed to the system in the active phase of their lives, since the aging and decline of the population will cause the decrease of the number of active contributors and the increase of pensioners in the future.

This injustice is explained, for example, in the latest report of the German Bertelsmann Foundation (Werding 2014). The author of the report counted the marginal revenue (for German state finances) of a German child born in the year 2000. The hypothetical child is considered average in all possible aspects, e.g. life expectancy, education, earnings, taxes, etc. On the one hand, all projected financial flows from the state to this child (cost of its education, child benefits, healthcare costs, its future pension, etc.) are counted; on the other hand all opposite flows (future payments of all kinds of taxes and contributions to the pension and healthcare systems) are counted. The balance is highly positive for the state, and negative for the child. The balance reached 50.500 €, and if we take into account also the hypothetical children of the child, the balance is as high as 103.400 €. It is evident, that this balance is caused by the existing pension and healthcare system, combined with the fertility decline. This balance "has the character of a net transfer to the previous, older, generation" (Werding 2014).

1.3 Weakened ties between parents and children

Social security systems weaken economic ties between parents and children. According to Gary Becker: "Social security payments financed by taxes on the working population reduce the amount spent by children to support retired parents, because children recoup for themselves some of their resources taxed away" (Becker 1981).

1.4 Social security as a cause of fertility decline

During the last two decades at the microeconomic level the discussion is held: Child—consumption goods or investment? There are two main competing theoretical models: In the Becker-Barro model, parents are altruistic, caring for their own, but also for their children utility (Becker & Barro 1986); In the alternative model (Boldrin et al. 2005), parents see children also as their "old age security". The results show, that generous social security systems are one of the reasons for the fertility decline. Therefore, the opposite logical implication exists to the one mentioned in Part 1.2. That is:

Generous social security systems

$\Rightarrow$

Fertility decline

This implication is caused by the price effect, known from the microeconomics. The introduction of the social security increases the price of children, since, as mentioned in Part 1.3, it decreases the old age benefits of a parent in the form of financial flows from the child to the parent. It is claimed in (Billari & Galasso 2009), that theoretical results of (Boldrin et al. 2005) were empirically confirmed for example after the Italian pension reform from the nineties, where the worsened prospects of high pensions slightly increased the fertility rates.

2 PRO-FERTILITY MECHANISM IN PENSION SYSTEMS

2.1 Systems independent of children

It is typical for the current pension systems ("pay-as-you-go", but also for saving pension systems), that the person's pension depends almost solely on the amount of the former contributions of the pensioner to the system. So, in the language of mathematics:

$$F = F(c_1, c_2, ..., c_k)$$

where F is the amount of the pension (per time unit, for example, month) and vector $c = (c_1, c_2, ..., c_k)$ is the vector of pensioner's contributions to the systems during the past time periods 1, 2, ..., k (for example months). One can expect, that F is positive and increasing in any variable c_i. Basically, F is independent of the number of pensioner's children, although it is clear, that the survival of the systems depends on the quantity of the future generations (and also quality in the sense of children quality and quantity discussion).

2.2 Parental pillar

A social security system will be called a social security system with built-in pro-fertility measure, if

$$F = F(c_1, c_2, ..., c_k, n, d_1, d_2, ..., d_n)$$

where n is the number of pensioner's children and d_j is the contribution of the j-th child to the system during some (unspecified) period. In a simpler form F can be defined as the sum of two mutually independent functions in the following way:

$$F = F_1(c_1, c_2, ..., c_k) + F_2(n, d_1, d_2, ..., d_n) \qquad (1)$$

where F_1 is the "classic" pension derived from pensioner's own contributions and F_2 depends on his children. F_2 is assumed to be positive and increasing in any variable.

Let us mention, that implementation of F in the form (1) means that price of the children decreases over the lifetime of the parent by the amount

$$F_2(n, d_1, d_2, ..., d_n) \qquad (2)$$

what can cause increasing demand for children in the frame of microeconomic price effect. Of course, to estimate the elasticity of this demand is another issue. We call F_2 the *parental pillar*.

The variable n represents the quantity of children and variables $d_1, d_2, ..., d_n$ the quality of children. In special cases F_2 can be of the form

$$F_2 = F_2(n) \qquad (3)$$

when the quality is ignored, or of the form

$$F_2 = F_2(d_1, d_2, ..., d_n) \qquad (4)$$

when the quality of each child is included.

3 ATTEMPTS TO PARENTAL PILLAR

3.1 *German discussion*

The idea of including children into the pension system is not a new one. We can find it, for example, in the discussion on the German pension reform in 1957 in the draft of Wilfrid Schreiber (Schreiber 2004). During the last fifteen years a revival of his ideas appeared.

In the German environment the idea of parental pillar is known under the name "Kinderrente". During the last decade the "Kinderrente" was supported by the German economist H.W. Sinn (Sinn 2005). In 2005 a commission led by the former Saxon prime minister Kurt Biedenkopf prepared a report "Strong family" for the Robert Bosch Foundation (Biedenkopf et al. 2005). The report states that "a stronger differentiation of pensions according to the number of children" is necessary, and that "besides a general pension, parents should also receive another portion of pension, financed by taxes".

In 2014, the Werding's report mentioned above (Werding 2014) contains a detailed quantified proposal of "Kinderrente". The whole pension should consist of two sums, as was defined in (1). The first is the basic pension, financed within the present pay-go system, the second one is the "Kinderrente", financed by pay-go system or by taxes. The proposal envisages an additional sum also for childless people. However, those must pay for it via higher contributions during their active period of life.

Let us note, that in Germany the definition of parental pillar F_2 ("Kinderrente") takes into account only the quantity of children, so F_2 is of the form (3), i.e.

$$F_2 = F_2(n)$$

3.2 *Slovak discussion*

In 2007 V. Palko and M. Krajniak published a (more political) document "Economics of conservative solidarity" (Palko & Krajniak 2007), proposing the pension F in the form (1), where

117

F_1 was based on the (then very popular saving pillar) and F_2 was based on direct contribution of working children to their parents.

In 2010 four members of the Slovak parliament for Conservative Democrats of Slovakia (KDS) V. Palko, R. Bauer, F. Mikloško and P. Minárik submitted a bill on the establishment of the parental pillar (BILLPROPOSAL 2010). The bill proposed the establishment of a fund, from which the part F_2 is paid to the pensioners with working children according to the amount contributed to the fund by their children. Every working person would contribute during the period of 30 years to this fund with 4% of the total amount of 18% of contributions to the social security system. In average case, the ratio of F_1 and F_2 would be approximately 6: 1. The bill was not adopted.

In 2014 members of the Slovak parliament for Christian Democratic Movement (KDH) J. Figeľ, P. Hudacký and J. Brocka submitted a bill establishing a miniature parental pillar (BILLPROPOSAL 2014). According to the bill, a pension of the parent should be increased by 0.5% of the gross wage of each working child. Due this proposal, for average pensioner with two children the ratio of F_1 and F_2 would be approximately 50:1.

Let us note, that all proposals in the Slovak discussion during the last decade considered the parental pillar of the form (4), i.e.

$$F_2 = F_2(d_1, d_2, \ldots, d_n)$$

Since F_2 was derived from the amount of children contributions, we can say, that also the "quality" of children was incorporated. So, unlike the German proposals, the Slovak proposals guarrantee a parental pillar only in the case of a "well brought up" working child. A non-working child guarrantees no parental pillar for the parent.

3.3 *Russian discussion*

The following case is an example of a succesfully implemented parental pillar. In 2006, the State Duma of Russian Federation adopted a bill "on further measures for the state support of families with children (BILLPROPOSAL 2006). The bill guarranteed the sum of 250 000 Rubles to the parents with the second, third or further child. One of three possibilities of how they can use this sum, is the increase of their pension.

Apparently the time of social security systems with built-in pro-fertility measures is coming.

REFERENCES

Becker, G.S. 1981. *A Treatise on the Family*. Harvard University Press, 1981.
Becker, G.S. & Barro, R.J. 1986. A Reformulation of the Economic Theory of Fertility. *New working paper Series, National Bureau of economic research,* 1986.
Biedenkopf, K., Bertram, H., Kässmann, M., Kirchhof, P., Niejahr., E., Sinn, H.W. & Willekens, F. 2005. *Starke Familie. Bericht der Kommission "Familie und demographischer Wandel".* Im Auftrag von Robert Bosch Stiftung, 2005.
Billari, F. & Galasso, V. What explains fertility? Evidence from Italian pension reforms. *CESifo Working Paper Series No. 2646,* 2009.
BILLPROPOSAL 2006. Federalnyi zakon o dopolnitelnyh merah gosudarstvennoy podderzhki semey imeushshih detey, prinyat Gosudarstvennoy Dumoy 22. Dekabrya 2006 goda, odobren Sovetom Federatsii 27. Dekabrya 2006 goda.
BILLPROPOSAL 2010. Návrh poslancov Národnej rady Slovenskej republiky Vladimíra Palka, Františka Mikloška, Pavla Minárika a Rudolfa Bauera na vydanie zákona, ktorým sa mení a dopĺňa zákon č. 461/2003 Z. z. o sociálnom poistení v znení neskorších predpisov a o zmene a doplnení niektorých zákonov (zákon o starobnom rodičovskom poistení a výsluhovom rodičovskom zabezpečení)—tlač 1472—prvé čítanie, 48. schôdza, 4. volebné obdobie.
BILLPROPOSAL 2014. Návrh poslancov Národnej rady Slovenskej republiky Jána Hudackého, Jána Figeľa a Júliusa Brocku na vydanie zákona o podpore viacgeneračnej solidarity v rodinách a o zmene a doplnení niektorých zákonov v znení neskorších predpisov (tlač 986), 35. schôdza, 6. volebné obdobie.

Boldrin, M., De Nardi, M. & Jones, E. 2005. Fertility and social security. *National Bureau of economic research, NBER Working Paper Series,* 2005.

CIA 2014 The World Factbook. Central Intelligence Agency. https://www.cia.gov/library/publications/the-world-factbook/rankorder/2127rank.html

Eurostat 2014. Total fertility rate, 1960–2011. *European Commission. Eurostat.* http://epp.eurostat.ec.europa.eu/statistics_explained/index.php?title = File:Total_fertility_rate,_1960–2011_(live_births_per_woman).png&filetimestamp = 20

Palko, V. & Krajniak, M. Ekonomika konzervatívnej solidarity. *Pravé spektrum,* 22.2.2007.

Schreiber, W. 2004. *Existenzsicherheit in der industriellen Gesellschaft.* Uveränderter Nachdruck des "Schreiber- Planes" zur dynamischen Rente aus dem Jahr 1955. Bund Katholischer Unternehmer e.V., 2004.

Sinn H.W. 2005. Führt die Kinderrente ein! *Frankfurter allgemeine Zeitung,* 8.6.2005.

Werding, M. 2014: Familien in der gesetzlichen Rentenversicherung: Das Umlageverfahren auf dem Prüfstand. *Bertelsmann Stiftung,* 2014. http://www.bertelsmann-stiftung.de/cps/rde/xbcr/SID-3C3F7D36–1F449E9 A/bst/xcms_bst_dms_39223_39224_2.pdf

Current Issues of Science and Research in the Global World – Kunova & Dolinsky (Eds)
© 2015 Taylor & Francis Group, London, ISBN 978-1-138-02739-8

Overview of the use of traditional and new financial resources for SMEs and access to finance in Slovakia and European Union

M. Sobeková Majková

The Faculty of Economics and Business, Pan-European University, Bratislava, Slovakia

ABSTRACT: According to many researches lack of the financial resources is one of the biggest problems of SMEs in Slovakia. In our paper we analyze access to finance of Slovak SMEs compared to the European Union and other European countries by numbers of regular EU statistics. We present actual numbers in using traditional own and foreign financial resources (bank loan, trade loan etc.) and new funds as private equity investments, venture capital and business angel investments. Our aim is to bring final overview of the biggest barriers in obtaining finance for SMEs in Slovakia.

1   INTRODUCTION

Problems with financing of Small and Medium-sized Enterprises (SMEs) were still on the top of all obstacles of business. Lack of finance is considered according to the authors Pissarides (1999) and Steinerowska-Streb, Steiner (2014) for the main obstacle to the growth of SMEs. Currently many authors and institutions declare the same problems of SMEs. Financing difficulties of SMEs is the world's problem (Xie, Chen, Xuan, Vlachynský, Berger, Udell, Sobeková, Solík, Fetisovová, World Bank, European Commision, etc.). The authors Chen and Xuan (2011) inform that financing difficulties of SMEs is not only a hot problem in European Union but also in the whole world and we agree with this opinion. It was one of the reasons of making this comparative study for presenting better overview of using different types of finance by SMEs. The aim of this paper is to show actual overview in using traditional and new financial resources used by SMEs in European Union and also in Slovakia. Finally we would like to present in conclusion actual barriers in obtaining finance for SMEs in Slovakia.

SMEs represent the most dynamic firms in an emerging economy. They have many comparative advantage and high value added. By the author Sobeková (2001) *"in Slovakia SMEs are 99.9% of all companies. They employ two thirds of population and they create 50% of added value in economy."* Problem is they often face many obstacles—finance, legal or institutional. (Pissarides, 1999) Basic problem of small company arises from the size. Self employers often start to do business only with minimum capital what intensively influences their further investments. According to the authors Jakubec, Sobeková Majková and Solík (2012) two thirds of young entrepreneurs start to do business with their own savings, 20% of them did not need any starting capital and one fifth borrowed starting finance from the family. What is interesting 19 from 20 young entrepreneurs have started to do business with capital less than 15 000 euro.

2   ACCESS TO FINANCE OF SMES IN LITERATURE

In the introduction of the paper we tried to show that lack of finance for SMEs is worldwide problem. In this part we would like to present possibilities of SME financing in literature and the views of authors to the access of finance for SMEs. Proper capital or equity is the most important financial resources not only for small but for big sized companies too. Small companies usually have problems to obtain needed finance (usually loans) because of their weak capital power in comparison with big sized companies. Weak capital power is closely connected with the lack of property to guarantee. The authors Dong Yan and Men Chao (2014) inform about this findings about financing of SMEs. They confirm that *"relatively small, young firms in nonmanufacturing sectors consistently face more severe financing obstacles/constraints and rely heavily on internal financing. The availability of credit information and the bank concentration ratio have a significant impact on SME financing."*

2.1 *Possibilities of SME financing and access to the finance in literature*

As we wrote in previous part of the paper according to researches Slovak small companies and entrepreneurs but also SMEs worldwide have problems with financing. The Slovak researches declare they are insufficiently informed about possibilities of financing. In this part we would like to present literature review of the financial resources suitable for SMEs. We could divide financial resource for SMEs according the author Sobeková (2011) to two basic groups as: traditional financial resources—in this group there are financial resources with long history of using. Entrepreneurs know them. They also know conditions of their obtaining and they are often used—the best example is bank loan or leasing. Bank loan is traditional external finance. According to the author Sufi (2009) lack of access to a traditional external finance bank loan or credit is a statistically powerful measure classified as financial obstacle. Second group of finance are new or alternative financial resources. They are used (especially in Slovakia) very rarely. Small companies and entrepreneurs are not informed about them. We consider for the alternative financial resource as private equity, venture capital, business angel investments, and finance from funds.

Bauchet and Morduch (2013) present thought of Vandenberg (2003) that *"SMEs have alternative financial resources even when formal-sector funds are limited."* For alternative sources of financing we consider private equity investment and also venture capital as a part of private equity, business angels, finance from structural funds, state support, finance from European Investment Bank and Fund and other. The author Brzozowska (2008) informs about using venture capital in European Union. She declares that *"especially in the new member's countries venture capital market is not so developed. Published data provide that using of venture capital in these parts of the world is increasing with growing level of innovation business activities."* This is the reason why we consider them for alternative or new financial resources. They are used especially in Slovakia very rarely because majority of entrepreneurs or SMEs don't know them or they are not informed about them. Especially private equity investments are not often used in Slovakia. Small entrepreneurs are afraid of new investor in company. They are very conservative and don't see the main advantage of new investor in company—he brings needed capital and also knowledge and needed experience and contacts.

There are lots of views on the division of the financial resources. According to the author Vlachynský (2009) they could be divided by many aspects as purpose, time, source and ownership. Next author Lupták (1998) presented different point of view. He divided finance only to two groups as proper finance and foreign finance as loans and other liabilities.

By the authors Cressy, Olofsson (1997) the specifications of SME financing results from different capital power of SMEs in comparison with big sized companies. Small companies have different structure of assets (different share of current assets and fixed assets). SMEs have lower share from fixed assets to total assets and share from current liabilities to their greater assets. These indicators show their greater financial vulnerability. Small and medium enterprises tend to have less financial strength, they don't have sufficient assets to liability which is usually the main reason why the bank denied a bank loan obtained by such companies with difficulties. Smaller businesses and enterprises with shorter history with—banks have only short-term contacts and therefore pay higher interest as well as higher guarantee is required from them (Berger, Udell, 1995). This fact is also true in Slovakia—companies with shorter existence and small businesses have stricter conditions for granting credit than older and larger firms.

Narrower topics—financing of small and medium enterprises—are engaged in the publication authors Finance of SMEs (Fetisovová, Vlachynský, Sirotka, 2004). Usually the different options for obtaining financial resources but do not analyze in more details specific barriers that small and medium-sized enterprises in Slovakia meet in the process or formulate proposals to support fund-raising.

SMEs are one of the most important parts of the market economy. (Sobeková, 2011) But they face many difficulties. It is caused because of its characteristics. They have weak capital power and credit degree. (Cheng, Tang, Shi 2012) According to the authors Mercieca, Schaeck and Wolfe (2009) access to finance is necessary for growth and next developing of SMEs. At the base of many studies we know that lack of finance is dangerous for future development of companies. The same fact is provided by authors Steinerowska-Streb, Steiner (2014) in their study An Analysis of External Finance Availability on SMEs' Decision Making. They present the thought at the Poland companies as one of the transforming economies like is Slovakia. Researchers of World Bank Ardic, Mylenko and Saltane (2012) presented topic access to finance of SMEs as a cross-country analysis and they agree that access to finance of SMEs is connected with their economic growth. The authors Mueller and Zimmermann (2009) declare in their study that lack of access to finance is great constraint special for smaller companies. Especially small companies face to regulatory and tax constraints, administrative burdens and these factors decline their growth.

Access of SMEs to alternative financial resources as private equity or venture capital is more difficult in Central and Eastern Europe. The authors Brzozowska, Sobeková, Solík, Jakubec, Fetisovová, etc. and also European Venture Capital Association (EVCA) at the base of their researches inform that using venture capital and private equity is in these parts of the world rare than in the Western Europe.

3   METHODS OF THE RESEARCH

The aim of this paper is to show actual overview in using traditional and new financial resources used by SMEs in European Union and also in Slovakia. We try to do it at the base of Eurostat, EVCA, EBAN and World Bank data. The result of this comparative study is to identify actual barriers in obtaining finance for SMEs in Slovakia.

To fulfil this goal we decided to use methodology of comparison. Methodology of this paper lies in comparing different statistical data and their following analysis. We used aggregated statistical data from Enterprise Survey of World Bank and also from European statistics of: European Commission survey Access to Finance, European Venture Capital Association (EVCA), European Business Angel Network (EBAN).

To see the perspective of finance in Slovakia we analyze in the paper statistical data of the World Bank from available from its web page www.enterprisesurveys.org. Our aim was to compare situation in Slovak republic with other member states in EU. This survey offers for comparison suitable group of states—Eastern Europe and Central Asia. However we don't have access to STATA or SPSS we processed aggregated data in Excel. We were able to obtain from the survey not only average, but also standard error and number of observations for years 2009 and 2013.

Statistical data from the first source are used to analyze traditional financial resources and EVCA and EBAN data are used for analysis of alternative finance used in business.

SMEs as an important building block of developed market economies, the European Union pays due attention. The European Central Bank carries out a survey on access for those enterprises to finance Access to Finance, which assesses the development of SMEs' access to finance. The survey was performed in 2009, 2011 and 2013. In this part of our work we would like to try to compare the development in the financing of SMEs.

The survey examines SMEs (European Commission, 2013):

– Financial situation, growth, innovative activities and need for external financing
– Use of internal funds and external sources of finance

Table 1. Regional structure of the sample and basic data set according national statistics.

| Location | Sample | | Basic dataset | Difference |
	Numbers	%	%	%
Bratislava	92	28,40%	34,70%	−6,30%
Prešov (Eastern Slovakia)	38	11,73%	9,20%	+2,53%
Košice (Eastern Slovakia)	37	11,42%	10,30%	+1,12%
Trenčín (Middle Slovakia)	34	10,49%	8,30%	+2,29%
Žilina (Middle Slovakia)	52	16,05%	9,40%	+6,65%
Banská Bystrica (Middle Slovakia)	31	9,57%	8,80%	+0,77%
Nitria (Western Slovakia)	21	6,48%	10,00%	−3,52%
Trnava (Western Slovakia)	19	5,86%	9,20%	−3,34%

– Experiences when applying for different types of external financing
– Use of loans, the size and reasons behind taking out specific loans
– Views about the extent to which different types of financing are available to them.

In this paper we bring also results of our research based on individual data of Slovak companies. It was done in 2012 in cooperation with Association of Young Entrepreneurs in Slovakia. There were 324 respondents (all of them were SMEs) chosen by random selection (Jakubec, Sobeková, Solík, 2012). More information about sample and basic data set are in Table 1. The results were statistical processed in Excel and software R. Because of length of this paper we present only chosen information from research.

4 DISCUSSION ABOUT THE CURRENT ACCESS OF THE SME TO THE TRADITIONAL AND NEW FINANCIAL RESOURCES IN SLOVAKIA AND EUROPEAN UNION

When the Slovak Republic became a member of EU country started to be influenced by its policy also in obtaining financial resources. For example Slovak companies and entrepreneurs could start to use supporting finance and other financial product of European Investment Bank. European Union created document called Small Business Act in which there are collected all types of supports for small and medium-sized enterprises. One of the goals is to make access to finance for SMEs easier also through products as micro loans and venture capital.

4.1 Financing of SMEs in Slovakia and European Union

Possibilities of SME financing are specific in each country. In the report Trade Finance for SMEs in CIS countries (Economic Commission for Europe, 2003) is written that loans for private sectors were in average in Czech Republic 54% from GDP, 26% in Estonia, 23% in Poland and Hungary. Here we could see significant differences among chosen EU countries. According to the study Financing of SMEs (Fraser, 2010) realized in UK in each size group of companies there are used for financing credit cards (more than 55%), than overdraft (more than 50%). Term loans were used only in 20%, grants 6% and equity finance only 3%, but in value of 14.3 billion pounds. In comparison with Slovakia we can say that Slovak companies don't use credit cards so often because it is quite new bank product with short history in Slovakia.

In Slovakia there are companies financed usually from their own capital (by deposits of entrepreneurs), bank loans and leasing. These financial resources are mostly used because of their history. The main barrier of using bank loans is lack of guarantees, in case of structural funds it is bureaucracy and corruption. Many entrepreneurs don't use for financing venture capital because they have never heard about it. (Majková, 2008) Business Alliance of Slovakia (BAS) in its research also informs about problems of SMEs with obtaining capital (BAS, 2008) which has been updated a year (BAS, 2009). According the institution in 2008 55% of companies presented deterioration in the availability of financial resources. One year later the group was greater by 2%. In 2008 bank loans were unavailable for 8% of companies (in 2009 it increased to 12%).

At the base of results of World Bank survey we have analyzed situation in Slovakia. The access to finance was marked as an obstacle of business in 2013 more often as in 2009. In Table 2 we could see distribution by size of company and average, standard error and number of observations. The results in Table 2 show that the worst situation was in the group of small companies. Average in 2009 was 6.7 and in 2013 it was increased to 18. Different situation was in group of medium companies. In average 13.9 companies marked access to finance as an obstacle to business in 2009. In 2013 it was 4.1 less (9.8). In the group of large companies the situation is better as in the group of medium-sized companies. We could say that small companies sense access to finance during finance and economic crises worst than other size of companies.

We wanted to know if there are some differences in access to finance and locations in Table 3. According statistical data is Bratislava the most developed region. Bratislava is the capital city of Slovakia. When we compare these two years we could say, that companies in Bratislava didn't feel the change in access to finance between these two periods. But companies from other parts of Slovakia felt access to finance worse—in Middle Slovakia the averaged increased from 4.1 to 18, in Eastern Slovakia it was increased to 10.1 and in Western Slovakia there was small decrease.

Table 2. Access to finance as an obstacle to business according the size of company in Slovakia.

Size of company	Indicator	2009	2013
Small (5–19)	Average	6,7	18,0
	Standard error (SE)	3,6	6,1
	Number of observations (Nr)	76,0	81,0
Medium (20–99)	Average	13,9	9,8
	SE	4,7	2,9
	Nr	80,0	136,0
Large (100+)	Average	16,9	8,5
	SE	6,1	4,9
	Nr	69,0	36,0

Table 3. Access to finance as an obstacle to business according location in Slovakia.

Location	Indicator	2009	2013
Bratislava	Average	7,0	6,2
	SE	6,9	2,4
	Nr	39,0	67,0
Middle Slovakia	Average	4,1	18,0
	SE	2,4	6,9
	Nr	43,0	53,0
Eastern Slovakia	Average	5,5	15,6
	SE	2,2	7,5
	Nr	63,0	57,0
Western Slovakia	Average	14,7	12,3
	SE	5,5	5,0
	Nr	80,0	76,0

Table 4 brings answer on the question if young entrepreneurs in Slovakia had lack or sufficiency of starting capital dividing by gender. Presented data are results of research made by Association of Young Entrepreneurs in Slovakia with cooperation of Paneuropean University. Research was financed by Iuventa in 2012 and authors of the research are Jakubec, V., Sobeková Majková, M. and Solík, J. From the results it is obvious there are statistical significant differences between men and women entrepreneurs in Slovakia and lack of their starting capital ($\chi^2 = 8.53 > 6.0 = \chi^2_{0,05 \text{ with 2 dgf}}$[1]). Entrepreneurs from both groups use the most frequently starting capital saved money (63%), loans from families and friends (22%) and bank loans (8%).

Results from our research (Jakubec, Sobeková, Solík, 2012) we could compare with results of research by Majková (2008). She examined how companies see lack of starting capital in distribution by size of company. If we connect the group of micro and small companies we could say that numbers of companies which have enough starting capital are increasing by the increasing size of company. These results show that small entrepreneurs really feel lack of starting capital—this fact see truly 32% of small companies (micro included), 16.7% of middle sized companies and only 4.9% of big companies. The lack of starting capital—it is a fact in Slovakia, but there is a question if the lack of starting capital is a real problem. Entrepreneurs whose declared lack of starting capital, said it forced them to be more effective in using financial resources. It is obvious from many scientific researches that small companies in Slovakia don't have starting capital in needed amount. On the other hand they are insufficiently informed about different financial resources.

SMEs as an important building block of developed market economies, the European Union pays due attention. The European Central Bank carries out a survey on access for those enterprises to finance Access to Finance, which assesses the development of SMEs' access to finance. The survey was performed in 2009, 2011 and 2013. In this part of paper we would like to try compare the development in the financing of SMEs. Figure 1 evaluates problematic areas of SMEs for years 2009–2013. Results of the survey show that problems with financing are one of the biggest which SMEs in EU solve. The graph compares development for 4 years and from the results it is obvious that after starting financial crises more and more companies still fight with those problems. For example finding customers was problem in 2009 for 19% of SMEs, in 2011 and 2013 it was problem for more than 22%. Graph 1 shows that situation is still worsening. Access to finance was marked as the big problem in 40% SMEs in Cyprus, 32% in Greek, 23% in Spain and Croatia. In comparison, in Germany it was only 8% and in Luxemburg only 6% of SMEs.

As it is presented in Figure 2—in the survey there was a lot of variation across countries to evaluation of SMEs access to finance. On the scale of 1–10 where 10 means extremely pressing and 1 means not at all pressing, how pressing is access to finance that your firm is facing. 42% of Slovak SMEs marked access to finance as extremely complicated because they gave it 10 points from 10. Right behind them was Greek with 28% and Cyprus with 26%. What is interesting only 5% of SMEs in Czech Republic identified access to finance as an extremely complicated problem.

Table 4. Lack/sufficiency of starting capital of SMEs in Slovakia.

Lack/sufficient of starting capital	Men		Women		P-value
	Nr	%	Nr	%	
Sufficiency of starting capital	54	25,96	43	37,07	**0,04925**
Lack of starting capital but later I obtained finance	42	20,19	30	25,86	0,2995
Lack of starting capital pushed me to be more effective	112	53,85	43	37,07	**0,00539**
Total	208	100	116	100	

[1]Degrees of freedom.

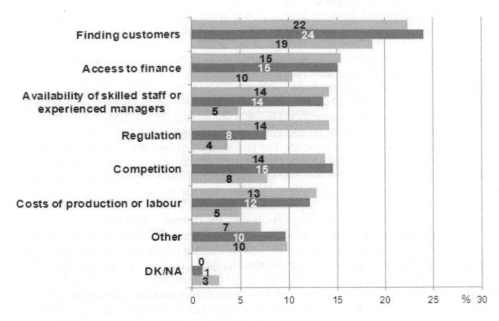

Figure 1.   Problematic fields of SMEs in EU in 2009–2013 in %.

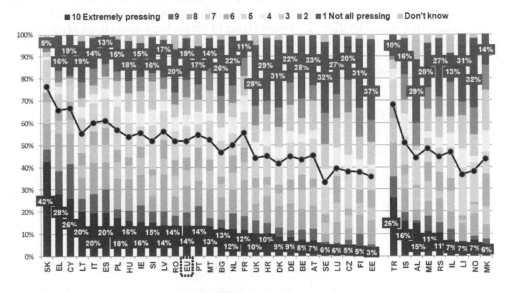

Figure 2.   Access to finance—country variations.

The survey also examined and compared kinds of external financing utilized for the last half year. In summary, it appears that in case of using bank loans there was some decline in 2011, but in 2013 again presents some recovery—increase in applications has increased by 2%. In total, the bank loan used about a fifth of businesses. The average in EU for the "extremely pressing" for access to finance was 14% and the average for reporting "not at all pressing" was around 19%. For the last six months SMEs requested loans most frequently in Luxembourg (32%), France and Slovenia (30%). Overall, for the last six months external financial resources used over 50% of EU SMEs. The most common reason why businesses did not apply for a loan or overdraft was sufficient internal funds.

4.2 Using a standard financial resources in Slovakia and European Union

We consider for standard financial resources that types of finance which have long history tradition in using for example bank loans. In Slovakia there are many types of capital which are quite new as venture capital or business angel investments, they are considered for new because many of entrepreneurs are not informed about them. In this chapter we would like to present using standard external finance in EU and in Slovakia.

Results at the Figure 3 show the most used source of external finance in 2013 was bank loan and overdraft—they were used by 39% of SMEs. Immediately beyond them there were leasing and factoring. Similar data were collected in 2011. On average 75% of SMEs used for the last six months some foreign form of financing. In the case of equity financing can be concluded that was benefited only 5% of businesses. The profits were funded 26% of SMEs. Larger companies with 50+ employees applied for bank loan more often than micro and their applications were also less frequently rejected, suggesting that the size of the company is closely associated with capital strength and sufficient assets to liability. The second factor that supports the thesis that larger businesses obtain borrowed capital more easily is that they staff is qualified and they are better informed about funding opportunities and credit conditions.

Graph 4 informs about rejected applications for bank loans. From companies applying for a loan 34% said that the interest rate for the last six months increased, 20% said that interest rate decreased and 40% were free of interest rate changes. Compared to the year 2011 52% of SMEs reported an increase in the interest rate. What was interesting: micro frequently showed an increase in the interest rate on the loan than larger companies with 50+ employees.

From the survey of the European Central Bank it is clear that bank loans are still the most commonly used source of finance for small and medium-sized enterprises. One of the factors of poor use of other forms of finance is according to Sobeková Majková (2011) low information ability about alternative financing sources. In examining the survey results from previous years, we find that Slovak SMEs in 2010 were trying to obtain bank loans most often, closely followed by Greece and Belgium. At least, only about 20% use bank loans to finance their SMEs in the Netherlands and Luxembourg.

Statistical data from the survey of World Bank in 2009 and 2013 bring following information: With the higher size of company using a bank loan was increasing in 2009 (35.5% of small companies, 55.8% of medium companies and 60.2% of large companies). Situation in 2013 was quite different. There was a great decrease in using bank loan in the group of large companies. Maybe because of financial crises they stopped some development activities.

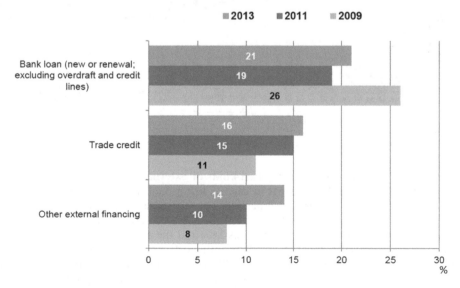

Figure 3.   Types of external finance (% of applications).

■ Applied but was rejected

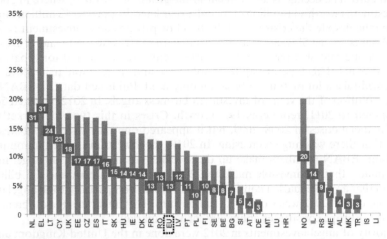

Figure 4. Rejected applied for bank loans.

Table 5. Development of private equity investment and venture capital in Slovakia and other V4 countries in thousands euro.

Country/type of investments and amount in thousands euro	2007		2008		2009		2010	
	VC	PE	VC	PE	VC	PE	VC	PE
Slovakia + Post Yugoslavia	38 150	43 150	49 050	111 560	42 700	42 700	9 000	9 000
Czech Republic	0	1 060	0	2 000	0	41400	10 450	10 450
Hungary	33 000	33 000	100 000	100 000	0	35120	133 690	133 690
Poland	11 120	823 780	8 860	760 460	13 360	145 350	7 660	114 760

It is interesting that medium companies have more rejected loan applications as small companies. It is probably because of medium companies applied for bank loan often than small companies. Next interesting fact is that number of companies identifying access to finance as a major constraint is decreasing with the increasing size of company. Small companies see access to finance often as a major constraint than middle and large companies. Thus the company is larger, the access to finance is perceived as less of a threat.

4.3 Using an alternative finance as venture capital and business angel investments in Slovakia and European Union

In this paper we consider for alternative finance non traditional financial resources as private equity investments—venture capital, mezzanine capital, business angel investments and structural and community funds. In this part we would like to present actual state in this area in European Union and compare this situation with Slovakia. There are two organizations which connected data for statistics—EVCA (European Venture Capital Association) and EBAN (European Business Angel Network). In this part of papers we would like to present their statistic data but corrected in our own processing.

Table 5 compares the development of Venture Capital (VC) and Private Equity investments (PE) in the V4 countries. We summarize the published statistics, the numbers that we compared the development of these investments even with the surrounding countries, the fact remains that venture capital is seen as a subcategory of private equity investments. Crisis had affected negatively development of private equity investments, not only in Slovakia but a significant decrease in 2009 was primarily in Poland, where the value of 760 million euro (2008) fell to

145 million euro. The decline is also evident in the group of Slovakia, where in 2008, private equity investments worth 111 million euro and a year later dropped to 42 million euro. For the Czech Republic despite the crisis period the trend of private equity investment is continually growing. What is special Slovakia is in one group with other states of former Yugoslavia.

Venture capital and also business angel investments are considered for a part of private equity investments. This type of capital is still more and more popular in Slovakia but we estimate it will take a lot of time to be commonly used. Published data of EBAN declare a sinusoidal evolution of the value of investment business angels. In 2012, invested 509 million euro, compared to 2011, represents 19% growth. Crises in this type of market weren't as standard financial crises in 2008–2009, but it appeared one year later. But it was only one year, after that there was huge increasing. In 2010, the lowest investment amount was only € 157 million. EBAN members make up the so-called visible market of angel investment, EBAN estimates that the invisible market in 2012 accounted for around € 5 billion, which would mean that a visible market consists only of 10%. In comparison, total angel investment market in the U.S. in 2012 was worth 17.4 billion euro. This illustrates the differences in various markets—Europe constitutes about 30% of U.S. market.

The majority of angel investments in 2012 were done in the United Kingdom and France. The highest value of the investments was in the United Kingdom in amount of 68.3 million euro. While watching the share of angel investments as% of GDP, it is clear that the largest share is represented in Finland to 0.0146% of GDP. The average amount invested by business angels in Europe was in 2012 1.1 million euro, with the highest value reached investments in Finland—well above average to 7.1 million euro in Luxembourg and at least 267,000 euro.

5 CONCLUSION

Finally we could say that the situation in using traditional and alternative finance in Slovakia is different in comparison with EU. This fact is declared by researches about access to finance of SMEs in Slovakia. We promised in the aim of the article to present final barriers of using standard and alternative financial resources in Slovakia. What obstacles have been identified on the basis of collected data? Small Slovak companies often don't use traditional external finance as bank loan because of their weak capital power and lack of guarantees. On the other hand they also don't use enough alternative financial resources. The reason is they usually don't know them. So the problem is low information ability. About this fact note more authors (Majková, 2008 and Jakubec et al. 2012) there is missing R&D infrastructure and weak legislative update of laws connecting with venture investments and state support of investment to innovative businesses. But there are also other problems.

Many Slovak institutions inform about high administrative burdens and fact that some of alternative financial resources as structural funds are often connected with corruption. There is a question what to do. At the base of our research of access to finance of SMEs in Slovakia we bring some proposals. Small companies need micro capital. From the state side it is quite missing there. The role of government or relevant state institutions for developing SMEs should be preparing supporting schemes with microfinance and also with guarantee schemes. Guarantee schemes could help solving special problem of SMEs with weak capital power.

An essential part of each market economy is knowledge and innovative economy. We identified lack of capital for innovative projects in Slovakia. There is a reason of low awareness about types of capital for financing innovative projects. Government should support founding of business angels nets and also spend money for publishing information about venture capital, business angels, mezzanine financing and other alternative financial resources.

What to do with administrative and legal burdens? What about corruption? It is too difficult to bring answers on these questions. We identified them as constraints but for the right solving it is needed to accept structural changes and also changes in people minds. In Slovakia there is still certain level of corruption acceptation. Government does not seem to liability for acts of corruption. There is missing criminal liability of politicians and other employees working in government. And for the change in this area it is needed to receive some important laws. But it is not in the certain of interest of this government unfortunately. So it requires the change of overall approach of relevant persons. We believe the change of overall approach will bring also changes in administrative and legal burdens, so they will not be set up for satisfying of interest of some politicians but Slovak people and also entrepreneurs.

Where is the right way? It is to show SMEs in Slovakia that in the market there are many possibilities of financing, not only bank loans, overdrafts and leasing. They could use many interesting sources, but the first step is to inform them about it. Entrepreneurs and also state institutions should be more active and not only to complain about the wrong system.

REFERENCES

Ardic, O.P., Mylenko, N., Saltane. V. 2012. Access to finance by small and medium enterprises: Across-country analysis with a new data set. In: *Pacific Economic Review* 17(4):491–513.

Bauchet, J., Morduch, J. 2013. Is Micro too Small? Microcredit vs. SME Finance. In: *World Development*, volume 3:288–297.

Berger, A., Udell. G. 1995. Relationship Lending and Lines of Credit in Small Firm Finance. In: *Journal to Business* 68(3):351–368.

Brzozowska, K. 2008. Business Angels in Poland in Comparison to Informal Venture Capital Market in European Union. In: *Inzinerine ekonomika—Engineering Economics* 2:7–14.

Cressy, R., Olofsson, CH. 1997. European SME Financing. An Overview. In: *Small Business Economics* 9(2):87–96.

Dong,Y., Men, Ch. 2014. SME financing in emerging markets. In: *Emerging Markets Finance & Trade*. Jan/Feb 2014, 50(1):120–149.

Cheng, ME, Tang, Y. 2012. Research on the Small and Medium-sized Enterprises Financing Problems. In: *Proceedings of the sixth international symposium—The development of small and medium-sized enterprises*. Dec 15–19, 2012:93–97.

Chen, X., Xuan, Y. 2011. The Influence of SME Financing Structure on Enterprise Growth. In: *Proceedings of the twelfth west lake international conference on small and medium business*: 237–246.

EBAN: Analysis of business angels investments. Online document: http://ec.europa.eu/enterprise/policies/finance/data/enterprise-finance-index/access-to-finance-indicators/business-angels/index_en.htm, 3.4.2014.

European Commission: Access to Finance 2013. Online document: http://ec.europa.eu/enterprise/policies/finance/files/2013-safe-analytical-report_en.pdf, 3.4.2014.

Fetisovová, E. & Nagy, L. 2007. Ciele a finančné nástroje modernej politiky rozvoja malých a stredných podnikov Európskej únie (Goals and financial instruments of modern policy of SME development). In *Ekonomické rozhľady* 36(4):502–512.

Fetisovová, E. & Vlachynský, K. & Sirotka, V. 2004. *Financie malých a stredných podnikov. (Finance of SMEs)*. Bratislava: Edícia Ekonómia.

Jakubec, V. & Sobeková Majková, M. & Solík, J. 2012. *Potreby mladých podnikateľov a prekážky v ich podnikaní. (Needs and obstacles of young entrepreneurs)*. Bratislava: ZMPS.

Majková, M. 2008. *Možnosti financovania malých a stredných podnikov v SR. (Possibilities of SMEs financing)*. Brno: Tribune.

Mercieca, S., Schaeck, K., Wolfe, S. 2009. Bank Market Structure Competition and SME Financing Relationship in European Regions. In: *Journal of Financial Services Research* 36(2–3):137–155.

Mueller, E., Zimmermann, V. 2009. The importance of equity finance for RandD acitivity. In: *Small Business Economics* 33(2):303–318.

Pissarides, F. 1997. Is lack of funds the main obstacle to growth? EBRD's experience with small- and medium-sized businesses in central and eastern Europe. In: *Journal of Business Venturing* 14(5–6):519–539.

Shi, L. 2012. The Factors of Affecting Financing and Countermeasures Based on Asymmetric Information and Marginal Information Cost. In: *Information and Business Intelligence*, PTI 267:672–677.

Sobeková Majková, M. 2011. Analýza bariér a faktorov financovania malých a stredných podnikov v SR (The analysis of barriers and factors of financing of SME). In: *Journal of Economics* 59(10): 1028–1032.

Sobeková Majková, M. 2011). *Ako financovať malé a stredné podniky (How to finance SMEs)*. Bratislava: Iura Edition.

Steinerowska-Streb, I.; Steiner, A. 2014. An Analysis of External Availability on SMEs Decision Making. In: *Thunderbird International Business Review*, July/August 2014, 56(4):373–386.

The World Bank. Enterprise Surveys (http://www.enterprisesurveys.org).

Aspects of green marketing: The current situation in the selected companies in the Slovak Republic

S. Supeková

Faculty of Economics and Business, Pan-European University, Bratislava, Slovakia

ABSTRACT: The aim of the article is to analyze the term of green marketing within Corporate Social Responsibility. Green marketing is one of the most important tools companies can use to highlight their activities in the field of environment and to get new customers and partners. Green marketing covers a wide range of activities, including product modifications, changes in the manufacturing process, packaging changes, as well as modification of the ads. It is not easy to define and specify the relationship between marketing and environmental protection. Sustainable marketing is for producers also a tool for improving competitive ability and position in the market. The situation behaviour of companies has changed in the Slovak Republic nowadays. Green marketing intends to promote, support and market products and services that are safe and friendly for the environment. One of the new projects to communicate green marketing of companies to the public is the "Step by step green project" started in 2011 in Slovakia. Promotion of businesses through green products could be the best way to a "Green Planet".

1 INTRODUCTION

Humans are inextricably linked to their environment. Currently, it is very important to ensure the balance of nature and the environment, as the effects of civilisation and modern times have disturbed this balance, ever more. Components of the environment could be considered resources and are used by most people to meet their basic physiological needs. All these elements form a global eco-system and people in this system maintain emotional, intellectual, or physical relations; and people provide the basis for a sense of meaning in life.

The rapid development of the period in which we now live, population growth, especially in developing countries, reducing the amount of non-renewable resources, and environmental degradation create the need for the rationalisation of economic processes to achieve sustainable development. This situation is increasingly clear to consumers. They are adapting their buying behaviour and increasing the demand for organic products. At the same time, consumers are well aware of how companies behave towards the environment. If we want to maintain the equilibrium of natural resources, we have to acquire not only organic products but also more businesses need to supply them, too. To increase this trend it is necessary for businesses to adapt and achieve improvement in their treatment of the environment. Environmental or green marketing is an effective tool for businesses to reach this objective. If we start from the fact that most natural resources are limited, we have no choice but to treat them sensitively and carefully. It follows that the importance of environmental or green marketing compared to traditional marketing approaches has steadily increased.

The companies in the past felt that activities leading to environmental protection caused additional cost increases and threatened the competitiveness of their companies. Recently, and increasingly, improving the environmental performance of a company can also lead to a better trade, economic and financial performance. Ministers and heads of NGOs, consumers, and companies are beginning to care about these environmental issues. Companies are a natural part of their surroundings as the natural environment becomes an integral component of their marketing environment.

1.1 *Methodology*

In conceptual methodology it is necessary to collect literature sources, first. Author presents the current system of evaluation of aspects to specify what is important from the consumers' behaviour and what from the producers'. Many papers have presented green marketing as a concept for marketing differently according to the country—e.g. China, USA, and the EU, green marketing success depends on the economic situation of customers. It comprises the theoretical analysis of the body of methods and principles associated with a branch of knowledge. The article examines what the term means, why it has come into favour, advantages, disadvantages and aspects of the execution of a mixed-method design. We start with a general look at the research methods associated with secondary data. It is important to examine the main types of data and look at how to incorporate the secondary data of research design.

1.2 *Structure of study*

The paper proceeds as follows: Section 2 presents briefly theoretical basis of green marketing and Corporate Social Responsibility. Section 3 provides examined green marketing mix, as product, price, communication and also distribution. Section 4 is focusing on Interbrand and practical skills.

2 THEORETICAL BASIS

Green marketing is part of the concept of Corporate Social Responsibility (CSR), which is defined by the European Commission in 2001 as "a concept where companies, on a voluntary basis, integrate social and environmental considerations into current business operations and interactions with stakeholders." At present ecological marketing is the mainstream of applied marketing. Ecological marketing could be called green marketing, environmental or sustainable marketing. Ecological marketing was introduced as a concept already in 1975 when it held the first seminar organized by the American Marketing Association (AMA). Experts from various field of ecology examined the impact of marketing on the human environment. Environmental marketing is understood as a full orientation of marketing and business concepts (i.e. corporate objectives, strategies and the entire marketing mix) to ecological aspects, respectively necessity and possibility. It follows general social responsibility and business activities. *(Nízka, H., Aplikovaný marketing. 2007).*

Castenow *(1993),* in his book, "New marketing" is describing a global trend defined as "natural environment with its influences on the future of marketing". This trend could be characterised by:

- Care about clean air, soil and water
- Nature conservancy before devastation
- Rational use of natural resources with an emphasis on recycling
- Popularisation of ecological orientation and a healthy lifestyle.

Realisation of the interconnection of the environment, hunger and overpopulation.

Marketing then generates call with two dimensions. In the short term point of view, there are the environmental and social issues that have become dominant in the external environment of a company. Functional substantiality of markets is directly affecting the daily operations of a company. Companies need to respond flexibly to the changing needs of consumers, new legislative measures and movements suggestive of negative social and environmental impacts of business. In the long term point of view, the pursuit of sustainability will require substantial turnover in management, which includes both marketing and other business functions. Modern marketing thus requires an environmentally—oriented approach that has to be understood as an element of environmental management. Individual elements of the marketing mix on a new "green" dimension. It naturally imposes new opportunities for competitive advantage. The aim of this paper is to define the concept of green marketing. This paper will explain how it is used in marketing companies operating in Slovakia.

Ecological marketing focuses on ecological aspects. The concept of ecological marketing can be named as environmental marketing, green marketing and sustainable marketing. "The aim is to carry out a business activity to prevent environmental pollution, or reduce the load on the environment and to preserve the natural living conditions for people, animals, plants or, at least, change the conditions to improve their lives". The movement "green marketing" began in the U.S. in 1990 following the Earth's Days activities. Companies have tried to capitalize on increased consumer sensitivity to environmental issues. In terms of brands, green marketing programs have had little success. Many marketers have tried to come up with organic products without success. Studies have shown that consumers as a whole are not willing to pay more because of ecological benefits at a higher price. Although some market segments have the answer—yes, it seems that most consumers are not going to give up the benefits of other alternatives and switch to "green products". Ecological marketing is based on the idea of environmental protection and maintenance of non-renewable resources. It is part of a new marketing approach that is not only aimed at modifying and improving the current marketing thinking and practice but it is looking for challenges and ways to provide a different, sustainable perspective. This refers to the process of implementation of products on the market based on their environmental benefit. Prerequisites for ecological marketing is that potential consumers accept and prefer the environmental attributes of the product for its contribution and to lay the foundation for their buying decision. In our case, however, it depends rather on how the business is environmentally reported by the manufacture of these products that should produce minimal adverse impact on the environment.

According to Zhang *(1999)*, there is a difference in the narrower and broader understanding of green marketing. We understand green marketing as a specific type of marketing based on the principles of traditional marketing, but focusing on organic products. More broadly, however, the term includes a variety of ideas, methods and processes that enable the implementation of the objectives of the enterprise. The form of green consumption leads the company to the tendency to protect the environment and make efficient use of natural resources. A company's activities as design, manufacturing, packaging, sale and recycling of green products are connected.

An even broader definition is offered by Dubey *(2008)*, which states that green (environmental) marketing includes all activities aimed at satisfying human needs and desires with minimal impact on the natural environment. Thus, the holistic and responsible management process identifies, anticipates and looks for opportunities to meet the needs of stakeholders while not negatively affecting society and the environment. That definition probably most accurately characterizes the mission of green marketing for a new millennium. This definition expands because the original concept of societal issues and environmental aspects of seeing potential sources for innovation and opportunities for companies to meet market needs more effectively.

John Grant described in his book, "The green marketing manifesto", five characteristics of green marketing, which states the acronym 5I. According to him, green marketing should:

- Be Intuitive—non-violent means should convince consumers that environmentally conscious behaviour started to be perceived as normal, which can become a normal part of everyday life. For most people seem to be living, working, travelling and enjoying life in a way friendly to the environment, very difficult. They do not want to give up their comfort and standards. The role of business is, therefore, to create products that are well accepted and easy to grasp. The role of marketers is subsequently arranged for people to begin to feel quite normal and it to be common to live an environmentally friendly manner.

- Be Integrated—have to combine business, technology, ecology and social impacts. They should seek to combine ecology with business and keep the idea of development. It is a new perspective on green marketing, which has nothing to do with anti-business ideas and radical movements of environmental activists.
- Be Innovative—have to create new products and new lifestyles.
- Be Inspiring—seek to involve as many people as possible. It is necessary to convince consumers that what is green is better, healthier, more affordable, efficient and, at the same time, in a friendly manner. It should create the concept of a green lifestyle that people would embrace and consider great and not abnormal and strange.

Be Informative—the less knowledge, the worse. If people lack information about something before it appears, some concerns, doubts and often perceptions can become hostile. Green marketing should not try to cheat people into buying a brand without their knowledge of the product, but rather consumers should be involved in the process and educated with all the necessary information to be able to assess the product, and clarity, truthfulness and sincerity should be promoted.

3 GREEN MARKETING MIX

3.1 Product

To create a significantly greener economy requires a number of new and mostly greener products and technologies—"Green Technology". Sustainable solutions are product changes, service changes or changes in systems that minimize negative and maximize positive impacts on sustainability (economic, environmental, social and ethical). One of the criteria for the production of new products is a concept design for the environment. Measurement and understanding of eco-product characteristics is important for all companies regardless of whether they promote green strategy or not.

Green products should consist of a number of attributes, which can be divided into two basic categories *(Baker, p. 747)*:

1. Attributes associated with social and biological waste of the product, such as the efficiency of utilization of energy consumption, safety and recyclability. Product, after its useful life, creates an interesting perspective on green product management. Use of this perspective requires the implementation of, at least, some or all of the "5R's":
   - Repair—designing the product to be easily and effectively fixed in the future without the need to purchase a new product
   - Reconditioning—repair parts from malfunctioning products and their subsequent sale,
   - Reuse—reuse parts of a product, such as returnable packaging
   - Recycling—necessary waste processing after use of the product and subsequent production of other goods or identical goods
   - Re-manufacture—used, old, worn, non-modern and otherwise unnecessary products collection, their subsequent use in the manufacture of new products
2. Attributes related to the process by which the product was produced and also attributes of the company that manufactures these products.

In the green products production, it is important to have regard for the product in addition to the green concerns, and comparable with competing products, in terms of functionality and price. This is not an easy task. However, it can reduce at least the difficulty of packaging the product with respect to the environment without costly changes to the properties of the product and the production process and also without risk of discouraging consumers.

3.2 Marketing communication

Many companies are already trying to communicate and present their products through associations with environmental or social problems. Environmental issues do provide good opportunities for informative but also emotional marketing communications. Few of them, however, were also successful. Marketing communications is one of the most controversial elements of the green marketing mix. Often the effort of green marketing communications, referred to as "green-washing", especially if the product advertised is only disguised as green, but in reality it is not. Therefore, consumer publicity campaigns on environmental scepticism. Correct marketing communication, promotion of environmental attributes of products should follow the following short manual *(Baker, p. 748)*:

- You have to ensure that the advertised product characteristics have a real impact on the environment
- Quote the data specific to a green product's characteristics
- You have to provide context, so that the consumer can form their own opinion and arguments to compare
- Define all technical terms
- Explain the benefits, because consumers have limited understanding of environmental issues

Problems with marketing communication are subject to the communication tool of choice. Advertising carries the risk of accusations of green-washing or trivializing the serious social and environmental problems since it is difficult to explain the complexities involved within such a narrow space. Sending emails, in turn, entails risk designation for spammers. Conversely, sales promotion or sponsorship appears to be an effective tool that needs only to pay attention to the synergy between the product being advertised and sponsored event. Also, public relations are a key communication channel.

3.3 *Price*

Green strategies affect the cost structure of the company. Development of new sustainable resources and materials, and technology increases costs for companies. This could be offset by savings due to lower consumption of materials and energy, packaging reduction, lower cost of waste disposal, and finding new markets for by-products.

Progress towards sustainability can be achieved by a focus on marketing off costs. Although low energy bulbs have a higher price, they have lower long-term operating costs.

3.4 *Distribution*

The environmental impact of products is largely determined by the amount of fuel consumed in transporting materials and goods to the consumer. The implementation of reverse logistics occurs within the green transport process, which provides rollback of unnecessary products, packaging, etc., back to the manufacturer for further processing. Labelling plays a large role in reverse logistics, as well as marketing communications.

Logistically, it is important for the labelling of products for consumers to encourage their recycling behaviour and habits.

In terms of marketing communications used, product labelling serves to inform consumers about the fulfilment of social and environmental standards.

4 INTERBRAND

Interbrand is rating, creating and managing brand value of the best green smart brands in the world. Last Best Global Green Brands report is the fourth in history and examines the gap that exists between a cooperation's environmental practices and the consumer's perceptions of those practices. In the year 2014, the automotive sector dominates again. When identifying the top 50 brands Best Global Green Brands each year, Interbrand starts with the 100 brands that make up its annual report. Interbrand then conducts extensive consumer research to capture public perception of the brand's sustainable or green practices and compares that to the environmental sustainability performance data provided by Deloitte; data that are based upon publically available information. Brands are measured against two sets of criteria: 1. Performance: organisations must demonstrate that they source, produce, and distribute products and services in an environmentally responsible manner. 2. Perception: organizations must work to build value amongst key audiences by credibly conveying the benefits of their environmental practices.

The first 12 of the Best Global Green Brands report are:

1. Ford. 2. Toyota, 3. Honda, 4. Nissan, 5. Panasonic, 6. Nokia, 7. Sony, 8. Adidas, 9. Danone, 10. DELL, 11. Samsung, 12. Johnson & Johnson.

The Slovak Republic provides real production for these brands from the Best Global Green Brands: Panasonic, DELL and Samsung. For the comparison we select these three companies. From this ranking list, Sony also had production in Slovakia till 2012 and Danone till 2006.

4.1 Panasonic

Mills of Japanese concern, Trstená and Stará Ľubovna, are starting new production and investing in automation. Panasonic Trstená is one of the largest electrical companies in Slovakia—Panasonic Electronic Devices Slovakia—and is focusing on the new form and plan an expansion. The company gave three million Euros last year for new production line voltage converters for cars. This will be delivered to the UK in October. Furthermore, the company is investing to increase the production of automation, especially in the manufacture of loudspeakers. This reinforces production for the automotive division and will constitute 20 percent of factory sales. Panasonic has received more ENERGY STAR award recognition than any other consumer electronics-manufacturer. Today, more than 300 Panasonic models across 9 product categories carry the ENERGY STAR logo, which indicates that a product uses less energy with no sacrifice in performance or features. Taking steps toward zero waste, Panasonic achieved a factory waste recycling rate of 99.3% in 2013. The brand also used approximately 12 thousand tons of recycled plastic for its products in 2013. In 2013, water used at factories per basic unit of production improved by 0.7% compared with 2012. Panasonic is also speeding up the development of industry-leading products that contribute to water saving. In 2009, Panasonic launched home appliances with ECONAVI function, which automatically control power and water consumption to cut losses using sensor and other technologies. The brand now offers 25 products with ECONAVI function, helping consumers across 88 countries conserve water and energy through its growing range of smart products. The company declares the using of efficient resources and green packaging, but just in its central company. We missed the information of their activities and direct green methods in Slovakia.

4.2 DELL

Dell was named a Green Power Leadership Award winner in 2013 by the U.S. Environmental Protection Agency for its longstanding commitment to purchasing renewable energy. More than 22.6% of Dell's electricity purchases were for renewably generated power, including wind and water. In 2013, Dell reached and exceeded its 2008 goal—a full year ahead of schedule—of collecting and recycling 1 billion pounds of used electronics. This is roughly the equivalent of 46 million average desktops—enough to fill London's Royal Albert Hall more than twice over. While Dell is not a large water consumer compared to other brands, it does operate in some water-stressed regions. It is committed to reducing water use in those areas by 20% by 2020 and continues to reduce water intensity through packaging innovations such as its wheat-straw-based boxes, which require less water and energy to produce than conventional cardboard and can be recycled just like any other box. Stressing "industry first," Dell has developed carbon-negative packaging bags made of AirCarbon, a material produced from air, not oil. The manufacturing process alone reportedly produces a net positive impact on the environment by sequestering more carbon from the air than it produces. DELL considers the environment at every stage of the product lifecycle starting with how a product is designed. Instead of one eco-friendly product, we consider the following when designing all of their products. DELL in Slovakia has mostly divisions for support of customers and delivery of goods. Therefore, the part of green-marketing activities is oriented mostly to packaging and product delivery and shipping. Packaging protects products and is a very important part of production. Shipping gets them where they need to go. Simple enough, but both can leave a substantial environmental footprint when you're shipping roughly a product per second around the world like we do. DELL follows a "3Cs" strategy around reducing our impact through the cube (the box itself), the content (what it's made of) and the curb (how easy it is to recycle). At Dell, we could see green packaging and product delivery as an opportunity for innovation—helping businesses and homes reduce their waste through the following important points: 1. Reducing the size of the box, so the customer gets more boxes in the same space for shipping. Put more products in one box—like DELL do using Multipack for some orders; there's also less waste overall. 2. DELL declares that they pioneered the use of natural materials in packaging, like bamboo and mushroom cushions, their new wheat straw initiative and new AirCarbon bags. 3. Better logistics. Shifting many international shipments from aircraft to ocean freight is just one example of how DELL works, to constantly refine its transportation network, delivering better ways for getting products to you safely and with a minimal footprint. The way of DELL's green marketing tools is very active and modern, but in the Slovak Republic the consumer could not find this information because DELL does not promote their green-marketing activities.

4.3 *Samsung*

In 2012, Samsung installed F-gas (fluorinated greenhouse gases) treatment equipment to reduce gases from the LCD and semiconductor manufacturing process, an equivalent of 1,030,000 tons of CO_2. As part of its Green Management framework, Samsung has recycled 325,545 tons globally and is running e-waste take-back programs in more than 60 countries. Samsung Electronics has a target to reduce 3% of its water use per production unit by 2015. The company collected water use data to identify plants with highest water use, established a monitoring structure, identified reduction measures, and implemented the most cost-effective measures to minimize business risks associated with water use and environmental impact. Samsung also achieved a recycling rate of 51% with ultra-pure water at semiconductor and LCD production plants (in 2011). Among Samsung's many eco-innovations, the brand developed reusable packaging technology for refrigerators by replacing paper and polystyrene with Expanded Polypropylene (EPP). The EPP packaging container can be reused up to 40 times for delivering refrigerators. Not only is EPP more durable and safe than paper, but by replacing paper packaging with EPP containers in Korea alone, an estimated 7,000-ton reduction (99.7%) of CO_2 could be achieved. Samsung Display Slovakia s.r.o. Voderady (SDSK) declare, that the aim of the production is to produce high quality LED panels while maintaining high quality and compliance and an adherence to the principles of health, safety and environmental protection. The company accepts commitment to ensure safe working conditions, with particular focus on prevention of injury and damage to health as well as a healthy environment with emphasis on pollution prevention. The company will also try to find a balance among customer focus, quality, environment, health and safety at work and economic performance. SDSK shall provide customers with environmentally friendly products that no hazardous substances are used to make. SDSK uses resources efficiently. SDSK shall minimize the amount of rare substances throughout all the processes. In addition, we concentrate on saving natural resources by effectively re-using and recycling energies, media, materials and raw materials. By continuous checking and controlling the processes and workplaces, SDSK tries to create and maintain a safe and zero-harm workplace and makes safety a routine part of work. Moreover, SDSK puts emphasis on health and hygiene-promotion activities and eliminating hazardous elements from the workplace. SDSK commits to complying with applicable legal and other require-ments in the area of the formation and protection of the environment and health and safety at work. The important point is that the company maintains and continually improves the effectiveness of its environmental-management system and health and safety-management system. In the field of environmental care, the company's goal is to reduce inputs and save natural resources, allow for careful protection of all environmental elements and optimize waste management. In terms of safety, in particular, they take care to minimize occupa-tional risks and the impact on the health of employees; they seek to protect their health, prevent occupational accidents and ensure consistent fire protection. The company man-agement provides the necessary sources and support to achieve goals by encouraging the programs of EHS management system based on ISO 14001 and OHSAS 18001. We sug-gest the company management to use these results in the marketing communication of the company. The company uses just the promotion in the general way of central Samsung. The results of green marketing put Samsung to 11th place in the Interbrand and it is not used in promotion of the brand.

4.4 *Step by Step*

This project can help companies, including in their social activities and operations staff, to behave ecologically responsibly. The project is a form of corporate responsibility of compa-nies called the Green Project Step by Step. The aim is to eliminate the use of plastic bottles, plastic cups and disposable packaging and help the environment. This project offers compa-nies to use a series of products that are made of organic material, which have multiple uses and can be also used as space/place for advertisement. The offer consists of stainless steel bottles, Green Bottle, Green Dutch and Roll'eat. These items can be a promotional gift, but also can help in educating and respecting a drinking regime for its employees. This project has been involved in Slovakia green company since 2011 in the companies such as Orange, Tatra Banka, Enviropak, Slovak Savings Bank and Synergy.

5 CONCLUSIONS

Sustainable development is a very controversial issue. There are different ways of inter-pretation, of measuring performance or putting it into the practice. We are currently wit-nessing continuous progress in technology, products, markets and marketing. Nevertheless, the environmental costs of production and consumption become increasingly part of the cost structure of companies and are not reflected in product prices. In fact, society and the environment are constantly subsidizing the excessive consumption and production of humankind. Marketing is often presented as part of the problem mainly due to the part used to support unsustainable consumption. Development as artificial needs of consumers or abusive use of public relations covers the negative impact on the business environment. On the other hand, green marketing is an essential tool in influencing consumption towards responsible environmental behaviour. Achieving stronger environmental performance, however, requires going beyond product orientation and eco-marks, and using all available tools of traditional marketing—price, communication and distribution. It can be only a green marketing source of competitive advantage. Green marketing has meaning only if it is accompanied by changes in corporate values, strategies, regulatory, investment, political systems, education, and, last but not least, trade and consumer behaviour. In the Slovak Republic, companies used mostly the way of green marketing dictate of their mother com-panies. The transnational corporations have the strategy of green marketing and the way of promoting these activities on a high level. In the Slovak Republic, the promotion of green products is just on the starting line.

REFERENCES

Baker, M.J. 2003. *The marketing book*. 5th edition. Burlington: Butterworth—Heinemann, 2003. 834 p. ISBN 0 7506 5536 4.

Castenow, D. 1993. New Marketing in der Praxis. ECON Executive Verlags, Düsseldorf.

Dubey, P. 2008. Recycling Businesses: Cases of Strategic Choice for Green Marketing in Japan. In. *IIMB Management Review*. Vol. 20, No. 3. 2008. pp. 263–278. ISSN: 0970-3896.

Filo, P. 2006. Spoločenské vedy a ekológia alebo čo je ekologické je aj ekonomické. In. *Prevádzka Národného registra emisných kvót skleníkových plynov SR: konferencia, 17.2.2006, Žilina*. S. 7–11— Bratislava: ERA OZ, 2006. ISBN 80-969367-2-7.

Ginsberg, J.M., Bloom, P.N. 2004. Choosing the right green marketing strategy. In *MIT Sloan management review*. Messachusetts Institute of Technology, 2004. p. 79–84.

http://www.interbrand.com/en/best-global-brands/Best-Global-Green-Brands/2014/best-global-green-brands-2014.aspx 20.6.2014.

http://panasonic.net/..20.6.2014 20.6.2014.

http://www.samsung.com/us/aboutsamsung/corporateprofile/history06.html 21.6.2014.

http://www.dell.com/learn/us/en/uscorp1/about-dell 21.6.2014.

Kotler, P., Keller, K.L. 2007. *A Framework for Marketing Management*. Klaipėda; Logitema.

Nízka, H. 2007, *Aplikovaný marketing*. 2007, Wolters Kluwer, Iura Edition, ISBN: 9788080781576.

Oyyman, J.A., Stafford, E.R., Hartman, C.L. 2006. Avoiding green marketing myopia. In *Environment*. Vol. 48. Heldref Publications, 2006. p. 22–36.

Paco, A.M.F., Rapaso, M.L.B. 2008. Determining the characteristics to profile the "green consumer": an exploratory approach. In *International review on public and nonprofit marketing*. Vol. 5. No. 2. Springer—Verlag, 2008. p. 129–140. ISSN: 1865-1984.

Rex, E., Bumann, H. 2006. Beyond ecolabels: what green marketing can learn from conventional marketing. In. *Journal of Cleaner Production*. Vol.15. 2007. pp. 567–576. ISSN 0959-6526.

Redfern, K., Crawford, J. 2010. Regional differences in business ethics in the People's Republic of China: A multi-dimensional approach to the effects of modernisation. *Asia Pacific Journal of Management*, 27: 215–235.

Shweder, R.A. 2008. The cultural psychology of suffering: The many meanings of health in Orissa, India (and elsewhere). *Ethos*, 36, 60–77, 2008.

Stevels, A. 2002. Five ways to be green and profitable. In. *The Journal of Sustainable Product Design*. Vol 1. 2001. pp. 81–89. ISSN 1367–6679.

Woolverton, A., Dimitri, C. 2010. Green marketing: Are environmental and social objectives compatible with profit maximization?. In *Renewable agriculture and food systems*. Cambridge: Cambridge university press, 2010. p. 90–98. ISSN: 1742–1705.

Zhang, X., Zhang T. 1999. Green Marketing: A Noticeable New Trend Of International Business. In. *Journal of Zhejiang University*. Vol.1. No. 1. pp. 99–104. 1999. ISSN: 1009–3095.

REFERENCES

Faculty of massmedia, section media-science-culture

Current Issues of Science and Research in the Global World – Kunova & Dolinsky (Eds)
© 2015 Taylor & Francis Group, London, ISBN 978-1-138-02739-8

Brno, starobrno and massmedia in the project partnership for local development

M. Foret
Faculty of Media, Pan-European University, Bratislava, Slovakia

I. Zdráhal
Faculty of Regional Development and International Studies, Mendel University in Brno, Czech Republic

ABSTRACT: The paper is devoted to the role of massmedia in the contemporary marketing project Partnership for local development as well as in the previous one Communicating Town. Some practical results and marketing methodology recommendations from the latest marketing research conducted in Brno in March 2013 are mentioned in the second part.

1 INTRODUCTION

Brno, the Czech Republic's second largest city, has a population of nearly 370,000 people. It lies in the central part of Europe and within its two hundred-kilometer radius there are other important European capitals: Prague, Vienna and Bratislava. There are a number of places that can attract tourists, not only from Europe, but surely from the whole world.

The city can be proud of its numerous landmarks, which are proof of its rich (800-year) history. The historical center of Brno is a listed urban preservation area. The city became prominent in European history during the Thirty Years' War, as it was the only town in Central Europe able to defend itself against siege by the Swedish army. Subsequently it underwent rapid economic development and during the 18th and 19th centuries it dominated the textile and engineering industries. Functionalist architecture contributed to giving the city its appearance as a modern metropolis. The most significant architectural landmarks include the Brno Exhibition Centre, founded in 1928, and the famous Villa Tugendhat, built in 1930, which is featured on the UNESCO World Heritage List. Today, Brno is a university town, home to the country's highest legal bodies as well as other important institutions. It is a center of business and trade, culture, sport, science, research and innovation. Brno is the capital of Moravia and a departure point for tourists who wish to explore the natural and cultural attractions of the South Moravian Region. To the north of Brno there is the protected landscape of the Moravian Karst (Moravský kras) and to the south one can find vineyards with stylish wine cellars. The city is surrounded by beautiful mixed forests offering numerous hiking and biking possibilities. The city hosts annual cultural events and festivals (e.g. "Brno—City in the Centre of Europe" featuring a firework festival known as Ignis Brunensis) as well as significant sporting events (motorbike Grand Prix). Modern shopping and entertainment centers can also be found here.

Accommodation data (ČSÚ, 2012) show that the number of visitors in Brno is growing constantly (486,318 visitors in 2007; 3,75% CZ). This indicator dropped in the following years due to a common decline in tourism during the crisis to around 400,000 visitors. New data point to stop in this decline and revival of visitors. Compared visits of South Moravian Region—39.9% of all visitors are heading to Brno. For foreign visitors, this share is higher at around 57%. The proportion (accommodated) non-residents to residents is generally about the same ratio. Foreign guests are mostly from Poland (but mainly transit), Germany, Slovakia, Italy and Austria. Among the reasons for attendance Brno are given the following: recreation (25%), visiting friends and relatives (26%), visiting cities (24%), tourism (23%), business trip (12%), sports holidays (9%), stop on vacation (8%) and shopping (6%). (Průzkum návštěvnosti města Brna, 2004).

South Moravian Region is famous across the Czech Republic because of its wine producers. But Czechs are also famous for they beer drinking. The average per capita beer consumption is about 160 l during every year gives the Czech Republic the first place in the world (the second is Germany with 138 l). It means the beer production and beer drinking are very typical for the way of life in the Czech Republic as well as for Brno and its inhabitants.

The paper is devoted Brno (Starobrno) brewery partnership and communication activities for tourism and local development in Brno.

The paper is devoted on local development, especially to its bottleneck that may arise as poor communication and cooperation among important stakeholders. The authors emphasize the role and importance mainly three stakeholder groups: 1) public administration, 2) entrepreneurs and 3) local public.

From methodology point of view, the representative research of Brno citizens older than 18 years was conducted jointly by the International Institute of Marketing, Communications and Business (IIMCE) together with the Faculty of Regional Development and International Studies, Mendel University in Brno. The main objective was to find out which company citizens of Brno considered as the most important for the further development of their city. Collection of data took place on 15th–28th march 2013, through personal structured interviews. Answers of 434 respondents residing in Brno were processed. Respondents were chosen by quota procedure and by socio-demographic characteristics (gender, highest educational attainment and age). The research sample can be considered as representative of all the inhabitants of Brno over 18 years. Interviews also included forms of open questions.

2 LOCAL DEVELOPMENT AND PROJECT PARTNERSHIP FOR LOCAL DEVELOPMENT

2.1 *Partnership and communication for local development*

The idea of local development issues from experiences and results obtained in previous international project Communicating Town in the second half of the 90's. (Foret—Foretová, 2001) and (Foret—Foretová, 2006). One of its general conclusions describes local development as a partnership and a communication among three main participants—the local publics (citizens/inhabitants, civic initiatives, politicians, journalists), the entrepreneurs and the public administration.

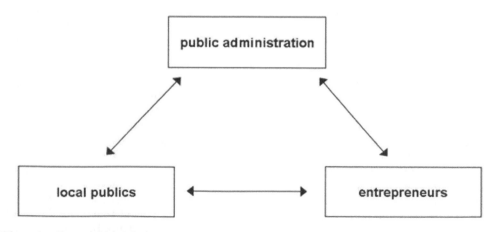

Figure 1.   Partnership and communication for local development.

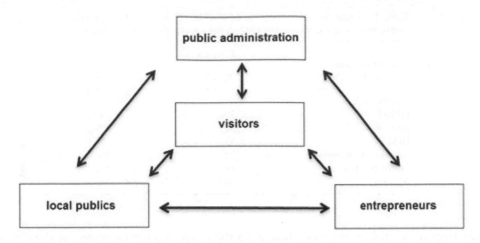

Figure 2.   Partnership and communication for local development.

Especially the relationship between the local publics and the entrepreneurs is full of conflicts. Both sides have very often different ideas about local development. The local publics think about pleasant life (with such values like nature, quiet, cleanness). But the entrepreneurs are thinking first of all about their economic goals (like profit). Therefore the public administration has the task to solve these conflicts and to find a compromise solution.

In the case of incoming visitors (tourists), then we should include other stakeholder group to the above diagram. This group is listed in Figure 1, also with its links to other stakeholders.

2.2   The research results—how citizens perceive different entrepreneurial subjects in region

In preparation for the practical application of the Partnership for local development have so far been carried out empirical research for the Municipality of Znojmo (in the south part of the region, approximately 50 km far away from Brno) in 2010 and 2011, and in 2013 in Brno. We will present the basic results in the next section.

As Table 1 shows, the central research question what firm or organization respondents consider most important in Brno for the further development of the city, the most frequently reported was Veletrhy Brno, a.s. (Fairs Brno). In second place was Dopravní podnik města Brna, a.s. (Transport Company in Brno, DPMB), with apparent distance took an imaginary third place Zetor, a.s. Next order consisted of IBM Czech Republic s.r.o., Masaryk University in Brno and Starobrno Brewery, which is part of Heineken Czech Republic, a.s. All less frequently nominated companies were included in the last category "other".

The results showed unusually rare consensus of respondents and there were minimal differences in responses according to socio-demographic characteristics such as gender, age and education. This conclusion was also confirmed by the statistical calculations of Pearson's coefficient of contingency, which was lower than 0,1 in all three cases.

The next analyses showed that as the main advantage of Veletrhy Brno people consider its contribution to the image and presentation of the city of Brno (32%). Fair events attract hundreds of thousands of domestic and foreign visitors, so they are also beneficial for the local tourism (26%).

As main contribution of Dopravni podnik mesta Brna to the local development was seen in the development of infrastructure, accessibility and transport as agreed 93% of its supporters. Most preferably represented contribution by the Zetor is a range of job opportunities (71%). Respondents pointed in particular to the fact that this company is one of the largest job provider in Brno. Another advantage was seen in the competitiveness of its products (12%). Similarly, IBM has been seen as the most cherished job provider (56%). The second

Table 1. Frequency distributions of answers to the most frequently mentioned the most important venture for the development of Brno.

Most important firms in Brno	Frequency	
	Absolute	Relative
Veletrhy Brno	69	16%
DPMB	59	14%
Zetor	34	8%
IBM	32	7%
Masaryk University	22	5%
Starobrno Brewery	18	4%
All other	200	46%

most frequently cited answer was a benefit for the image and representation of the city of Brno (16%). Respondents also pointed to the fact that this company is very modern, perspective and employs leaders in their field.

According to the respondents, the main asset of Masaryk University is the development in the field of education (82%). They regard it as one of the most important and best universities within the Czech Republic. Due to the fact that both provides quality education to students, as well as facilities top scientists who are successfully involved in research in various fields, contributes to the image and representation of the city of Brno (18%).

In the case of the brewery Starobrno, respondents praised the tradition and history of the brand, the availability and quality of its products. Equally (22%) have been appointed and offer advantages such as employment opportunities and benefits for the image and representation of the city of Brno.

On the other hand, the following paragraph shows the most commonly cited deficiencies of considered enterprise. In the case of Veletrhy Brno, the mostly cited deficiency was transportation (41%). This problem is clearly associate with the aforementioned influx of a large number of visitors during trade fairs, often causing traffic jams and columns that complicate the situation for the residents in the city. With the increased traffic related negative impacts on the environment, as reported by respondents in second place (10%). Dopravni podnik mesta Brna was criticized because of price increasing of fares (46%), but without at the same time improving the quality of public transport. As mentioned in the second place (36%), a delay occurs most connections and buses and trams are often crowded. Respondents blamed Zetor for environmental pollution (32%). The IBM is negatively perceived because of influx of foreigners (53%) and the associated loss of jobs for local residents. The citizens of Brno often associated this fact also with the rise in crime, prostitution and alcoholism. The main disadvantages of Masaryk University were similar to the previous mentioned with IBM, when its supporters as negatives also reported an influx of foreigners and foreign students or students from other towns of the Czech Republic (27%). The most negative view on Starobrno Brewery was featured negative impact on the environment (39%), particularly in the form of a smell from the brewery and drunken behavior of consumers.

If we look again at the Table 1, it is clear, how little citizens of Brno realize that the decisive role of engineering companies as Zetor, as well as nearly fifty-five tradition of organizing international trade fairs, are not a decisive for development of the city anymore. Especially in the case of Fairs Brno is more than obvious that the idea of Brno as fair city or town fairs is slowly becoming a thing of the past. Just a quick glance at their website (www.bvv. cz) and from the published annual reports show that for the years 1997 to 2011 the number of employees declined by half—from 763 in 19997 to 357 in 2011. Undergoing a similar development as well as gross turnover, which for the same years fell from 2 billion CZK to 1 billion CZK. Also the interest of key customers, which are undoubtedly exhibiting companies declines as evidenced by the following figures: whereas in 1999 were attended by nearly 14,000 from 63 countries, in 2011 it was 7,300 companies from 50 countries.

Similarly, earlier in that view, that Zetor is one of the largest employers in Brno is rather out of the realm of memories. Currently, it operates less than 900 people. In any case, however, remains in the results of research only representative of large engineering companies such as the Prvni brnenska strojirna, Kralovopolska strojirna, Zbrojovka Brno and others that played a determining role in the development of Brno for at least half a century—since the end of World War II until the early 90s.

In contrast, IBM and Masaryk University, represent the current and future direction of the city into areas of new technologies, education, science and research.

Following the previously mentioned negative perception of Starobrno Brewery in connection with its alleged negative impact on the environment, as 39% of its supporters, particularly in the form of odors from the brewery and drunken behavior of consumers would be appropriate to properly communicate and explain to the public through activities such as public relations, according to experts' smell from the brewery "really not in the least threat to the environment in the vicinity of Old Brno Mendel square, let alone the entire city. It is controlled by the administrative authority within the integrated prevention and complies with regulations on environmental protection. Furthermore, what some of the above mentioned supporters Starobrno labeled as "bad brewery" does not have to be perceived as "odor", but on the contrary as the Mendel Square and the nearest neighborhood for decades, even centuries, "a typical smell of malt" that accompanies the transport of malt and brewing.

As shown by the results of the representative survey of the population of Brno Brewery Starobrno is considered by residents of the city as one of the leading companies for the development of the city and for its tourism. In the following section, we will present some company's benefits in this area.

2.3 Starobrno brewery and local development and tourism

The Starobrno brewery was founded in 1325, as a part of Cistercian convent settled below the Špilberk Castle in old part of the town. Starobrno means in Czech "Old Brno". In 1994, Starobrno became a part of the Austrian concern BBAG (Österreichische BrauBeteiligungs AG). The merger between BBAG and Heineken Group in 2003 allowed for the establishment of the most important brewing company in Central Europe. HEINEKEN is the third largest brewing group in the world and Europe's number one.

Thanks to the modernization, brewery Starobrno belongs among the most modern of its kind in the Czech Republic today, even the beer is brewed according to traditional recipes. Starobrno produces more than one million hectoliters of beer per year and offers six different brands of beer in regular production: "Ležák", "Medium" and "Tradiční" (pale lagers at 5%, 4.5% and 4% ABV respectively), "Řezák" (a semi-pale lager with caramel flavouring), "Černé" (a dark lager) and "Fríí" (an Alcohol-free beer). Starobrno also produces special beer as "Zelené pivo" (green beer) or Modrá Kometa (blue Comet). Since 2012 Starobrno have replaced process of pasteurization and is using unique technology of micro-filtration, so beers longer retains its original properties and also fresh and natural taste.

Starobrno brewery offers for visitors a tour on which they will get to see the most important parts including the new brew house which was completed in 2008. They will be told about the nearly 700 year history of the brewery, will get acquainted with the Starobrno beer production process and will get a chance to taste the beer brewed in the center of Brno. The main aim of the excursion is to inform all visitors about all phases of beer production and especially procedures that make the Czech beer in the world famous. Visiting route starts outside the boiling room, where all the tourists heard all the attractions of the history of the brewery from a qualified guide and continue to boiling, fermentation and filling room. Beer is produced in the same place since 1872.

In 2012 visited the brewery 4,797 visitors from the Czech Republic and abroad. These figures relate to the period from January to the end of November—and it's more than in the whole year 2011 (3,382 visitors). Attendance at the brewery Starobrno increase the third year in a row. It is also partly due to newly introduced weekend excursions without prior order, which began with the year 2012. Guided tours on the weekend, was introduced because of the increasing number of candidates who would like to explore brewery during the free day. In 2013 is planned to extend the excursion route, so visitors can other parts of brewery and there will be add a new information panels. Visitors can also buy beer souvenirs, gifts and all different Starobrno beer brands and products in Brewery Shop.

Visitors can combine the Starobrno brewery tour with a tour of the city of Brno with its main attractions: Špilberk Castle, Petrov Cathedral, Brno Zoo, Villa Tugendhat (World Heritage Site UNESCO since 2001) or the reservoir—all within easy reach by public transportation or on foot).

Example of grand events organized within the traditional public relations are the Doors Open Days in the brewery Starobrno, the third September 2011. To the event came about 16 000 spectators and they drank about 220 hectoliters of beer Starobrno. In September 2012, Starobrno celebrated 140 Years Anniversary. In addition to tours in the brewery and tasting of beers, Starobrno also prepared a big concert on two stages. This action visited 14 000 visitors and they drank 500 hectoliters of beer Starobrno. Organizing of these events contributes significantly to promoting Starobrno and attracts many visitors from greater distances. But there should be also required cooperation with local public administration and good communication with the local inhabitants, because it is a type of action that significantly disrupts the peaceful life at the venue.

Starobrno also started to support local hockey club Kometa Brno. Kometa returned after 40 years to the First Czech hockey league and gradually became an urban sport phenomenon, with a large number of supporters and highest attendance of matches in all Czech cities. Linking the Kometa reached Starobrno strong anchoring of its products in Southern Moravian Region, raise awareness and promoted its beers and not only among the inhabitants of Brno, but also throughout the Czech Republic.

3 CONCLUSIONS

The current project Partnership for local development is based and builds on the results obtained in the second half of the 90s project Communicating in the city, especially on experiences that local development is dependent on partnership, cooperation and communication of three main actors:

First is the local community, where people also ranks alongside various civic initiatives, political parties, non-governmental organizations, local media, academic and research institutes, development, consulting information centers, etc, second are entrepreneurs, including local associations (Chamber of Commerce) and third is local public administration that local development controls and is responsible for it.

It is clear that the source of the problems of local development is primarily contradictory or conflict between the local community on the one hand and business on the other side.

Interests and ideas of both groups (stakeholders) on further development of the region tend to be quite different. The local communities should be represented generally as strengthening the conditions for a happy and enjoyable life, based on values such as unspoilt nature, tranquility, cleanliness, order, security, etc. In contrast, entrepreneurs usually goes primarily to its economic recovery, to their business benefit. This could in some cases contradict the above mentioned values of the public. The challenges for the local public administration than arise, if these difficulties can be prevented. In the worst case the local public administration must this problems addressed and prepare and adopt the concept (strategy) for local development.

Benefits for local public administration from enterprises could be certainly counted on tax payments, job creation, care and infrastructure, the creation of GDP contributes to the growth of the standard of living n region, etc. For the local community could be important in the case of organizing and sponsorship of popular events, contributing to the knowledge and fame areas, as well as activities within Social Responsibility (CSR).

In managing local problems could also be successfully used such marketing tools, that represent marketing research (sociological research or public opinion research) and marketing communications, especially public relations. With their help should first of all be objectively identified and present the views and ideas of local public on the proposed and existing initiatives, business and public administration. Consequently, there must be searching for the partners and consensual implementation of these plans. Here, then exits the partnership for the local community, business and public administration as a basic assumption of local development.

The aim of the project Partnership for local development, as was already the case in communicating the city, is to assist local public administration. The whole principle of methodological procedure based on mutual sequence, consistency and constant repetition of two basic activities:

First monitoring specific situation in the area—the views of citizens, second the design and implementation of measures to improve the situation. The results of both activities are always properly communicated to all parties—public government, the public and businesses. To do this, there are used the aforementioned tools of marketing communication, especially public relations. The next step would have to devote a specific local business entity.

In any case, the activity of the business entity (organization) in given place always benefits (jobs created in the form of taxes are paid funds, increasing the attractiveness of the place) and losses (worsening living conditions, safety and health). Among the organizations are significant differences from what and to what extent brings to a particular place. It depends on their size (large versus small), the focus of the (chemical production compared to educational facilities), etc. For existing organization is already possible to identify how local residents perceive it (value), what are the benefits and the losses. What are the possible improvements (corrective measures). It is important the local public properly and openly inform about the intended business plan, its local benefits and losses all in the spirit of public relations. Only then is possible to ask what is opinion of local community and based on this knowledge of the situation, use the marketing communication tools to improve.

REFERENCES

ČSÚ—Český statistický úřad. Cestovní ruch—časové řady. [online]. [qtd. 2013–12–01]. http://www.czso.cz/csu/redakce.nsf/i/cru_cr.

Foret, M.,—Foretová, V. 2006. Marketing Communication in the Czech Republic and Slovakian Localities: Ten Years of the International Project Communicating Town. International Review on Public and Non Profit Marketing. Vol. 3, Number 1, pp. 81–92, ISSN 1812–0970.

Foret, M., Foretová, V. 2001. Jak rozvíjet místní cestovní ruch (How to Develop Local Tourism). Grada, Praha, 1st edition, 178 p., ISBN 80–247–0207 – X.

Město Brno [online]. [qtd. 2014-03-01]. http://www.brno.cz/turista-volny-cas/informacni-materialy-turista/.

Průzkum návštěvnosti města Brna. 2004. Centrála cestovního ruchu Jižní Morava [online]. [qtd. 2014-15-01]. http:/www.ccjm.cz/doc/pruzkumnavstevnostimestabrna-2004.com.

STAROBRNO website. [online] [qtd. 2014-02-01]. http://www.starobrno.cz/cs-CZ/uvod.html.

Current Issues of Science and Research in the Global World – Kunova & Dolinsky (Eds)
© 2015 Taylor & Francis Group, London, ISBN 978-1-138-02739-8

A problem of responsibility in media studies

Ľudovit Hajduk
Faculty of Mass Media, Pan-European University, Bratislava, Slovakia

ABSTRACT: The current media debate generalization primary responsibility for the form of media discourse attributed to the media. Derive from it serious cultural implications to the unwavering certainty. The focus of responsibility for the state a shift to media products, media text as their product or media as such. How critical scientists, we need to ask in whose interest may be carried out by this center of gravity shifting responsibility to the media, which results in masking the true origin of the discussions?

1 THEORETICAL AND PHILOSOPHICAL ASPECTS OF RESPONSIBILITY

I decided a long time, as my paper will contribute to the analysis of very serious of responsibility issues in media studies. The question is, which is very general and theoretical concepts are inspirational and can form the basis for serious empirical scrutiny of this issue in the context of media studies? Responsibility belongs to determining the professional and ethical categories primarily in the media and especially in journalism (Remišová, 2010). From this perspective, ideal candidates for the analysis of accountability in media studies now represent two particular philosophical and theoretical concepts of responsibility, and this: the concept of Hans Jonas and Jurgen Habermas.

Jonas writes about the ethics of duty to make the future, writes that it is a search for "fear" in the context of today's world and to test the possibilities of application of certain ethical principles (within, say, the thought experiment). Similarly, writes about the obligation to exercise reasonable idea of feeling distant effects. Fear according to Jonas does not by itself nor by itself does not afraid of what is to be afraid. In the real world we used to get the experience of fear. As to future generations, we need help here with reason and analysis of our actions in the present. The fate of future mankind us today is basically indifferent, but the idea of what may await our successors, should help us to make sure we were allowed to reach—the future Thinking—luck or misfortune's followers (future generations). It is the responsibility of today's man, realizing the threat of possible orientations of human behavior and actions. Uncertainty of future projections, technology and science work with short-term forecasts. You give them the opportunity and strategy as well have some mathematical sequence. However, if you realize that human behavior and actions are not governed by mathematical formulas, and it is difficult to make predictions of human behavior is almost impossible, and therefore uncertain as to design the image of man and mankind in the future. The only thing we have and what we can use the analysis of past optionally analogy of past and current behavior of the companies, countries and people. Thus in a sense, the media can distinguish between individual and collective responsibility.

"Almost all of those now writing about collective responsibility agree that collective responsibility would make sense if it were merely an aggregative phenomenon. But they disagree markedly about whether collective responsibility makes sense as a non-distributive phenomenon, i.e., as a phenomenon that transcends the contributions of particular group members. In this context, as in many others, skeptics set the agenda. Two claims become crucial. The first is that groups, unlike individuals, cannot form intentions and hence cannot be understood to act or to cause harm *qua* groups. The second is that groups, as distinct from their individual members, cannot be understood as morally blameworthy in the sense required by moral responsibility." (Smiley, 2011). Timeliness and importance of Habermas discourse ethics contributes despite all the critical comments their concept of responsibility: a) The nature of moral reasoning intersubjective phenomena and the possibility of confirmation of the social dimension of morality; b) to reveal discursive procedures as a democratic way of reasoning and validation of moral norms, c) The deepening of normative criteria of criticism. Discourse ethics is not ethics which deals with some moral man project. Habermas has not a problem. His interest in morality concerns only questions of the possibility of moral reasoning standard. Power that shows us who we are and leads to the realization of our inherent nature, is neither God nor face each other, it's just a force of rationality are creating our living world. The founder and creator of discourse ethics as generally recognized is Jürgen Habermas. Habermas used to describe an ethical concept connection Diskursethik that into Slovak or Czech Translation different. In its later interpretations he makes a clear distinction between the moral and ethical discourse emphasizes that his efforts are directed in particular to the analysis of discourse theory of morality notes that will stick to the established terminology connection Diskursethik discursive ethics (Habermas, 1991, p. 7).

2   A DISCOURSE-ETHICAL CHALLENGE TO THE RESPONSIBILITY

Habermas goes back to Kant and analyzes the function of reason in different areas of our world. His inspiration European, Kantian tradition embodying ambition to grasp the development of critical thinking it leads to finding and uncovering the relationship between the functioning of theoretical and practical reason. His efforts in the field of ethics is aimed at transforming Kant's moral theory through communication and pragmatic means.

Habermas bases his approach on the theory of communicative action. Trying to redefine Kant's categorical imperative monologue from a position of moral character principle in the form of public dialogue. Habermas sees the main problem right categorical imperative in character monologues moral principle which leaves unanswered questions possibility of intersubjective validity of moral courts. Kant's categorical imperative is a moral law which reason makes a wish to act like moral imperative to obey the call. But what sense moral law commands are not specified standard. Categorical imperative can be seen as a formal procedure for the determination of the will "Act only according to that maxim which can also want to become a general law ' ... resp. "Watch out as would be the case of your maximum on the basis of your will to become universal law of nature." (Kant, 1976, p. 62) categorical imperative requires Take on the role of the legislature for oneself and others and to act in accordance with such rules I want such laws universally valid maxims costs. But how to ensure that what one person may want as a valid norm will want a different and that therefore the standard in question actually valid intersubjective maxim procedure? In the case of Kant's theory can only rely on the order of reason. The controversy with Horkheimer and Adorno,who through their interpretation of Kant radically question the possibility of fair and universal functionality Kantian categorical standards and interpret them from a whole traditionalist position where actually dissolves otherness, as the premise where there is a kind of ideal, but that is not demonstrated (Habermas's referred to as a myth), Habermas is back to Kant. Believes that "cleansing" of Kant's ethics of metaphysical residue can be moral justification of normative assumptions based on theoretical and communicative means. Deontological dimension of moral phenomena are in fact already is derived as in Kant from the operation of goodwill but rational subordination inherent in the communication process into which the all participants as equal partners. In his work The Theory of communicative procedure Habermas emphasizes that communicative reason is directly contained in the processes of social life since acts of mutual communication play a role in the mechanism of coordination costs. Habermas sees the specific competence of the human race is not in the ability of knowledge but the ability to take language to participate in communication (Habermas, 1998, p. 156–157). According to T. Machalová "man we are no longer depicted as being monologues blessed with good free will but as a being whose essence is formed in the process of communication. Communication ... ceases to be the only medium of information transfer and becomes as a means of interaction that is mediated by the language of the proceedings. Structure of the case as a form of human being." (Machalová 1998, p. 117).

If we sum up: Habermas' critique of Kant's understanding of morality we learn: 1. Minimal source of morality is not good will but the communication process respectively communicative action. 2. Moral reflection is a demonstration application of communicative competence. 3. The synthesis takes place as the transcendental Self discursive embodiment of a consensus and finally. 4. Categorical imperative as a moral principle is transformed into the principle of universalisation, i.e., principle of formation of generally applicable rules of communication.

3 COMMUNICATION AS THE SITUATION UNDERSTANDING

Habermas has created a discursive ethics within his theory of communicative procedure and communicative rationality morality presents us as everyday matter. Moral action according to him creates in everyday communication in solving problems. Its aim was to "justify" philosophical ethics as a special form of argumentation theory" (Habermas, 1983, p. 54). Habermas could imagine that communication as a source of morality minimum have to justify its normative dimension. This problem started in the 70s already paid 20 century, when it developed a theory of communicative competence, which is known as the "universal pragmatics" (Habermas, 1984, p. 354). Habermas finds that if it explains the use of language as a course of action in particular to clarify the normative conditions of the relationship that exists between speaker and listener. What actually happens when talking to another person when we talk to him any sentence? Such information is subject to the rules of speaking?

Habermas defines the necessary conditions for communications and show that transfer information to which here there is not just a normal transfer of information but that situation understanding. The understanding of the situation in relation distinguishes the speaker and the listener are two basic levels: 1. Plane of intersubjectivity—in which the speaker and listener talk to communicate, 2. Plane of substance, i.e. what is to be understood.

Communication and poses as a situation where there is a possible understanding of that sentence here bind to extraverbal context. Transforming sentence to notice here takes place by the means pragmatic universals. Habermas refers to these universal and universal constitutive dialogue. Their function is to design rules, enabling the creation or generation of speech situation as understanding the situation.

Understanding of the constitution is characterized as a situation that is generated by idealising power of its participants. Terms of performance when speaking in its communication must master grammar respectively linguistic and logical rules. Should these rules not make it he could not communicate to other people. What would tell them would be incomprehensible to them and would lead to not understanding and not understanding. Act of communication should ever could occur. Based on this finding Habermas distinguishes idealizations embodied in language and idealizations embodied in communication.

Idealizations embodied in the language they are idealizations: 1. Generally, the language respectively concepts of thinking that is enshrined in the language for example vocabulary grammar, guaranteeing its general use and invariance; 2. force which scenes in the operation of speech acts. Equation for the statements receipts for valid and accurate, respectively real. In communicative action then these idealizations bind to another type of idealizations that are associated, 3. with 'delivering the performative—constructive opinion. These idealizations rise in the use of language normative pressure which is manifested by the participants of communication not only control the grammar vocabulary language in which they say but also voicing their performative attitude. This means that their communications producing activity which gives rise to mutual interaction. Habermas believes that the parties enter into the conversation of communication as speaking the truth responsible single people to understand you and to speak. This leads him to believe that the speakers and listeners occurs only correct lead in the coordination of their actions but also the effective functioning interaction. Otherwise the act of communicating undergoing.

According to Habermas person when using language anticipates the counterfactual i.e., unconditionally accepts the universal language of force requirements. These requirements are clarity, truthfulness, reliability and accuracy (Habermas, 1984, p. 354). These requirements derived from different forms of speech acts such as: communicative constitutes of representative of regulations. Force requirements guarantee that the speaker performs its information positively and constructively as elected understandable words and expressions i.e., linguistically and grammatically correct language use. Its expression thus giving the listener something should understand and he assumes that these statements are true. But speaking listener "credit" only when it is satisfied about its credibility. Only then will consider the recommendations and calls for taking certain attitudes and actions correct. A person who is not expressed clearly say it once then again says something else ceases to be a trusted partner communication. His statements are ambiguous signal that it seeks not understanding and understanding but getting hold of handling listener.

Habermas shows that the force requirements are: a) The term contingention will and their adoption party communication is not empirically motivated, b) Have a universal nature. c) Have only a coordinating effect but lead to the fact that the situation of interaction between actors communication is considered true.

Claims force is equipped with every language and its application in speech is unproblematic. But Habermas considering communication as a form of action, i.e., the as a situation where there is a real fruition. This also is a possibility that there appear testimony of various truth—content that problematic validity claims. Habermas wants to be paid to how these problematic conditions removed. Assumes that communication is the unity of the two planes when the first plane connected with the functioning of linguistic terms associated with context i.e., enable interaction. Second plane then creates its own argumentation language—discourse. In this plane the language statements are only thematically. Although the procedure of communication and expression participants accompanying discourse are not part of it. So the only question argumentation speech situation. Habermas about it literally says: " Under the slogan procedure introduces sphere of communication in which information is exchanged (i.e., the action drawing on experience) where implied in the speeches (even arguments) estimates and also recognized the validity of claims. Under the slogan discourse I introduced a form of communication characterized by argument the theme of which is to achieve eligibility problematic validity claims. If we keep the discourse, the same way we need to step out of the context of the case and experience, there does not occur the exchange of information, but the arguments that are used to justify (or reject) a problematic validity claims (Habermas, 1984, p. 138).

4   BASIC PRINCIPLES OF DISCOURSE RESPONSIBILITY

Habermas does not remain only in terms of validity of the reasoning, but also wants to explain why it may be consensual discourse led not only normative condition of the case, but also the criterion of correctness. He assumed this role will explain the "logic" of discourse. Argument does not consist only of a series of sentences but also speech acts. However there is a transition from one to another? How acquires character argument to support or weaken the validity of claims? How can we the arguments confirm anything disable evaluate or reject? This relation can not be explained on the basis of formal logic. Habermas in this regard based on the concept of justification put forward by American philosopher Stanley British origin Toulmin (1922). There is argument against analytical method based on formal logic in substance of reasoning that is not based on the premise already contained in the summary, but counts the reasons for this conclusion. The argument is in substance conclusion stronger "suported". This finding Habermas used to justify so principle of universalization which indicates a large letter "U". Until its definition to proceed to the formulation of the concept of discursive ethics. Habermas assumes that the theoretical and practical discourse is a fundamental difference

Arguments practical discourse relates mainly to justify social phenomena. It is thus an argument according to which there is a rulemaking proceeding—standards. There have been reasons for the inductive nature but based on generally accepted needs and interests. Among the material content of the standards and expectations of the addressee standard is completely different than the relationship between the existence of subject matter and their expression of astertorical sentences. Here according to Habermas needed so bridging principle the principle of universalisation performs the function U (Universalisierungsansatz). This principle is: "sufficient condition for each applicable standard is that all participants will discourse voluntarily, without coercion to accept direct and side effects of their joint efforts to satisfy the interests of each individual." (Habermas, 1983, p. 75–76).

Function principle of universalisation. Arguments in practical discourse is under Habermamsa always based on the Functioning of the generally applicable rules. This rule is no longer the result of favorable recognition of the interests of all participants in the discourse and allows you to take a neutral perspective. Given standard thanks to this principle becomes valid and is generally recognized when all the participants to achieve the consent of discourse that this standard should apply.

At least the basics of morality. Acceptance and recognition of the principle of universalisation process of argumentation presupposes the existence of a knowledge requirement which, according to Habermas always a case of moral obligation. General normative validity claims embedded in the rules of discursive communication (clarity, truthfulness, veracity, accuracy) may be seen not only in normative pressures but also ethical relevance. Habermas them as a basis for a minimum of morality. Defining principle of universalisation was found a general normative framework needed to justify the discourse ethics (Habermas, 1983, p. 75–76).

Award of ethical relevance of the discourse Habermas still while accepting question of how to become standards in the discursive situation of current moral standards. What is going on here that all participants discourse not only recognize a certain standard as valid, but in its place? Initially addressed this issue by reference to its concept of universal pragmatics. However, it was only possible to the moment when he ceases to be observable difference between ethics and morality. This leads to the definition of Habermas moral discourse principle which refers to the letter "D" (Diskurssprinzip). This principle is: "The discursive ethics is attributed to the validity of the standard only when those affected by this standard be any way as participants in a practical discourse agree that this standard is applied" (Habermas, 1983, p. 76).

1 What is the principle of discursive D? This principle is often compared with Kant's categorical imperative which leads to the "cleansing" of our will from its subjective wishes and purposes. In contrast, a discursive principle has led to the formation of a collective will—figuratively speaking—to the "cleansing "of strategic leadership communication. This means that each participant in adopting normative validity claims the use of their language also undertakes to comply with the conditions that allow a certain standard as a valid moral norms. Participants entering the discourse has certain moral ideas. The procedure argumentation speech then place their correction in a way when will the consensual agreement of the parties shaped by adopting the views and interests of others as their own. The ethical discourse is only about whether and how this will be substantiated and confirmed the accuracy/validity of existing moral norms (Habermas, 1983, p.76).

The first correction discursive principle. Habermas considers the level of awareness of moral discourse participants sufficient reason to act in recognition of the validity of an act also saw its acceptance i.e., act in which the standard becomes immediately recognized the motive of the case. This undifferentiated definition of discourse as situations justify the validity and application of standards also led to a very large wave of criticism and under the influence of Habermas came to a certain correction of their views. The first correction made to work Explanation discursive ethics. It admits that the discourse may include a discussion of ethical and pragmatic issues in general so the theoretical problem here is the logic state of the application. This shift seeks to show that discourse performs the role of suppression mechanism of absolutizations specific content of moral values and uncritical promotion of only one idea of the good and the good life (Habermas, 1991, p. 100–118). Formalism and universalism of ethical discourse are presented to us here: a) As legitimate forms of reasoning where the meaning and importance must be re-interpreted; b) Paradoxically not as opposites pluralism but as necessary conditions for its preservation.

The second correction discursive principle. The more radical correction came Habermas work facticity and validity which is already covered by the formulation of the principle of pragmatic D into shape which did not use conditional, but only talks about the possibility that "force those just standard procedure in which can be identical as possible all those concerned as participants in rational discourse" (Habermas, 1992, p. 138). Rewording actually looking for ways in which the discursive reasoning made available by any standard procedure morally. This is confirmed by the finding that the discourse is open to different topics, information reason, etc. This gives it universality principle of universalization U. Habermas does not want to talk about the pragmatic dimension of discourse just as metalevel, but wants to deal with specific types of reasoning (legal, political), and thus show that they also own a discursive principle. In this context, the so-called talking, ethical—political discourse. To distinguish the various forms of discourse chose the easy way. Discourse principle in a situation of moral reason simply left to the principle of universalisation as functions as rules of argumentation. Morality had to be left universal form otherwise they would be called into question the foundations of discourse ethics. For other forms of discursive reasoning acquires the form of the so-called principle, democratic principle. His interpretation of Habermas completes their ethical approach as proof of eligibility critical theory.

The crisis of the traditional way of legitimization. Democratic principle works for other reasons not the ones that lead to the validation of moral standards. Its operations forces legitimacy crisis law. This means that appeared legitimation claims that cannot be it resolved only by reference to legal form. The traditional way of legitimizing does not guarantee the democratic nature of the law because they are strongly reflected the interests of the state and the entire remoralisation of positive (statutory) rights in his regime. This way is not instrumentalised only the right but also morality. Advancing the growing influence of law in modern society and causes them to not only expand the radius of normative regulation, but in this hidden form by the control of power, even where social issues should be addressed and regulated morality. This creates the possibility of abuse of moral arguments in the name of democracy. The whole process of legitimization is literally becoming "immune" to some external correction and concluded in selflegitimated procedures i.e., produce their own reasons its own rules of legitimization. For these reasons, the legitimization of norms by reference to their legal form becomes the current unsustainable pluralistic society (Habermas, 1992, p. 49). Habermas justification democratic principle is an attempt to break the cycle of power selflegitimated through the inclusion of all stakeholders—citizens, both as creators of moral norms in force.

5 CONCLUSIONS

Current discourse—an ethical challenge to the responsibility of media based primarily on defining the concepts of Jonas and Habermas. In terms of conceptualization of responsibility in the media plays an important role collective responsibility of management of the media, the problem of shared responsibility and responsibility procedures that can build on discourse. Ethical ethical discourse of this aspect as a new form of communicative rationality. Ethical discourse as a form of argumentation practice is characterized by empirical and normative beliefs of its participants as well as Their interests are subject to change under the influence of the arguments raised. Discourse represents a new type of communicative rationality and its basic approaches to the interpretation of Habermas structure of modern society. Functioning mechanisms of communication According to him leads to various ways of social integration for both social integration Which embodies the life world (Lebenswelt) while the integration which Represents the world system (Systemwelt). Organization of society on the basis of systemic relations according to him however does not reflect the natural sense of social integration. Their continual expansion Considered a major threat to the

Living World. In this context speaks of "colonization of the life world" where the growing influence of the world system leads to degradation of its foundations, i.e., cultural value, consensual understanding and solidarity.

REFERENCES

Habermas, J. 1981. *Theorie des kommunikatives Handelns*. Bd. 1. Frankfurt, Suhkamp.

Habermas, J. 1983. *Moralbewusstsein und kommunikatives Handeln*. Frankfurt, Suhrkamp.

Habermas, J. 1984. Vorstudien und Ergänzungen zur Theorie des komunikativen Handelns. Frankfurt am Main: Suhrkamp.

Habermas, J. 1988. *Nachmetaphysisches Denken. Philosophische Aufsätze*. Frankfurt: Suhrkamp.

Habermas, J. 1992. *Faktizität und Geltung*. Frankfurt: Suhrkamp.

Habermas, J. 1996. *Moralbewusstsein und kommunikatives Handeln*. Frankfurt: Suhrkamp.

Habermas, J. 1998. Teória komunikatívneho konania. In Buraj, I. et al. 1998. *Sociálna filozofia*. Bratislava: Filozofická fakulta Univerzity Komenského.

Habermas, J. 1999. *Dobiehajúca revolúcia*. Bratislava: Kalligram.

Honneth, A. 1994. *Kampf um Anerkennug zur moralischen Grammatik sozialer Komflikte*. Frankfurt: Suhrkamp.

Honneth, A. 1996. *Sociální filosofie a postmoderní etika*. Praha: Filosofia.

Horster, D. 1995. *Jűrgen Habermas*. Praha: Svoboda.

Kant, I. 1976. *Základy metafyziky mravů*. Z nemeckého originálu Grundlegung zur Metaphysik der Sitten preložil L. Menzel. Praha: Svoboda.

Machalová, T. 1998. Diskurzívna etika. In: Gluchman, V.—Dokulil, M. (Ed.) 1998. *Súčasné etické teórie*. Prešov: LIM, 1998. p. 111–121.

Remišová, A. 2010. *Etika médií*. Bratislava: Kalligram.

Smiley, M. 2011. Collective Responsibility. In Zalta, E.N. (ed.) 2011. *Stanford Encyclopedia of Philosophy*. Stanford: Stanford University.

Current Issues of Science and Research in the Global World – Kunova & Dolinsky (Eds)
© *2015 Taylor & Francis Group, London, ISBN 978-1-138-02739-8*

Emotions as a key element of news reporting

Z. Hudíková
Faculty of Mass Media, Pan-European University, Bratislava, Slovakia

ABSTRACT: Television news still ranks among the most watched TV programmes. Slovak nationwide television channels synchronized their times of broadcasting, as well as the length of the main news programmes. They compete for viewers not only by setting the content of the main daily events and by including the input of interesting editors' and viewers, but also by the form of processing the topics. Apart from the informational function, the entertainment function of the news broadcasting too rises in prominence. Emotion and story are becoming dominant means of attracting and maintaining attention, as well as communicating information. In this paper the author presents how these two functions—informational and entertainment—blend in news reporting, and how are the emotions of respondents and recipients are used in news production. The author arrives at the results based on qualitative content analysis of television communicants—main news programmes on the television channels RTVS, TV Markíza and TV JOJ in the span of three months (January, April, June 2014).

1 INTRODUCTION

Television news is still one of the most important programmes on television channels and main news programmes are one of the most viewed programmes of the programme structure (Department of Research Analysis RTVS). Viewers of the middle and older generation, but often also of younger the generation, conform to their every day stereotype and turn on the TV in the evening to watch the selection of the main events of the day to gain the information about domestic and world affairs. Even though the viewer has already received the information about the events during the day, he/she expects to see the summary of the most important news in a consistent form: an analysis of reasons and consequences of the events, a prioritization of the events according to their significance, a prediction of possible consequences for society, as well as for the individual life of the viewer. Apart from all this, the viewer expects to receive it in a clear and comprehensible form. News reporting provides the viewer with meanings and various interpretations of events (Trampota, 2006, p. 104).

Every day, through various channels and media, a recipient obtains and processes a great amount of information necessary for their lives. Quality, amount and character of this information creates high demands on its journalistic processing, as well as on the quality of its reception. Nowadays, two key functions of media predominate television production—to inform and to entertain. People watch TV to gain information and knowledge, and to fulfill their needs, dreams and desires through emotions, that are triggered by the programmes. Television entertainment provides specific kinds of secularized ritual, as well as the specific emotional and mental frame of mind (Pravdová, 2011, p. 9). Both functions—informational and entertaining—blend together not only within the programme structure, but significantly influence the content and form of news—a journalistic genre, which was until recently based on mere transmission, analysis and interpretation of facts. Apart from the need of being informed, the demand for entertainment is gradually growing (Nordström, Riddesträle, 2005, p. 15). H. Pravdová states, that the "current media production is overloaded by genres that are to entertain and please the 'recipient's soul'" (Pravdová, 2009, p. 294). In this paper, the author points out the way of how informational and entertaining functions blend, and how the emotions of respondents and recipients are used in the television news production. This paper is based on her long-term three-month research (January, April, June 2014) of the main news of RTVS (Rozhlas a televízia Slovenska—Slovak Radio and Television), TV Markíza and TV JOJ. The author arrives at the results based on qualitative content analysis of television communicants—main news programmes on the television channels RTVS, TV Markíza and TV JOJ in the span of three months (January, April, June 2014).

2 THE EMOTION—A PART OF AN EVERYDAY LIFE

To attract someone's attention, it is necessary to offer an experience, which is spontaneous, immediate and intense (Nordström, Riddesträle, 2005, p. 86). Therefore television production is more and more focused on the close connection between the informational and entertaining aspects.

Emotions form a fundamental part of a person's life. They are triggered by various impulses and situations; they reflect relationships with the outer environment, and with actions that happen in the environment. Emotions are of various kinds and intensity; its experiencing is solely individual. Emotions can be defined as the "state of increased activity of the organism, when experiencing the reality (pleasant—unpleasant), and are manifested by physiological changes and by characteristic behavior (posture, gesture, facial expressions)" (Cakirpaloglu, 2012, p. 204).

Emotions mobilize a person's energy and power, galvanize him/her into action (sthenic emotions), or attenuate and paralyze one's activities (asthenic emotions). D. Krech and R. S. Crutfield divided emotions into: 1. elementary emotions, that are congenital, 2. emotions triggered by sensoric stimuli (external or internal) 3. emotions of self-evaluation (shame, pride, etc.) 4. emotions connected to others (love, hatred, envy, etc.), 5. emotions of appreciation (humor, beauty, admiration, etc.) (Czako, Seemanová, Bratská, 1982, p. 24).

Emotions are experienced in three basic aspects:

- the aspect of quality, as pleasant or unpleasant, and are determined by the intensity of the stimulus,
- the aspect of the quality of the stimulus, as excitement or satisfaction,
- the aspect of the duration of the stimulus, as tension—relief (tension is followed by relief).

Experiencing a particular emotion can be identified by the verbal description, as well as by some of the physiological responses. Physiological responses are automatically triggered by a particular emotional situation (without any possibility to control it consciously), and are demonstrated by a universally recognizable facial expression (Ekman, Izatrd In Slaměník, 2011, p. 20). Some of the physiological responses are visible (e.g. blushing, growing pale, chin trembling), others are unconscious (e.g. change of the blood substances ratio, hypertension or hypotension). The best observable demonstration of experiencing a particular emotion is behavior. This is observable for the other participants of the situation, while the one experiencing an emotion does not have to be aware of his/her behaving, or he/she is aware of it and can even be able to control the emotion.

Most psychologists distinguish higher and lower emotions. Lower emotions are determined by the activity of the basic signal system, and occur as unconditioned responses. They are connected with the vitality of the organism, with the satisfying of the most basic life needs, and are specific by their simplicity. They are linked with instincts. Among the lower emotions belong elementary experiences of pleasant and unpleasant, which are triggered by various sounds, smells, colours, etc., or by complex somatic perceptions, by sensing one's own self. Among these are also experiences of the basic instincts (hunger, thirst, sexual needs), somatic states (pain, fatigue), or instinct of security (defensive or offensive responses of a person—fear, cry). Higher emotions occur on the basis of conditioned responses; they are particular for a human being and are socially determined. They are closely connected with satisfying the secondary needs, and therefore are also called feelings. According to E. Hradiská, these emotions express a person's relationship to his/her surroundings, to their own actions, to others, and to one's own self (Hradiská, Brečka, Vybíral, 2009, p. 316). Feelings can differ in intensity and length. They are determined mostly by particular kinds of actions; they depend on the actual person's relationships and create permanent emotional relations.

J. Boroš [In Czako, Seemanová, Bratská, 1982, p. 34) categorized higher emotions into five following groups:

- emotional relations to non-human events—relationships with non-living things and with living and nonliving natural phenomenons,
- interpersonal emotional relations:
  o relationship between a man and a woman,
  o relationship between parents and children,
  o emotional relations to others,
- relationship to one's own self,
- aesthetical feelings,
- ethical feelings.

Human emotions are of evaluation, communication and adaptation importance. Z. Milivojević (In Cakirpaloglu, 2012, p. 204) stated that "emotions are responses of a subject to stimuli, which are regarded by the subject as important and which mobilize the subject to an adaptive action in visceral, motoric, incentive and mental way." The Object of emotions can be formed by various contents closely connected with needs, efforts and interests of a person (Hradiská, Brečka, Vybíra, 2009, p. 315). If we are experiencing intense emotions, we also often realize that memories of other situations. When we were experiencing similar emotions in the past, these memories of intense emotions come spontaneously to our minds. The state of adaptation signalizes the achieved balance of the internal and external environment.

Generally, there are 6 basic emotions distinguished that are of specific adaptive importance (Stuchlíková, 2007, pp. 54–55):

Joy—is a universally recognized signal of a good disposition, an openness to friendly interaction; or can be eventually recognized as a signal of presence of the important sources that help a person to overcome stress.

Sadness—is noticeable mainly by the decrease of mental and motoric action which has an eventual adaptive effect. The decreasing of this action enables a person to persue a more detailed and punctuated examination of actions and circumstances that led to the loss causing the sadness. This kind of analysis and recognition can help in gaining the new perspective and the different point of view to understand and plan one's behavior. Expressions of sadness also communicate that a person is experiencing a difficult situation which can eventually lead to emphatic feelings or responses from observers.

Anger—incents to mobilize and maintain the required amount of energy necessary for coping with the obstacles while reaching the target. It can accelerate mental and motoric functions, and keep them in that state for relatively a long time. This eventually leads to their effective functioning. Research proved that it is a far more effective way of motivating action than emotions of happiness and interest.

Disgust—motivates a person to secure his/her environment. Responses of disgust are determined by all potential threats; every threat can become by the way of learning a stimulus that develop disgust, not only in its direct biological meaning, but also in its indirect psychological meaning.

Shame—is probably the most noticeable when connected with self-evaluation. Shame is an effective means of directing attention on flaws of one's own self. While experiencing this emotion, self-evaluation grows and thus a person may discover in what aspects he/she does not match with the conditions of the environment. Shame holds an important role in developing a balanced self-perception.

Fear—its basic role is to help a person to avoid danger. The endangering of self-perception, one's own integrity, or a good mental disposition can cause fear and therefore the escape of such a situation. Fear has influence on perceptive and cognitive processes, e.g. fear causes so-called "tunnel vision", which focuses all attention on the source of potential danger and reduces the ability of an unfavorable action.

Adaptation to a situation is accompanied by feelings of pleasure and joy; failure causes negative feelings and frustration.

According to the length of an emotional state, the following categories are defined:

Affect—is defined as an intense and sudden emotional response, which can be caused by various stimuli. Numerous emotions such as anger, rage, joy, fright, shame, or sadness lead to affect if their intensity is high. The course of affect is turbulent and a person loses his/hers rational thinking temporarily. The degree of controlling affect depends on a person's character, educational influences and on life experiences; controlling affect is more successful in its initial stage when it is not fully developed.

Mood—is long-term, less intense than affect and lacks particular stimulus. Mood is subjective and has various forms—joyful, euphoric, depressive, anxious, optimistic, irritated, sarcastic, ironic, etc. To a high degree, mood influences memory, motivation, thinking, attitudes, behaviour and other mental functions and creates a disposition of more agreeing attitudes.

Passion—is an extremely strong, deep and long-term emotion that determines a person's thinking and action. It is embodied in the overall approach of reaching one's target and also in their single steps to fulfill one's desires. Passion is always concrete and particular. As a very intensive emotion, it can be directed towards other people (love for a child), groups (fanatical following of a political movement, ethnic group), objects (obsession with garden, car), activities (charity, sport), etc. Comparing with passion, enthusiasm is linked to a particular idea that strongly attracts a person's attention (e.g. enthusiasm about ideas—e.g. building an ideal society) (Czako, Seemanová, Bratská, 1982, p. 33). Excitement, as passion, is characteristic by a strong affection felt for an object. The difference between excitement and passion is that excitement lasts for short time; it can disappear quickly; it is considered to be the first stage of passion.

Emotions are a result of man's action—including the mental action. However, at the same time they influence the process of that action—its organization and dynamics, as well as the perceptual processes, information processing, memory, or the level and result of cognitive processes. Emotions and related physiological and cognitive processes prepare an individual for the right reactions to the opportunities and problems that occur in the course of social interactions, of which a person does not have to be aware. The basic function of a subjective emotional experience is to trigger a particular intended behavior. After the initial emotional response (which is followed by spontaneous behavior), the subsiding subjective emotions help a person to clarify what he/she feels, to reflect and talk about the circumstances that led to the particular emotion, and to create future plans for the potential similar situations (e.g. to avoid such situation, or to develop an effective solution for it). Subjective emotional experience plays an important role in learning—not only classical, but also in operational. Communicating emotions (the ability to express and identify them) serves for mediating the information about objects in the environment. Emotional response directs attention and intended behavior of others, and therefore plays an important role in learning social behavior. Sharing emotions with others often serves as stimulus for others to provide support and understanding of one's reasons for one's behavior. Emotions also form a part of socializing practices, and help children to learn the standards and values of the culture and society they live in (Stuchlíková, 2007, p. 95).

Emotions can be triggered by the real situations in which a person directly participates, but they may also be triggered by indirect stimuli, i.e. the person is an observant of the situation, or has heard about it. This second mechanism is also the one that works within media communication.

3   TELEVISION NEWS REPORTING AND EMOTIONS

The main aim of news reporting is to provide information. TV news reporting provides viewers with information via various journalism genres (TV news, TV report, interview). However, the current situation is, as we have already stated, that due to the aim to maintain viewers' attention for as long as possible, the entertainment function of these factual genres is becoming dominant. Over the plain factual information stories, emotions predominate and the form of information processing is less formal.

In Slovakia, the nationwide broadcasters of news programmes are RTVS (Rozhlas a televízia Slovenska—Slovak Radio and Television), TV Markíza, TV JOJ, and by news television TA3. Shorter news is also broadcasted throughout whole day in 5–20 minute long programmes. News reporting is also provided by all regional or local televisions and is mainly focused on local affairs.

The main news programme of the Slovak Radio and Television is News RTVS; TV Markíza—TV News, and TV JOJ—Main News, which all start at 19:00. In September 2001, news TV TA3 launched its broadcast and broadcasts news almost 18 hours a day (6–24 h) in 30-minute programmes in addition to the Main News at 18:30. Apart from this, the TV produces programmes of various journalism genres—political (Debata [Debate], V politike [In Politics], Téma dňa [Theme of the Day]), educational (Cestopis: sú..., [Travelogue] Cestovanie [Traveling], Cestujeme bez hraníc [Traveling without frontiers], Check in), economics, sport (Bago, Bez dresu [No Jersey]), cultural (Bravo, Portrét [Portrait]), showbiz (Showbiz). RTVS broadcasts main news titled as News RTVS, presented by Janette Štefánková, Viliam Stankay, Jana Košíková, Ľubomír Bajaník; RTVS also broadcasts short news—News RTVS at 12:00 and at 16:00 (presented by Miroslav Frindt, Simona Simanová), programme News and Commentaries (21:30) with Marta Jančárová or Miroslav Frindt.

TV News is the main news reporting programme of TV Markíza. It brings current and dynamic political, as well as socio-cultural information from home and abroad. As it is stated at the TV's webpage, every topic is processed according to the rule of respecting the universal viewer and of optimal informing without commenting. The programme is presented by Zlatica Puškárová, Mária Chreneková Pietrová, Patrik Švajda and Jaroslav Zápala. Among the other news programmes are 5-minute long news, forecast and sport news every 30 minutes in the morning show Teleráno, and First TV News—a 25-minute long programme at 17:00.

TV JOJ broadcasts its main news TV JOJ Main News, subtitled as Professional, Objective and Clear News, every evening at 19:30 to 20:00. It is presented by Lucia Barmošová, Adriana Kmotríková, Lubo Sarnovský and Ján Mečiar. The other 40-minute news programme is broadcasted at 12:00. It is titled News at 12: The Freshest News at 12 (presented by Dana Strculová, Andrea Pálffy Belányi), and 20-minut News at 17:00: The Freshest News at 17 (presented by Hana Gallová, Aneta Parišková), broadcasted between 16:55 and 17:15. At 19:00, 30-minute news named Crime News, presented by Monika Bruteničová, Stanislava Jakubíková, Michal Farkašovský and Pavol Michalka are broadcasted.

The popularity of TV news is also influenced by the emotional bond of the recipients to it. This is also confirmed by the RTVS survey of popularity of all the nationwide television news.

Graph: The popularity of the main news of the three Slovak nationwide television channels, source: Department of programme research and development, December 2013.

The graph of the news' popularity in 2013 shows, that the most viewed TV news were TV News at TV Markíza, second was TV JOJ News followed by Crime News at TV JOJ, and third with the growing popularity was News RTVS. The fact remains that televieweres have their habits, influencing the bond between them and the programme, which in case of being broken would result in loosening the popularity, as it happened in September 2013, when TV JOJ adjusted the broadcasting times of the main news. TV JOJ News were renamed to TV JOJ Main News and its broadcasting time was adjusted to 19:00; Crime News was moved to JOJ PLUS channel with broadcasting time at 20:00 (RTVS, Department of programme research and development, December 2013).

3.1   *Results*

Television news is supposed to trigger neutral feelings so it does not divert attention from the content; but even the apparently neutral factual content may stimulate certain feelings and emotions. This response is linked to a potential relation between the recipient and the content, e.g. if a news concerning an increase of taxes which would impact the self-employed is broadcasted, it may trigger disgust and fear and in less assertive individuals even worries about one's own existence. News programmes may also contain various emotional news. Since news reporting that fulfills its entertainment function (infotainment) is becoming more common, the news structure and news content are formed in a way, which is not only informational, but also entertaining. Producers use narration, which means that the news is presented as a problem of one of the viewers. This means of presenting, and if the topic is familiar to a viewer, can stimulate feelings similar to the presented ones, especially if it is case of a tragedy, or sufferingwhere the viewer may feel in solidarity with the victims. A specific kind of news is that which presents violence and disasters. TV JOJ produces a particular programme which is focused on this kind of negative emotions.

Differences exist between state and commercial television channels. State television usually presents political and socio-political news first in the structure of the news programme; disasters and tragedies are presented first only if it is a serious domestic case (e.g. bus crash of high school students coming back from school trip), or global (conflict in Ukraine) importance. The common structure of news is as following: important domestic affairs (political and social), international affairs, civic topics, important cultural events. Topics are presented in a neutral, informational manner, which is based on presenting the facts and connection between them. On the contrary, commercial television channels first present tragedies, which are followed by important political and social events. It is a common practice that civic problems are prioritized, or are presented as a scandal of a minister, an official, or government. A positive presentation of a topic is rare, or is produced only if an affair of great importance was either solved with the help of the particular television or if it is attractive and pleasant information about animals. Most of the topics broadcasted by commercial televisions are presented as long-term issues that are not being solved, and are even presented as unsolvable—be it social, foreign or civic issues. The majority of topics are presented in a negative, not neutral manner, which may trigger emotions of frustration, negativism, melancholy and helplessness, and therefore eventually traumatize the whole society.

News broadcast provides information about events of the day as well as of socio-cultural themes. Whether an event becomes broadcasted news depends on a value which is based on the event's frequency, placement, clarity, meaningfulness, urgency, continuity, possibility of further development, relation to elite nations, states, people, personalization, negativity, consonance, surprise, predictability, and variations. From the perspective of presenting emotions and triggering potential responses, the important values are: proximity, relation to people, negativity and personalization. Events happening in the recipient's surroundings are more identifiable to him since they are happening in the locality in which he/she lives, and therefore a recipient perceives more intensely that the presented events could have happened to him/her. If a positive, but mainly negative event is happening in the recipient's country, city, village, or the street he/she lives in, the emotional response is far stronger than if the event were \happening on the other side of the world. Deviations, whether it be violation or destruction of the environment (natural disasters), breaking the social and cultural rules (sexual, moral, ethical errors of people, mainly of those holding important positions), deviations from the socially recognized lifestyle and other eccentricities, are becoming the attractive objects of news programmes. There are numerous examples broadcasted by commercial television channels. According to E. Hradiská, the reliable way of attracting viewers' interest is the presenter's announcement in which he/she states that the following news is not suitable for children, adolescents or sensitive people (Hradiská In Magál, Mistrík, Solík (eds.) 2009, p. 109).

Various journalistic genres are used for presenting particular topics and events (TV news, TV report, interview). Emotions are, more than with a genre, connected with the form of the report and interviewed respondents.

TV news is a journalistic genre with dominant informational and educational functions. TV news is broadcasted in the following forms (Lokšík In Osvaldová, 2001, pp. 73–87):
- read news
- video news
- combined news

Read news—is the basic information about an important social event (Tušet, 1999, p. 93). The presenter reads it from the prompt or from the printed pages; it is short; it contains only the basic information about the event formulated into a few sentences. This news is usually obtained from the agencies and do not contain any visual material. Read news is nowadays a less common genre; in the case of special breaking news, a presenter obtains a texted message and reads it from the printed paper.

Short video news (so-called flash news) is a part of a video news section; it is a short and eloquent video without original sound and with a journalist's pre-recorded commentary. This news is usually of lower importance and only complements the daily mosaic of events. Its duration is several seconds and it is a part of the video news section formed by several pieces of video news.

Combined news (extended news) is presented in the following form: a presenter announces the core of the news which is subsequently explicated in following text and accompanied by illustrative pictures. This type of news is used if there is a sufficient amount of material sources.

TV report (or shot; report) is the most complex genre of the TV news broadcasting, which, by an effective usage of the richness of the audiovisual language, portrays the natural character of the TV presentation in the most adequate way. It is a journalistic genre characteristic by its informational and analytical function, and by using documentary principles. It is based on the presenter's or respondent's authentic witness. This report is also characteristic by its formal diversity and by a variety of the means and techniques of expression. The aim of the report is not only to report, but also to reveal the causes and consequences of the affair. The content of the report is based on facts; it is formed by the wide spectra of political and socio-cultural affairs and events. A specific feature of report is its energy and dynamism; it captures the atmosphere of the action and presents the most attractive parts of the event. Report combines factual text with the real sounds, and with authentic emotions of the participants.

News interview is a dialogized report. The interview forms a part of a news report. Sometimes it is broadcasted separately as a whole programme. The interview provides the viewer with only a short part of the interview lasting for a few seconds, with or without the presenter posing the question. This report usually contains interviews with several respondents.

News broadcasting, as well as other television production, uses picture, subtitles, spoken word, authentic noises, and background sounds and music. Verbal and visual components are variously combined. The composition of news report is determined by the overall connection between the picture and the text and it can be arranged according to chronological, logical, or dynamical principle. In practice, all three principles are used.

Picture visually captures objects and events. The author captures not only the facts, but also the surroundings and atmosphere; the author provides a visual description of an event. To emphasize the emotional aspect of the event, sometimes the important and interesting details are used (e.g. the detail of the instrument that caused the tragedy, or details of victims' or their relatives' faces), arranging of the recorded objects that were discovered, seen and recorded; making connections between events—these all form the intended emotion. In news broadcasting, it is admissible to break the rules of composition and of montage, e.g. quick adjusting of camera focus, presenting events in their real duration, montages without inter-cuts, cut off panorama, etc. By combining picture with spoken word, viewers are provided with the answers of the basic journalistic questions. Picture and text does not have to answer them simultaneously.

Spoken word/text (background stand-up and commentary) is to announce and present the major facts which are further explained, localized and generalized by the visual picture. Comprehensibility is determined by the clarity of the text and by logical arrangement of the facts. The text of report is sometimes more personal; the viewer may notice the presenter's subjective approach, by which the viewer is led throughout the topic; the presenter explains and points out the important facts, and sometimes express his/hers personal attitudes, judgments, impressions and conclusions (commentary, polemic, sketch). The reporter presents his/her own subjective point of view with the hint of generalization.

The presenter provides the viewer with facts, through the personal experience transformed the information. To emphasize the emotions, presenters use emotional expressions, which can also connote the evaluated attitude.

To explain the event, one intensifiesone's own speech, or approximates the authenticity of the situation. The presenter uses the *respondents' statements*. Respondents are competent and reliable representatives of organizations, institutions and corporations whose statements tend to be neutral; respondents are also participants or witnesses of the events—the emotional character of their statements is determined by the character of the event (enjoyable cultural event, floods, bus crash), or victims or their relatives—whose statements are extremely emotional, often they are not able to fully answer the questions. Emotional responses can also be triggered by neutral, impersonal statement of a government representative or an official if the expected and the real statement are in contradiction (e.g. presented solutions will not be applicable according to the recipient of the information—i.e. "empty" promises, or if the attitude of the representant/official is inadequately indifferent.)

Part of the reports is also formed by *authentic noises and sounds*. These are often used at the beginning of the report to trigger certain emotions—e.g. when reporting on war conflict (shooting, scream of the crowd, etc.). Its atmosphere is strongly emotional and thus influence the report.

A component that is becoming of great significance in reports is *music*. Its function is not only to link a picture with a text (e.g. background music in news broadcasting), but it also contributes to the overall aesthetic and emotional character of the report. It is used to stimulate positive emotions of celebrations, or negative emotions of tragedies when it is combined with slow-motion and detailed pictures. In the journalistic genres such as feature or sketch, music emphasizes the grotesque character of the picture.

What is also attractive is the doing of the members of the social elite of various spheres—politics, social life, showbusiness or sport. People, who are attractive for the positions they hold (members of government, opposition, mayors, celebrities, etc.), attract media attention, even though nothing has to be going on with them. These topics are processed in an over-emotional way mainly in the pre-election period (general, presidential or regional elections, etc.); celebrities are spotted on mainly during the summer "silly season" (e.g. auction of Mrs. Obama's dresses). The personification of information, i.e. processing the topic through a personal story, is nowadays a very common way of presenting mainly for two reasons: it provides the topic with attractive content and the dynamics of the report is becoming more familiar to the recipient. Participants of the stories are two types of people—common people who reinforce the authenticity of the report, or publicly known people, who increase the credibility of the report. Another attractive theme is ethnic harmony or antagonism.

4 CONCLUSIONS

Negativity is nowadays a dominant feature of news reporting—to attract and to shock often form a significant percentage of daily news (Ruß-Mohl, Bakičová, 2005, p. 104). This criteria of selection has become dominant, because people respond more strongly to threat than to positive information. When feeling threatened, the recipient's attention grows and perceives the information with higher focus, and therefore it is easier to remember. On the other hand, it can trigger feelings of fear and pose a threat which can consequently lead to preventive action against the presented threat. The attention of the viewer is also captured by news that is surprising, unexpected, quaint or sensational—i.e. television brings the information as the first, or it is presented from an alternative point of view, or a journalist was able to interview respondents unspotted by the other channels. Among the significant criteria are also poignancy and sentimentality of the news, i.e. it can stimulate feelings in the majority of recipients. Producers of TV news bring in "live broadcasting" of the despair of victims and of their relatives, with crying, hysterical screaming, and a detailed description of the tragedy. In picturing such event, producers use video record without any commentary, only with authentic sounds, detailed pictures of wounds and with blurred faces or places of tragedy.

In case of confrontation, the contrast of emotions—anger of victims versus unemotional attitude of the competent ones—can be seen. Apart from the negative emotions, the positive ones also have their place in news reporting. These are presented at the end of the programme and are usually less serious. Small kittens or ducklings stimulate positive, pleasant emotions, and ought to relieve the potential tension from all the news. Positive, but mainly negative emotions attract the attention of viewers, and the presentation of these emotions in news reporting creates and provides models and patterns, shapes attitudes, creates the picture of "real relationships" in the "real" world created by the media. The greatest threat of this trend is that it lowers the threshold of sensitivity towards negative events and emotions; the overall social sensitivity is lowering and indifference is growing, the exceeded presence of negativity is becoming a standard. This can create great impact on the development of the whole society, but mainly of the younger generation. Enormous presentation of emotions in connection with the "objective" news that reflect the real world may potentionally lead to apathy on one hand, and cause neurosis to a part of the society on the other hand.

REFERENCES

Boroš, J. In Czako, M., Seemanová, M., Bratská, M. 1982. *Kapitoly zo všeobecnej psychológie: Emócie.* Bratislava: Slovenské pedagogické nakladateľstvo. p. 34.

Cakirpaloglu, P. 2012. *Úvod do psychologie osobnosti.* Praha: Grada Publishing, a.s. p. 204.

Czako, M., Seemanová, M., Bratská, M. 1982. *Kapitoly zo všeobecnej psychológie: Emócie.* Bratislava: Slovenské pedagogické nakladateľstvo. pp. 24, 33.

Ekman, P., Izard, C.E. In Slaměník, I. 2011. *Emoce a interpersonální vztahy.* Praha: Grada Publishing, a. s. p. 20.

Hradiská, E. 2009. *Emócie v spravodajstve a konštruovanie reality* In Magál., S., Mistrík, M., Solík., M. (eds.) *Masmediálna komunikáci a realita I.* Trnava: Fakulta masmediálnej komunikácie UCM. p. 109.

Hradiská, E., Brečka, S., Vybíral, Z. 2009. *Psychológia médií.* Bratislava: Eurokódex. pp. 315, 316.

Lokšík, M. In Osvaldová, B. 2001. *Zpravodajství v médiích.* Praha: Karolinum. pp. 73–87.

Milivojević, Z. In Cakirpaloglu, P. 2012. *Úvod do psychologie osobnos*ti. Praha: Grada Publishing, a.s., p. 204.

Nordström, K., Riddersträle, J. 2005. *Karaoke kapitalismus.* Praha: Grada Publishing, a.s. pp. 15, 86.

Pravdová, H. 2009. *Determinanty kreovania mediálnej kultúry.* Fakulta masmediálnej komunikácie UCM, Trnava, p. 294.

Pravdová, H. 2011. *Fenomén zábavy a úloha stereotypov v produkcii a recepcii mediálnej kultúry.* Commmunication Today, 1/2011, II. Ročník. p. 9.

RTVS, Department of programme research and development, December 2013.

Ruß-Mohl, S., Bakičová, H. 2005. *Žurnalistika.* Praha: Grada Publishing, a.s. p. 104.

Stuchlíková, I. 2007. *Základy psychologie emocí.* Praha: Portál. pp. 54–55, 95.

Trampota, T. 2006. *Zpravodajství.* Praha: Portál. p. 104.

Tušer, A. 1999. *Ako sa robia noviny.* Bratislava: SOFA. p. 93.

Current Issues of Science and Research in the Global World – Kunova & Dolinsky (Eds)
© 2015 Taylor & Francis Group, London, ISBN 978-1-138-02739-8

Communication intersection—journalist and spokesperson

E. Chudinová
Faculty of Mass Media, Institute of Media Studies, Pan-European University, Bratislava, Slovakia

ABSTRACT: Communication intersection—journalist and spokesperson. [Contribution to conference]—International Scientific and Research Conference. Current Issues of Science and Research in the Global World. Vienna, Bratislava: Faculty of Mass Media, Pan-European University in Bratislava, May 27-th and 28-th 2014, 8 p.

Journalism cannot work without public relations specialists. Public relations staff is not qualified to work without good contacts with the media. Spokesman is a person who is authorized to communicate with the media and act as an interface between the institution and the media. Journalist and spokesperson both have common objective—to achieve to make their product published in the media. Despite the various communication goals, some of communication products of a journalist and a spokesman have common features.

1 INTRODUCTION

Media environment and the world of media are certainly the most dynamically developing sectors. Their main object is the information and it is clear that the current globalized world is based on fast and dynamic transfer of information across continents, countries, various subjects and especially the media. Rapid information flow through available technology, new technologies, but also the ability of people to handle with technologies is self-evident today. It can often happen that the topic of the day is audiovisual information that is captured by an amateur on his smart phone. Within a few minutes such information reaches millions of people around the world. Globalisation has brought in new opportunities for public relations particularly in the form of communication opportunities for globalized players. According to L'Etang, globalization affects PR for several reasons:

- Globalisation offers the opportunity to formulate new ideas (including political and religious ideas) and newly presented goods and services in various markets.
- Globalization increases communication opportunities and challenges.
- Globalization changes organizations, key players, the public, key issues and relationships.
- More and more organizations are being globalized and use international labour and markets.
- Public relations are closely related to capital (economic power), its maintenance (searching for new markets and cheaper materials, markets and knowledge levels.
- Public relations cohere with global power (governments and international organizations such as the World Bank and World Health Organization).
- Resistance to globalization requires diplomacy and public relations for organizations that are involved in globalization (L'Etang, 2009, p. 282–283).

The ability of media to quickly move information to the consumer is motivating for all producers and sellers who can assert their interests and forward information quickly to their customers. The communication in marketing is the exchange of information between buyer and seller through communication tools: information (message) and media. Features of communication are carried by these components: coding, decoding, response and feedback, whereby marketing communication system itself consists of advertising, sales promotion, public relations, personal selling and direct selling (Labská, Foret, Tajtáková, 2009 p. 14).

Researches in recent years showed that journalists cannot exist without information from the public relations workers. Based on the elements of the system of marketing communications when searching communication intersections between the spokesman and the journalist, public relations are the most important component. B. Baernsová states that: "42% of all posts in the newspapers relating to the Coca-Cola concern are fully based on the PR texts" (In: Serafínová, Duchkowitsch 2005, p. 14)., R. Burkart points out a research of Torsten Rossmann, who observed the impact of all activities of the Hamburg Greenpeace printing places within two months and the result was: 84% of all, approximately 900 articles, which appeared in the press during these two months, were based on the ideas or events that were initiated by Greenpeace for press in the form of activities, press conferences and announcements, which had an impact on Media (In: Serafínová, Duchkowitsch 2005, p. 15). This means that journalists compose their articles also upon available PR information, but that does not mean that journalism has lost its status.

PR is a permanent process, set of activities, actions that are deliberately aimed at beneficiation of relationships of a subject. The aim of public relations is to obtain the public on their side and favourably gain the public. Each firm or a company that wants to be successful in getting the public to his side must use PR activities consistently. An important person in public relations area is a spokesman. The spokesman is the person responsible for the communication of company or organization.

2   COMMUNICATION GOALS

Journalism as a science is being examined at several levels. On the one hand, journalism is conceived as a social system which fulfils an exclusive function for the company and it is characterized by specific communication mechanisms, professional observation of topics from different social areas that are new and relevant in order to provide them to the public communication. In this way, the journalism differs from other forms of social communication (literature, advertising, PR, etc.). On the other hand, journalism may be examined at the organizational level—which is represented by the media businesses and products of journalistic institutions which upon specific rules and routine processes continuously provide journalistic communication and thus operationally fulfil the journalism function. Using forming and contextual criteria at this level journalist and non journalist organizations and media products can be distinguished. Finally, at the level of professional actors, journalistic working tasks are conceived as part of journalism system, where they apply to permanent workers and co-workers who deal with creating journalistic content (products) in their main profession. Journalism is the main profession which provides a journalist more than half of the income from journalism, or more precisely, if the journalist dedicates more than half of his working time to this activity (Weisenberg, Maliková, Scholl 2006, p. 346 In: Brečka, Ondrášik, Keklak 2010, p. 77).

When searching for the intersections of communication, it is necessary to emanate from communication goals. Common and fundamental goals of the spokesperson and the journalist are to achieve that their product will be published in the media. The aims of the spokesperson and journalist arise from their role which they play. The spokesperson is keen to provide such information that can draw an attention to him mostly in a positive way or at least in a neutral way.

Aims of spokesperson within media relations:

– Building an image of company and products among journalists
– Building an image among the public through the media
– Enhancing awareness of the company and products among journalists
– Increase publicity (positive and neutral) about the company and products
– Support objectivity of journalists when creating articles and reports
– Building the image field in which the company operates (Věrčák, Girgašová, Liškařová 2004, p. 28).

In researches on the role of the journalist, the most often following roles appear in the research.

– Neutral and fast reporter, moderator of information
– Advocate, lawyer
– Advisor and helper, service provider
– Educator, teacher
– Critic (social, political, critic of economic deficiencies)
– Investigative journalist—revealer od scandals
– Spokesman of citizens
– Civil and political activist
– Populist
– Entertainer (Brečka, Ondrášik, Keklak 2010, p. 100–101).

According to the abovementioned authors, the media not only influence public opinion, but also use it to support their intentions. Journalists many times raise the agenda for public discussion, but also the initiative often comes from the public. Compared to the past, thanks to the Internet, much closer connection between the journalists and the audience is established and the media are more likely to speak on behalf of the public. This option is often used by the media to spread banalities that are part of tabloid intentions, and only rarely the real public issues, quality of justice, health care or the tax system find their way into the content. Of course, the statutory media, according to their aim, have different status in this context.

Communication goal of journalist is to gather information, create a media product, which will be released (published) and it does not matter whether its form will be positive, negative or neutral. The journalist wants his product to be the best or the most attractive, the most interesting, the most charming or the most touching. He needs information and without it he does not create any media product.

3 COMPETENCES OF SPOKESPERSON AND COMPETENCES OF JOURNALIST

In Slovak republic the spokesperson position belongs to very attractive ones. Mostly important publicists working in full covering or regional media become spokesmen of both private and public institutions. Up to 2/3 of the spokesmen worked as journalists in the past. (Chudinová, Tušer, 2013). The motivation of a journalist to become a spokesperson was shown by the poll in 2011 (Chudinová, Tušer, 2013, p. 102).

The necessary competences of spokesperson are prerequisite for high quality communication. The most important competences are considered to be:

– Skills—to communicate, to think, to listen
– Expertise—experience, information equipment
– Skills—communicate well, to act professionally in the public
– Personal characteristics—credibility and reliability (Chudinová, Tušer 2013, p. 127–128).

In the same poll the journalists' view of the spokesperson in Slovakia was examined. When listing facts use either the style tag List summary signs or the style tag List number signs.

Table 1. Why do you work as a spokesperson?

Other reasons	38%
It is a job like journalism	33%
I got an offer	26%
Popularity is attractive to me	2%
I love camera	0%
The camera loves me	0%

Source: Chudinová, Tušer 2013.

Table 2.   What is your view on the spokesperson in Slovakia?

Some time positive, some time negative	63%
It is positive	21%
It is negative	13%
I have no opinion	4%

Source: Chudinová, Tušer 2013.

If a journalist wants to follow his journalistic goals, it is necessary to possess prerequisites—competences:

– Skills—assertiveness, empathy, ability to persuade
– Knowledge of the theory of media, forms of media work, knowledge of audience
– Skills—communication, organizational, mastering the required technology,
– Personality features—moral stability, regulation by conscience.

It was shown in the poll (Chudinová, Tušer, 2013, 127–128) that the spokesmen themselves consider the communication skills to be the most important. The need for schools aimed at mass media has to be associated by the ability of school to improve the communication skills of their students. The poll showed that the graduates from other universities work at the spokesperson positions as well. It may mean that the individual is interested in working at this position and amend the knowledge in the field of communication by self-study or natural communication skills that can a graduate of a different university or technical university use to enforce his/her personality in different profession.

4   NEWS AND PRESS RELEASE

The most frequent product created by journalist or spokesperson is a press release or news. The news answers the basic questions: Who? What? When? Where? How? Why? (Sometimes: Where from?). The same questions are answered by press release created by spokesperson as well. The press release is official news referring to an important event at the institution which can be current or new or both current and unknown, same as it is created by journalist. The spokesperson preparing the press release has to proceed the same way as a journalist, the text has to be:

– high quality language
– grammatically correct
– use clear statements
– the language used is the same as in a traditional news article (Věrčák, Girgašová, Liškařová 2004, p.49, 55)
Language style in news media is:
– fast, striking, informational
– simple short expressions
– titles rich in information, less emotional
– standard grammatically correct language (Kasarda 2010, p. 74, In: Tušer et al.)

It is clear that both journalists and spokesmen use the same language and stylistic tools for creating the news and press releases. The group, at which the news is aimed, plays a crucial role. It is necessary to study the topics while looking for intersections. The spokesperson will take into account the seasonal demands based on his/her knowledge of the media market, informational vacuum at the media market, new products and services, education in the field, target audience (Věrčák, Girgašová, Liškařová 2004, p. 29).

The topics of the spokesperson within media relations

What does the institution do and what people work for it, How is the institution extraordinary and what are its new products, How is the institution extraordinary in its field and whether it is environmentally friendly, What is its attitude to the region and labour union,

174

Preparation of some campaign and social event, New branch, new markets, presentation at the trade fair, Sponsored events and humanitarian activities, Awards in contests, Gain of new important partners or clients.

The journalists take into account several factors while choosing the content, Importance of the event, Effect of the event, Reaction—human interest, Recency of the event, Newness (Chudinová, Lehoczká 2005, p. 72–74).

Topics and sources of the journalist

Public institutions—state administration and municipality, Conflicts and disputes, Lobbies, Planning and development, Industry, economics, agriculture, environment, Science, technology, research, Health service, education, social services, culture, free time activities, Human factor, celebrities, Accidents, crime, transportation, Sport, weather, seasonal news (Chudinová, Lehoczká 2005, s. 70).

The attention needs to be paid to text composition and language tools as well. These are very similar because both spokesperson and journalist must take into account mostly the targeted audience.

In the poll (Chudinová, Tušer, 2013, s. 133–134) the respondents from the groups of spokesmen and journalists answered some similar questions. How did the communication intersections show?

The respondents from the spokesmen group consider the creating the relationship to media to be the most important (90.4%) and the cooperation with media they see as good to average (86%).

The respondents think that they are perceived by the journalists well or very well. Apart from that 69% of the responding spokesmen were working as journalists before so it is possible to assume that the opinion is based on their own experience as well.

Journalist respondents confirmed (87.5%) that the public institutions need spokesmen, they are in contact with them (87.5%) and almost all (95.8%) of them use the work of spokesmen mostly as background information source.

The journalist respondents consider the spokesmen sometimes positively, sometimes negatively (63%) but more positively than negatively. The negative perception needs more study. We can only assume that spokesmen's responses might be dubious or unconvincing what can raise doubts that the spokesmen try to hide the true information from them.

Regarding the comparison in the personal competences we have compared the answers to questions related to abilities, personal characteristics and expertise.

The aforementioned findings can actually mean that the journalists are unable to imagine their work without cooperation with spokesmen because they are often the only source of information from the public institution. On the other hand, the spokesmen consider the good relationship and cooperation with media to be the most important. Based on our approximate poll it is possible to assume that the relationship between journalist and spokesperson is affected by the joint need and necessity.

The spokesperson respondents chose the characteristics in this order: communicate, think and listen. From the personal characteristics they put trustworthiness and reliability and within professional characteristics they chose experiences and information equipment.

The journalist respondents chose ability to form competent answers. Regarding the personal characteristics of the spokesperson they chose trustworthiness, intelligence and emotional quotient. Then from the professional characteristics group they chose specialized knowledge and information equipment.

Such comparison can help especially spokesmen to adapt according to journalists' demands because they are the audience for the press release for their media. The journalist perception affects the production of media outputs and the public audience—consumers of these outputs as well.

5 CONCLUSIONS

Both the spokesperson and the journalist pursue the same aim—publicity, which means they are interested in getting the information to media and through them to the audience. The spokesman cannot do that without the journalist because he has no media available if we do not take the institutional website or personal website or social networks into the account. The journalist needs the spokesperson to create his media product because high portion (42–84%) of journalistic products is based on the PR news of different subjects. The PR news preparation methodology is slightly different for spokesperson in comparison to the journalist even though their aim is the same—to inform about the event, phenomenon or fact and to reach publicity. The most significant differences in news preparation are in content selectivity. Language style is almost the same; both spokesperson and journalist use grammatically correct language with high quality, simple, short and clear expressions. From the composition point of view, the journalist uses any style suitable for his medium. The spokesperson cannot use chronological method as he would minimize the chance of publication of his press release. There are distinct differences even in the expression of the press release in the medium. The journalist wants his product to be most interesting, dramatic, the best, etc. while it is not important whether the final expression will be positive or negative. On the other hand the spokesperson needs the press release to be positive, at most neutral and definitely not negative.

REFERENCES

Brečka, S., Ondrášik, B., Keklak, R. (2010). Médiá a novinári na Slovensku 2010. Bratislava: Eurokódex, 2010. 232 p. ISBN 9788089447329.

Chudinová, E., Lehoczká, V. (2005). Fenomén rozhlasu v systéme masmédií, Základy teórie rozhlasovej žurnalistiky v systéme masmediálnej komunikácie, Trnava: Univerzita Sv. Cyrila a Metoda v Trnave, Fakulta masmediálnej komunikácie, 2005. 230 s. ISBN: 80-89220-04-5.

Chudinová, E., Tušer, A. (2013). Kompetentný hovorca. 1. vydanie. Žilina: Eurokódex, 2013. 168 p., ISBN 978-80-8155-019-5.

Labská, H., Tajtáková, M., Foret, M. (2009). Základy marketingovej komunikácie. Bratislava: Eurokódex, s.r.o., 2009. 232 p. ISBN 978-80-89447-11-4.

L'Etang, J. (2009). Public relations. Praha: Portál. 2009. 344 p. ISBN 978-80-7367-596-7.

Serafínová, D. (zost.) (2005). Vedeckovýskumné dimenzie žurnalistickej komunikácie. Zborník z medzinárodnej konferencie Katedry žurnalistiky FF KU 11. novembra 2005. Ružomberok: Katolícka univerzita v Ružomberku, 2005. 124 p. ISBN 80-8084-052-0.

Tušer, A. a kol. (2010). Praktikum mediálnej tvorby. Bratislava: EUROKÓDEX, s.r.o., 2010. 368 p. ISBN 978-80-89447-16-9.

Věrčák, V., Girgašová, J., Liškařová, R. (2004). Media Relations není manipulace. Praha: Ekopress, s.r.o., 2004. 136 p. ISBN 80-86119-43-2.

Current Issues of Science and Research in the Global World – Kunova & Dolinsky (Eds)
© *2015 Taylor & Francis Group, London, ISBN 978-1-138-02739-8*

Playwright Peter Scherhaufer (an essay from theater archives)

D. Inštitorisová
Pan-European University, Bratislava, Slovakia

ABSTRACT: The article deals with drama and directing creation of significant theatre director, artistic director, pedagogue, translator and theoretician of Slovak origin Peter Scherhaufer—founder of famous Goose on a String Theatre (Czech Republic). Our perspective will emerge mainly from available archivalia (drama texts, theatre performances, articles, studies, monographs, bulletins, etc.

1 INTRODUCTION

Peter Scherhaufer (*29 June 1942–†31 July 1999) is one of the best known Czechoslovak—as well as Czech and Slovak—playwrights both in Europe and in the world. He earned his greatest success as the director of the Goose on a String Theater in Brno, Czech Rep., which he had also co-founded. He was also a university lecturer, a translator, and a theorist/historian of direction. Above all, he was a drama director occasionally working for television. His style, whose chief characteristic was irregularity, was based on different principles than the style of era. This was because he had been educated in a different spirit, one which might have felt strange at the time. Very often he had to work in non-traditional environments, including a street, an exhibition hall, a classroom, and an area beneath the stage. Also the way he treated literary texts was rather untraditional: he did not make typical stage adaptations. He substituted dialogue for inscenation, often with very few changes, going so far that once he staged the train schedule. He also worked with dramatized reading and scene drafts and other open drama forms. He trained many directors, and not only professionals, as he often taught at workshops for amateur directors. A university drama festival organized by the Faculty of Education of Constantine the Philosopher University was named after him. After his death, a series of publications based on his not-yet dramatized texts and unfinished works was published. He stood at the beginning of direction theory. His work, both in ink and on tape, still remains an inspiration. Many contemporary Czech and Slovak playwrights proudly affiliate themselves to him.

His was also vice-dean at the Janáček Academy of Music and Performing Arts (JAMU) in Brno.

In this study I will attempt to explain his long-lasting influence on Czech and Slovak theater by analyzing the strangeness of his works.

2 RESEARCH QUESTION

What was—or is—the general essence of Scherhaufer's drama?

Was it theatrical?

Was/Is it un-theatrical? What do we know about the author today? Where do we find information about him? Does his theatricality come from outside theater? Why does his 'strange' theater inspire even today, giving way to new 'strange' approaches?

Or let's put it differently:

Is 'his'[1] strange theater—the theater of irregular dramaturgy, of irregular scene setting—still popular because of its 'strangeness'? What about the Goose on a String? Did the theater succeed from the very beginning (1967/68) due to Scherhaufer's authorial strangeness? Or—if you should fancy a metaphor—what is the right 'strangeness', the right 'strangeness' to overcome the strangeness of socialist and early capitalist, barbaric, Czechoslovakia, that has managed to survive it all?

How different is Peter Scherhaufer?

3 THE RESEARCH

First all, he had a mixed Slovak and Czech theatrical background. He was of Slovak origin, born in Bratislava. Even so, he became the co-founder and main director of the legendary Czech, Brno-based Goose on a String Theater (GOST) in 1967/68. In 1967 the theater, originally called the Mahen No-Theater, did not perform publicly in the Procházka hall, an empty exhibition room in the House of the Arts. GOST officially started on 15 March 1968. Those who co-founded it were: teacher Bořivoj Srba as the dramaturgist; JAMU direction students Eva Tálská, Zdeněk Pospíšil as directors; Jiří Pecha, Hana Tesařová, František Hromada, Jana Švandová doing the acting. In 1969–1990 it was forced to change its name to the Theater on a String for besmirching the name of Gustáv Husák, the Czechoslovak president at that time. During the Soviet occupation of Czechoslovakia, the people used to put a—k ending to the word husa (Eng. goose) in its name. Hence, a subversive Husák on a string pun was created.

Another strange thing about Peter Scherhaufer was his language. He had lived too many years in the Czech part of the country, so he spoke and wrote very strangely, Czechoslovak (not so much Slovak, nor Czech). Nobody cared much, though. How he got to theater was strange as well. In 1968 he finished his directing studies at JAMU, even though he had originally come from a non-theatrical field. He had studied the construction of aircraft engines at the military department of the Brno University of Technology. Before going to JAMU he had to work in a mine to recompense the state for his engineering studies, which he quit after two years. When he thought about GOST and its management, he implemented scientific principles such cybernetics or information theory, which was very novel and unprecedented in theater. He discussed those principles in detail in his diploma thesis *Theater on a String—Model 1970* (Inštitorisová, 2006: 143–150). The following factors influencing the degree of intelligence of a given system prior to processing new information were of primary importance to him:

a. source of information,
b. interference, malfunction, accidents,
c. memory, i.e. already stored data,
d. the recipient's background knowledge and his attitude to the source—theater.

Yet, the poetics of GOST was rather understandable. The new theater wanted to differentiate itself from the old, official one. The Zero Manifesto reads as follows: "Its members believe that (…) they will find inspiration for a new, poetical expression on stage, an expression which would uphold the living roots of human existence, a poetry internal and external which man, burden by daily routine, does not exercise—or even feel—anymore" (Retrospektiva: 32).

Peter Scherhaufer, as well as other GOST personalities, admired Jiří Mahen, Emil František Burian, and Vsevolod Meyerhold. Those people were famous reformers of theater poetics and esthetics. Mahen's screenplay collection, The Goose on a String, gave the ensemble their name.

Due to his admirable knowledge of many scientific fields, Scherhaufer had been able to come up with a unique director's interpretation method he called 'tanking'. He emphasized long,

---

[1]This was a concept shared by the rest of the co-founders of GOST.

178

never-ending reading and analyzing of large amounts of materials which would lead to fantastic stage results. When he decided to dramatize a particular topic, his insight in a given period, its atmosphere and problems, was immense. He studied historical documents and present-day material systematically, relentlessly, and intensively. Thanks to this he could see through life off and on stage. We might go as far as calling him the visionary of the post-1989 era:

"The circumstances have changed. That's the REVOLUTION, take it or leave it. The fastest to adapt to this fact (it doesn't matter what works depicting change they read and play, for example Tolstoy's *Resurrection* or Sholokhov's *And Quiet Flows the Don*) will have an advantage and a priceless know-how. You know, I had my students work on such a thing in January 1989. 'Theater in the Times of Social Upheaval' it was called. They sneered back when I was assigning it; when they handed it in, they were speechless at how I had got it right. I'm still collecting material for that study (I'll use yours as well, if I may). I have material from 1968, from Russia, from Poland, and Hungary. I have enough material from Portugal when Salazar was ousted, from Spain when Franco came down, etc." (Inštitorisová: 238).

Even though Scherhaufer was professionally active only for 31[2] (having died at 51), the body of his works is monumental both in its vastness and its scope. He was not just a drama director, he worked for television too; he was a playwright, a translator, a teacher, and a drama theorist. He has earned his place in the annals of Czech and Slovak theater as a director, co-director, or co-author of great European or world international theater projects, such as *Hopes* (Wrocław, 1978)[3], *The Spring of Nations—Wiosna ludów* (Łódź, 1979; Brno, 1980)[4], Together—Společně (Copenhagen, 1983)[5], *Routes (Crossings—Train Schedules—Meetings)* (Brno, 1984)[6], *PROJECT 1985 – A Dramatized Reading of the Soviet Nations' Contemporary Literature* (Brno, 1985)[7]. His acclaimed projects also include the biggest international theatrical tour to this time, *Mir Caravane* (Moscow—Paris, 1989)[8], which GOST decided to

---

[2]If we take the moment he finished his JAMU studies as the start of his career.

[3]A joint theater project on hope. Participants: Communa Baires (Italy), Le Temps Fort (France), Teater 9 (Sweden), Teatr 77 (Poland), Theater on a String, Esperanza (the USA), Orchestr Teatra Ósmego dnia (Poland), Katka Manolidaki (Greece), Liliana Duca (Argentina), Pavel Büchler (Czechoslovakia). Location: Olešnica, Wrocław, 21 Sept.–8 Oct. 1978.

[4]A joint performance by the Theater on a String and Teatr 77 (Łódź)
Script: Paweł Chmielewski, dramatized reading project by Zdzisław Hejduk, Petr Oslzlý, Peter Scherhaufer, Andrzej Podgórski, dramaturgy: Petr Oslzlý, Andrzej Podgórski, setting: Boris Mysliveček, Krzysztof Rynkiewicz, music: Miloš Štědroň, Julius Wacławski, choreography: Alena Ambrová, directed by Zdzisław Hejduk, Peter Scherhaufer, premiered: Łódź 22 Oct. 1979, Brno 26 Jan. 1980.

[5]Together—Společně. Labyrinth of the World and Paradise of the Heart—Labyrint světa a Ráj srdce (An international inscenation—happening).
Script, artistic director: Richard Gough, Zdzisław Hejduk, Alexander Jochwed, Petr Oslzlý, Krzysztof Rynkiewicz, Peter Scherhaufer, performed by: GOST, Cardiff Laboratory Theatre (UK), Den Bla Hest—Arthus a Group of The Copenhagen International Theatre Festival (Denmark), Teatr 77 – Łódź (Poland), Location: Denmark, Valseverket Copenhagen, 6–17 July 1983.

[6]Created as a cooperation between: Divadlo na okraji, GOST, HaDivadlo and Studio Y. Creative team: Arnošt Goldflam, Zdeněk Hořínek, Miki Jelínek, Jan Kolář, Josef Kovalčuk, Petr Oslzlý, Zdeněk Potužil, Peter Scherhaufer, Jan Schmid, vVýprava: Ján Zavarský, Jiří Benda, Jan Konečný, Miroslav Melena, Jana Preková, music: Iva Bittová, Jiří Bulis, Marek Eben, Miki Jelínek, Miroslav Kořínek, Jaroslav Pokorný, Miloš Štědroň, premiered at GOST, 29 Oct. 1984.

[7]More on work with short stories, novellas, and novels such as Alexander Gelman: *Replika zrozená životem (polemika)*, Arkadij a Boris Strugackí: *Z deníku poctivě smýšlejícího občana* (sci-fi), Andris Jakubáns: *Sněhobílý kufr* (burleska), Valentin Krasnogorov: *Tři patrony z jedné sumky* (malé tragédie), Enn Vetemaa: *Pomník*, Vil Lipatov: *Šedivá myš*, Jefim Zozulja: *Zkáza hlavního města* (fantazie) OSLZLÝ, 2006: 49–65.

[8]*Mir caravane – Karavana mir* (An international theatrical tour). Participants: Akademie Ruchu (Poland), DNP, Cirk Perillos (Spain), Dog Troep (the Netherlands), Footsbarn Travelling Theatre (UK), La Compagnie du Hasard (France), Licedei (the USSR), Svoja igra (the USSR), Teatr Ósmego dnia (Poland), Teatro Nucleo (Italy).
Performed in: Moscow, Leningrad, Warsaw, Prague, West Berlin, Copenhagen, Basel, Blois, Lausanne, Paris, president: Nicolas Peskine, coordinator: Pierre Lauôanné, period: 10 May–17 Sept. 1989.

're-run' this year. The triptych *Shakespearomania* (1988–1992)[9] and plays like *The Commedia dell'arte* (GOST Brno, 1974)[10], *The Wedding* (GOST Brno, 1978), and *The Brothers Karamazov* (GOST Brno, 1981) are also part of the endless list. Most notable of Scherhaufer's Slovak projects include *Who Needs You Anyway '91* (Prešov, 1990), *Wherein All Our Misery Lies* (Košice, 1996), Bocatius '98 (open environment, Košice, 1998), *Slovak Classics Remade* (ZŤS Martin Theater, 1976), and *The Geometry of Deadlock* (Bratislava, 1999).

For Scherhaufer, working on a drama, meant going through a lot of materials and making a lot of drawings. As GOST takes pains to documents its history, many of Scherhaufer's sources, such as bulletins, scene drafts, annotated scripts, and photos, can be found in their library.

Scherhaufer was also very active and influential as a teacher of professional and amateur actors, the latter of which he felt had greater possibilities. He led many workshops for amateurs, he taught at JAMU (since 1990) and at the Academy of Performing Arts in Bratislava (since 1998). For both amateurs and professionals he drafted syllabi and various lesson work materials, which are still being used today (Inštitorisová, 2006: 239–241, 246–257 and others)[11]. In his lesson plans, here and there, Scherhaufer scribbled remarks on directing theory. Sadly enough, since they are just unfinished sketches, they cannot be published.

Scherhaufer also worked as a director and dramaturgist for amateur theaters. This collaboration started shortly after he finished his JAMU studies in 1968. In Slovakia he authored a number of memorable dramas[12]. Many amateur dramatists treasure their personal correspondence with him, in which they discussed theater issues, possible inscenations, individual plays, etc.

Scherhaufer's publications are equally admirable. Along with Ľubomír Vajdička, he still remains the only Slovak drama director who ventured into theory.

In 1984 Chapters on Directing came out, followed by *Inscenation in an Irregular Space* in 1988, *Theater Projects of the Goose on a String Theater* in 1996, *A Chronology of Drama Directing and two Drama Directing History Readers* in 1998, and in 1999 the third *Drama Directing History Reader*. After his death, his students helped to get some of his works published. Thanks to Claudia Francisci *The So-Called Street Theater* was published in 2002. In 2007 the fourth *Reader* came out, with the fifth issue being prepared for printing. The Readers contain large amounts of translated material dating from the beginnings of theater to the end of 20th century. Scherhaufer collected the materials on his travels around the world, and, in many cases, he was the first in Czechoslovakia to come up with certain findings. His diploma thesis, titled *Theater Model* 1970, and habilitation thesis, thesis *Directing as a Study Field*, and professorial lecture, titled *Come See the Lions, or Contemporary Street Theater*, have all been published as well. He was also the first one in Czechoslovakia to translate a part of Eisenstein's *The Art of the Mise-en-Scène*, published in his *Chapters on Directing*. One of his writings, known under the working title *A Breviary for the Lazy Director*, is 2,500 pages long. Another manuscript which he left unfinished in electronic form is, on the other hand, too fragmentary.

The current members of the GOST ensemble continue to document Scherhaufer's projects and inscenations. Supported by the Czech Science Foundation, GOST was able to *have The Commedia dell'arte* reconstructed in book form (by Petr Oslzlý[13] and Jozef Kovalčuk from FAMU's Faculty of Theater). The same happened to *Routes (Crossings—Train Schedules—Meetings)*.

---

[9]More later on.

[10]Due to space limitations, the rest of the projects and inscenations will not be mentioned. (They can be seen in: Inštitorisová, 2006). In total Scherhaufer made 236 drama and 23 TV directions.

[11]He prepared thorough compulsory reading lists, tests, questionnaires, analytic materials, etc.

[12]Notable examples include plays with Detský divadelný súbor Lúč (Martion), Divadelný súbor pri Dome kultúry ROH Závodov ťažkého strojárstva Martin, Divadelný súbor Jána Chalupku Brezno, Kremnické divadlo v podzemí z Kremnice.

[13]One of Scherhaufer's closest colaborators, a dramaturgists with many of his GOST inscenations, currently the chair of the Centrum experimentálního divadla.

Scherhaufer's impact on Slovakia has been significant. The year 2010 saw a reconstruction of his inscenations of Valentin Krasnogorov's Pelicans of the Wilderness[14] at the Divadlo úsmev theatre in Bratislava. In the fall of 2010, this inscenation was the grand finale at the new festival dedicated to Scherhaufer. Former directors from 1992, Alena Michalidesová, Marián Labuda, Jr., and Erik Peťovský, helped with the re-inscenation.

Teacher Peter Scherhaufer was appalled when theatrical problems were left to be solved by shear 'theatrical sense' or supposed talent. By publishing so much and teaching students of theater in a deep and systematic manner he wanted to challenge such views. The urge to enlighten can be seen in many of his studies and articles. Scherhaufer was a teacher who hated complacency—and he permanently questioned the state of theatre in his country. Even in his professorial lecture in 1997 he asked why Eisenstein's *The Art of the Mise-en-Scène* and other milestone works of world and Russian theater, such as the ones by Vsevolod Meyer-hold, Constantin Stanislavski, Sergei Diaghilev, Louis Jouvet, Max Reinhard, and Rudolf Steiner, had not already been translated by that time. To him, they had brought a completely different 'landscape' to theater. Scherhaufer and other GOST members used the untraditional approaches from these works to create the concepts of scene drafts and dramatized readings, which they—according to Petr Oslzlý—understood as follows:

- genres between literature and theater;
- specific forms incorporating the scenic articulation of situations and reading aloud;
- a result of the search for a specific epic subject in all scene drafts;
- a possibility for the actor to creatively interpret and present literature without reduction and the need to memorize a text (Oslzlý—Scherhaufer: 1985).

Scherhaufer's unorthodox views on the essence and meaning of theater presumed intertextuality as a common trait of his works. In many of his inscenations he used metatextual references to other semiotic systems (of art and reality), to variant translations, and to all sorts of parallel texts, quotations, references to other plays or modes of representation (allusions, persiflage, palimpsest, travesty, pastiche, caricature, etc.).

By using these technique he was ahead of time, foreshadowing postmodernism and postdramatic poetics.

The *Shakespearomania I—III* (GOST, Brno, 1988–1993) and *Bocatius '98* projects are great examples of intertextual compositions.

Shakespearomania consisted of three inscenations compiled from William Shakespeare's plays:

1988: *Their Highnesses Fools*, a montage of *A Midsummer Night's Dream, King Lear, Henry IV, King John, and Richard II*;

1990: *The Hamlets*—a montage of all Czech translations of Hamlet, including quotations from Hodek's rendering of *Much Ado About Nothing* and other plays;

1992: *The Tempest Man*—a montage of King Lear, Macbeth, The Winter's Tale, and *The Tempest*.

In the given plays Scherhaufer used various kinds of intertextuality not only in the text itself but also in its on-stage representation. For example, a certain Shakespeare excerpt was parodied in one context only to illustrate tension in another. In most cases he used quotations and historical references (costumes, props, on-scene music, etc.).

In the very first scene of *The Hamlets*[15] the Prince actually dies, having said just six sentences altogether. His replicas are a montage of replicas from all throughout the original play, uttered under different circumstances. The murder of Polonius is another typical example of the play's intertextuality—he is murdered four times. Not only is he murdered by Hamlet, by Hamlet's friends, but also by the Queen (lead by Hamlet's hand). Thus, Polonius' death

---

[14]First premiere: 11 Oct 1992: Divadlo úsmev Bratislava, directed by Mr. and Mrs. Michalides; reconstruction: 3 March 2010: students of the Department of Music CPU Nitra under the tutorage of Mr. and Mrs. Michalides).

[15]Apart from Scherhaufer, Karel Král worked on the script.

is caricatured. *The Hamlets* is also interesting not only because of the sheer montage nature of the text, but also due to its narration. At the beginning, we see Polonius who is about to tell the story of Hamlet. However, he can only do this in translations, namely those of Milan Lukeš a Zdeněk Urbánek[16]:

Excerpt from *The Hamlets* (p. 2):
2 Horatio sets the scene
Horatio divides up the scene space with duct tape. There is nothing else on stage.
[Translator's Note: Two versions of the Czech rendering of Hamlet follow.]
Possible text:
"Jak k tomu všemu došlo. Uslyšíte
o cizoložství, krvi, zvrácenostech,
o náhlých soudech, letmém zabíjení,
o smrti chystané a vynucené
i o lsti, která selhala a padla
na hlavy strůjců. O tom všem jsem schopen
říci vám pravdu." (p. 476)[17]
The same in Urbánek's translation:
"Svět dosud neví, jak ke všemu došlo -
Dovolte, abych já to pověděl.
Tak uslyšíte mnoho o neřestech,
O krvavých a sprostých zločinech,
O smrti zaviněné přehmatem,
O lsti, jež donutila zabíjet,
A nakonec o zrádných úkladech,
které se vymstily svým průvodcům.
O tomhle všem vám povím celou pravdu." (p. 168)[18]
Possible action:
While Polonius is reciting, the Hamlets appear and Horatio shows each where to stand. The dialogue starts sooner than Horatio finishes dividing the scene into rectangles.

In the project *Bocatius '98* two texts merged in the scene visuals. First of all, there were the realities of Košice and, secondly, there was the context of history in which meanings of different historical periods intersect.

On the textual layer of the play, various text types merged (with P. Scherhaufer penning the script):

– the project came about as an edition/montage of Rudolf Schuster's *Johannes Bocatius* 1992 radio drama;
– parts of the radio drama could be heard during the play;
– actors or cars carried banners with various inscriptions, the mob scribbled exclamations like 'This city belongs to us!', 'Get out!' on shop windows.

The publication of Schuster's book *Ján Bocatius* (1998) was also part of the project. This can also be considered a case of intertextuality. The book also contains an excerpt of the *Bocatius '98* script and Schuster's re-telling of Bocatius' life.

Irregular drama compositions are a significant part of Scherhaufer's works. Many of these texts are calques of literary texts. The play *A Replica from Real-Life*[19] is, for example,

---

[16]His morality play *Slovak Sayings*, based entirely on Slovak sayings, is made under the same principle. The play was banned in 1978. The main characters include Lie, Drinker, Babble, Charity, etc. (Inštitorisová, 2006: 329–337).
[17]Here I mean the Odeon anthology Pět her [Five Plays] from 1980.
[18]The edition number is not stated in the text.
[19]Alexander Gelman: Replika (polemika). Translation, script, and direction: Peter Scherhaufer, scene: Miloň Kališ. Premiere: 27 Oct 1985. The short story was published in the magazine Tear 3/1984.

modeled after a short story of the same name (by Alexander Gelman). The actual play was the short story was read onstage by an actor playing the author who talked about his professional experiences. This particular story has more often been inscened.

Excerpt from *A Replica from Real-Life* (p. 2)
"... and that was the 1st act. After a smoking break we went to see the 2nd act. It was a good play, so full of life. All I wanted to say was clear and impressive enough. The reactions of the audience, numbering almost a hundred people, were astonishing. (A short applause interrupted by live applause from the actors.)[20]
"So? What do you think? – Vera Vladimirovna took hold of my arm on the way to the artistic director's office.
"Did you like it? Was it better than in the Moscow theater?

The Slovak project *Who Needs You Anyway* is similar. It is based on the stories by Eastern Slovak authors, such as Ján Patarák's A Man Walking Too Quickly, Peter Juščák's They Killed My Friend, Štefan Oľha's A Drunkard's First Battle Cry, Karol Horák's A List of Predators, Stanislav Rakús' The Song About The Water from the Well, Milka Zimková's Bonsua, My Artist, and Pavla Sabolová's Lumpy Brackets. Also the short story A Unique Saint by Slovak dissident author Pavol Taussig was used. This all was complemented by a montage of Vasil Biľak's texts, along with Dezider Banga's gipsy tales *A Black Hair*, episodes from Andy Warhol's life, etc.

Scherhaufer liked working with open text structures, allowing actors (or other performers) to implement other texts into his script if needed. A great example of such a this method is the Slovak project *A Provisional Audit of the Revolution*, which Scherhaufer started to work on right after November 1989 in cooperation with the Kremnické divadlo v podzemí from Kremnica[21]. The structure of the play, however, had remained open for political and social struggle of the day to be implemented. (A 1988–1989 GOST and HaDivadlo joint project, called *Rozrazil*, had the same concept.)

Another typical feature of Scherhaufer's style is collective authorship (as far as it was possible within the framework of GOST). Cooperation on scripts or the creation of his own scripts was for him always a matter of context[22]. When creating a script, he always had in mind how it would or might look like onstage. This presupposition was for him the very semantic base of a text. The possible structure of a text onstage was its context. Scherhaufer felt the need to collaborate and was able to inspire his collaborators to look for common ground and do their best. This was also the case of his last work. Being ill, Scherhaufer was mostly able just to supervise. With some of his last works, like *The Geometry of Deadlock*[23], the last project realized in Slovakia, he would neither live to see them finalized nor oversee their development. Graduate and PhD. students, along with some alumni of the Slovak direction workshop, namely Michal Hatina, Claudia Francisci, Pavel Baďura, and Tomáš Svoboda, premiered his concept. They called it *Dramatized Reading of The Geometry of Deadlock, the Selection of Works from the Short Story '99 Literary Competition*. The dramatized reading featured their own selection of short stories from the Short Story '99 collection (published by L.C.A. Levice). Having to stage a dramatized reading which would preserve the essence of the original is one of the greatest challenges for a drama director. The reading of Rado Olos' To Leopoldov and Back, directed by M. Hatina, was the most interesting one. The director managed to interconnect the theme of the story with the act of reading it aloud (done by Dano Dangl). The short story Tiso, by Boris Filan, directed by C. Francisci, brought a great performance by Mária Kráľovičová who used her great charm and sense for atmosphere to convey the life story of a woman who attacked a famous actress in the street.

---

[20]A handwritten note.
[21]Premiere: 9 Jan. 1990.
[22]More in: Scherhaufer, 1988b.
[23]Ludus theater, Bratislava, premiere: 29. June 1999.

Scherhaufer's last directly made inscenation ever took place in his home theater, and it was even his last one in the Czech Republic. It was an inscenation of a play by the Flemish playwright Michel de Ghelderode called *The School of Clowns*[24]. It is a story of an elderly Clown Master and knight Folian (Pavel Zatloukal) who is about to—as always—see his oldest apprentices off to the world. Their last task is to prepare a school leaving feast under the supervision of valet Galgüt (Matěj Dadák). It is here Folian is said to relate to them the secret of his art. Yet, none of the apprentices has freely chosen to become a clown. They were unwillingly proclaimed 'clowns' right after birth. Their physical deformities lead their parents to condemn them to die. This story about the costs of doing theater was not unlike Scherhaufer's own life story. By making this play Scherhaufer also managed to stage Artaud's[25] famous concept of theater as a plague from which there is no cure and which forces us to obey its rules ...

4   CONCLUSIONS

Scherhaufer's 'strange' background brought forward techniques which permanently altered the imagery of strangeness in their time, and it has done so at present as well.

Scherhaufer's strangeness—seen and read—in drama or dramatized texts has lost the stigma of strangeness in the eyes of playwrights and other professionals.

This is because Scherhaufer is essentially individual. Not even the ordinary montage of evolutionary unrelated parts seems mechanistic in his work. Repetition served him for poetic purposes—to illustrate a feeling which one needs to communicate at once, in a particular manner, and—above all—ALOUD. His drama, the way he treated texts, crosses borders between landscapes which we thought were unrelated.

Above all that, the way Scherhaufer treated texts has helped revive the spirit of theater—in times of censorship and in times of freedom.

ANNOTATION

The article deals with the drama and directing creation of the significant theater director, artistic director, teacher, translator and theorist of Slovak origin, Peter Scherhaufer—the founder of the famous Goose on a String Theatre (Czech Republic). Our perspective will emerge mainly from available archivalia (drama texts, theatre performances, articles, studies, monographs, bulletins, etc.).

ARCHIVAL SOURCES

A letter from P. Scherhaufer to Viera Pelikánová from Dec. 1990.

SCRIPTS AND LIBRETTOS BY PETER SCHERHAUFER[26]:

1968. Anton Pavlovich Chekhov: *Tři sestry* (Ruský muzikál podle A. P. Čechova) [Three Sisters (A Russian musical based on Chekhov)] [Incomplete.]

1972. *11 dní křižníku Kníže Potěmkin Tauričeský* [The 11 Days of the Battleship Potemkin]

---

[24]This title inspired me to name a monograph I co-authored and lead.—Peter Scherhaufer—Učiteľ "šašků" [The Teacher of Clowns] in the year 2006. It is about his work in Slovakia.
[25]Antonin Artaud—a renowned French theater visionary and experiment master.
[26]Co-authors stated separately.

1974. Mikhail Bulgakov: *Divadelní román* [A Theatrical Novel]

1975. Jozef Ignác Bajza: *René mládenca príhody a skúsenosti* [The Adventures and Experiences of the Young Rene]

1978. *Slovenské porekadlá* [Slovak Sayings]

1981. Fyodor Dostoyevsky: *Karamazovci* [The Brothers Karamazov]
Also scripts (based on Scherhaufer) by Zdeněk Petrželka, Petr Oslzlý, Alena Ambrová

1981. Peter Scherhaufer—Miloš Pospíšil: *Faustiáda* (Montáž z diel slovenskej klasiky) [The Faustiad (A Montage of Slovak Classics)]

1984a. Spoločný projekt na tému *CESTY (křižovatky—jízdní řády—setkání)* [The Joint Project Routes (Crossings—Train Schedules—Meetings)] [Incomplete.]

1985a Alexander Gelman: Replika zrozená životem (polemika) [A Replica from Real-Life (A Polemic)]

1986. *Balet makábr* [Ballet Macabre] [Incomplete.] Libretto also by Petr Oslzlý

1988a. William Shakespeare: *Veličenstva Blázni—Shakespearománie I* [Their Highnesses Fools—Shakespearomania I] Script also by Petr Oslzlý

1990. William Shakespeare: *Lidé Hamleti—Shakespearománie II* [The Hamlets – Shakespearomania II]
(A Montage of Hamlet Translations) Script also by Karel Král

1991. *"KEMU CE TREBA '91" V.—Čierny vlas* (Performance) [Who Needs You Anyway '91 V.—A Black Hair (A Performance)]
(Based on Dezider Banga's Gipsy Stories A Black Hair, Helena Rudlová's Gipsy Laughter, and a selection from the Kale Rose anthology by Milena Hübschmann)

1991. *"KEMU CE TREBA '91" VI.—Ľubojsc, bože, ľubojsc, jaka ty presladka, alebo Ňichto ňema take gamby* [Who Needs You Anyway '91 VI.—Love, oh, Love, Thy Sweet Love, or Nobody's Lips Are as Sweet as Yours]
(A Dramatized Reading Project – a selection from a Zemplín love poetry Anthology)

1991. *"KEMU CE TREBA '91" VII.—Na stred Ameriki karčma murovana* [Who Needs You Anyway '91 VII.—In the Middle of America, There's an Inn]
(A Dramatized Reading Project—a scenic proposition based on the play by Kazimierz Braun Saint Wood and information from the books by Konštantín Čulen—The History of Slovak in the USA, Ján Sirácky—Slovaks in the World, Ladislav Klíma—Czechs and Slovaks Abroad)

1992. William Shakespeare: *Člověk Bouře—SHAKESPEAROMÁNIE III* [The Tempest Man—Shakespearomania III]
(Scripts from *King Lear, Macbeth, The Winter's Tale, The Tempest*)

1996. Michal Hatina—Milena Hurajová—Peter Scherhaufer—Alžbeta Verešpejová: *Kde leží naša bieda* [Wherein All Our Misery Lies]
(Third Part of the Wherein All Our Misery Lies Project—A Fairy Tale Environment)

SCHERHAUFER, P.

1970. *Divadlo na provázku—model 1970*. [Theater on a String—Model 1970]. Divadlo, no. 2: 71–77.

1976. *Divadlo väčších možností*. [A Theater of Greater Possibilities] Javisko, 8, 1976, no. 8: 240.

1982a. *Paňáca v kanále* [A Clown in the Sewer] In Program '82 Státního divadla v Brně. [Zvláštní číslo k patnáctému výročí činnosti Divadla na provázku.] Brno: Státní divadlo: 8–17.

1984b. *Kapitolky z réžie.* [Chapters on Directing] Bratislava: Osvetový ústav.

1988b. *Inscenování v nepravidelném prostoru* [Inscenation in an Irregular Space]. 1st edition. Brno: Krajské kulturní středisko, Ostrava: Okresní kulturní středisko Karviná.

1996. *Divadelné projekty Divadla Husa na provázku* [The Theater Projects of the Goose on a String Theater]. 1st edition. Bratislava: Národné osvetové centrum.

1998a. *Kalendárium dejín divadelnej réžie* [A Chronology of Drama Directing]. 1st edition. Bratislava: Tália-press.

1998b. *Čítanka z dejín divadelnej réžie* [Drama Directing History Reader]. [1.] Od neandertálca po Meiningenčanov. 1st edition. Bratislava: Národné divadelné centrum.

1998c. *Čítanka z dejín divadelnej réžie*. [Drama Directing History Reader] [2.] Od Goetheho a Schillera po Reinhardta. 1st edition. Bratislava: Národné divadelné centrum.

1999. *Čítanka z dejín divadelnej réžie*. [Drama Directing History Reader] [3.] Od futuristov po Ejzenštejna. 1st edition. Bratislava: Divadelný ústav.

2002. *Takzvané pouliční divadlo* [The So-Called Street Theater] (a fragment study). Brno: JAMU.

2007. *Čítanka z dejín divadelnej réžie*. [Drama Directing History Reader] [IV.] Od Artauda po Brooka. 1st edition. Bratislava: Divadelný ústav.

REFERENCES

Inštitorisová, D. a kol. 2006. *Peter Scherhaufer—Učiteľ "šašků"* [Peter Scherhaufer, The Teacher of Clowns]. Bratislava: Eleonóra Noterská—NM Code with Asociácia Corpus.

Kovalčuk, J.—Oslzlý, P. 2013. *Společný projekt Cesty (křižovatky—jízdní řády—setkání)* [The Joint Project Routes (Crossings—Train Schedules—Meetings)]. Brno: Masarykova univerzita.

Oslzlý, P. 2006. *Na hranici mezi literaturou a divadlem* [Between Literature and Theater] In Inštitorisová, D. a kol. 2006. *Peter Scherhaufer—Učiteľ "šašků"* [Peter Scherhaufer, The Teacher of Clowns]: 25–42. Bratislava: Eleonóra Noterská—NM Code with Asociácia Corpus.

Oslzlý, P. 1982b. Retrospektiva Divadla na provázku ke dni 18. 2. 1971 [Theater on a String in Retrospect up to 18 February 1971] In *Program Státního divadla v Brne*: 12–21. Brno: Tisk.

Oslzlý, P. 2010. *Commedia dell'arte Divadla na provázku* [The Commedia dell'arte in the Theater on a String] (1974–1985). Brno: Janáčkova akademie múzických umění.

Oslzlý, P. & Scherhaufer, P. 1985b. *Projekt 1985—Scénické čtení.* [Project 1985—stage readings]. Program. Brno: Divadlo na provázku. [Pages not numbered.]

Introduction of intercultural models used in global marketing

Z. Ihnátová
Faculty of Mass Media, Pan-European University, Bratislava, Slovakia

ABSTRACT: The factors of globalization strongly influence the field of business and marketing. The companies that choose to run their business outside their home country need to make decisions about an appropriate choice of the marketing strategies. Both, the globalization and the adaptation approach in global marketing have advantages and disadvantages. Although the globalization approach is more widely used today, its effectiveness across cultures is questioned. The main goal of this paper is to introduce selected intercultural models that are mostly used in the field of global marketing and serve as the main platform for making decision about how to adapt marketing strategies to local markets in order to reach more effectively the target audience.

1 INTRODUCTION

The basic discussion in global marketing should not only be about the efficiency of standardization but about the effectiveness of cultural segmentation (De Mooij, Hofstede, 2010). Although, a global, i.e. standardized approach in global marketing is preferred in the current global scale and has substantial advantages, such as benefits of economies of scale in procurement, logistics, production, marketing and in the transfer of management expertise, factors that all should lead to cost reduction and its effectiveness on different target markets is often questionable.

If we look at the classical marketing mix as defined by Kotler's 4P's (product, place, price and promotion), we can argue that all of the P's are to some extend influenced by cultural aspects. The product itself, including its product category, usage of the product and also life cycle stage it exists in, is strongly influenced by cultural values and preferred habits of the consumers, even more if we look at the cultural-bonded products. The distribution channels used by the company reflect and should respect the culture of the selected market. For example, in the USA, the main factors that are considered in choosing the distribution channels are mainly the aspects of economic profit. On the other hand, in the Arabic world or in Asia, more than the profit is considered the long term relationship and trust between the partners/suppliers. The factor of price is perceived more sensitive in some countries than in others. In fact, the Czech and Slovak market is a good example of the consumers who are very responsive to how much they have to pay for the product or service. On the other hand, German consumers considered the price in strong correlation with the quality and are willing to pay more if their expectations are met. Finally, we can argue that the last P from the marketing mix—promotion (i.e. marketing communication) is an element that is for the most part influenced by the culture of the consumers. In fact, the strategy of appealing to a single global market by a single global product, and by a single global communication in recent years appears to be less effective than adaptation to local markets. More insightful knowledge and understanding of cultural specificities of each country are getting to the forefront of today. Moreover, they become the basis for the selection of effective marketing communication strategy of a company, which has ambitions to effectively reach international markets.

Although, there are several models that are used in global marketing, we chose to focus in this paper on those that have numerical data available, because it allows to compare and contrast results from different countries involved in the adaptation process of marketing strategies (i.e. country of origin vs. county/countries of entry).

Therefore, the aim of this paper is to introduce three selected cultural models that are used most frequently in the field of global marketing and explain the cultural differences in the consumer behavior supported by the numerical data. Firstly, each of the models is briefly introduced. Then, we describe the collection of data: sample, respondents and questions used. Finally, we evaluate the application of each model and its weaknesses. In conclusion, we propose the selection process in making decision about which intercultural model is the most appropriate in a given situation from the marketer's point of view.

2 HOFSTEDE MODEL

2.1 Brief introduction

Geert Hofstede, the Dutch researcher, while working for the large multinational organization IBM realized that even though the company operated under its own set of corporate culture, the cultural differences on the individual level have also highly influenced the work of its employees. Therefore, he started to explore existing differences in thinking and social action among company's members by asking questions about their values. Hofstede has introduced based on the results of his research a model that identifies five primary dimensions that help to understand the intercultural differences. For each of the five dimensions the model provides a scale from 0 to 100. Each country has a position somewhere between the scale, relative to other countries. These five dimensions have been empirically verified, are statistically independent and arise in all possible combinations. They reflect basic problems of any society that need to be managed although the way of coping differs. Hofstede model has been used and cited widely since its introduction by the international scholars and practitioners (Wang, Shi, 2011). Lately, Hofstede added a sixth dimensions called Indulgence vs. Restrain.

2.2 Cultural dimensions

- Power Distance (PDI)—to what extent the society expects the power to be distributed unequally
- Uncertainty Avoidance (UAI)—to what extent people feel threatened by uncertainty and ambiguity in their lives
- Individualism (IDV)—To what extent the individual should look for himself or to stay integrated within his/her social group (opposite is Collectivism)
- Masculinity (MAS)—to what extent the typical masculine values such as achievement, success, etc. (opposite is feminity, caring for others, quality of life, etc.) prevail in the society
- Long-Term Orientation (LTO)—if the main focus in the functioning of the society is placed on today or to the future
- Indulgence vs. Restrain (IND)—Indulgence is typical for a society that allows fairly free gratification of basic and natural human drives associated to enjoying life and having fun. On the other hand, Restraint is typical for a society that suppresses gratification of needs and controls it by means of rigorous social norms.

2.3 Sample/respondents

The original sample was completed from the questionnaires of 116,000 IBM employees from 7 occupational categories in 66 countries (De Mooij, 2010).

2.4 Type of questions

The questions asked in the survey were designed to ask about individual behavioral preferences, preferred or actual state of being (the desired). For example questions were asked about the time one has available for family life, about the job diversity, or how often one feels tense or nervous.

2.5 *Applications*

The Hofstede model has the strength in its predicting power and thus is suggested to be used in the area of predicting behavior (De Mooij, 2010). Also, the model has been used to clarify the intercultural differences in the field of self-concept, personality and identity. All of this in turn helps to explain the variations in branding strategy in global marketing (De Mooij, Hofstede, 2010). The model serves as a base for explaining the area of information processing, such as differences in perception and categorization which in turn influence interpersonal and mass communication as well as how advertising works (De Mooij, Hofstede, 2010). To summarize, when marketing managers need to make an approximate about cultural differences between the company's home market and the foreign market, this model is most used. It helps to segment the world on a country level and allows adapting similar strategies in a country segment (De Mooij, 2011).

2.6 *Critique*

Firstly, many questions in the survey were asked about work related behavior and preferences that can be truly answered only by people who are familiar with the relevant work situation (De Mooij, 2011). Secondly, Hofstede did not measure feminine scores directly. In fact, while measuring the MAS index he considers the lack of masculinity directly to be feminine, which does not need to be always true (Wang, Shi, 2011). Thirdly, results for some countries have not been measured directly. For example, the data for China has been only estimated. Hofstede derived its results from other Asian countries, Taiwan and Hong Kong (Wang, Shi, 2011). Finally, the composition of Hofstede model sample has been widely criticized in the past.

3 GLOBE MODEL

3.1 *Brief introduction*

The Global Leadership and Organizational Behavior Effectiveness (GLOBE) research program was introduced by US University professor Robert J. House in 2004 in the publication *"Culture, Leadership, and Organizations: The GLOBE Study of 62 Societies"*. The major argument behind this model is that leader effectiveness is contextual, that means it is rooted in the societal and organizational norms, values, and beliefs of the people being led. In fact, if the leader wants to be seen as effective by others, he needs to respect and act according to the cultural rules preferred in the country he functions. This model empirically established nine cultural dimensions that describe the similarities and differences in norms, values, beliefs and practices between examined countries of the world. Moreover, based on the quantifiable results it allowed researchers to place 60 of the surveyed countries into clusters in which cultural similarity is greatest among societies that represent a cluster, on the other hand, cultural difference increases the farther clusters are apart (Hoppe, 2007).

3.2 *Cultural dimensions*

- *Power Distance*—the extent to which members of society expect power to be dispersed equally.
- *Uncertainty Avoidance*—the degree to which a society relies on social norms, rules and procedures to ease unpredictability of future events and actions.
- *Human Orientation*—the degree to which a society encourages and rewards individuals for being fair, altruistic, generous, caring, and kind to others.

- *Collectivism I. (institutional)*—the degree to which institutional and societal institutional practices encourage and reward collective distribution of resources and collective action.
- *Collectivism II. (In-Group)*—the degree to which individuals express pride, loyalty, and cohesiveness within their organizations and families.
- Assertiveness—the degree to which individuals are assertive, confrontational, and aggressive in their relationships with others.
- *Gender Egalitarianism*—the degree to which the society reduces gender inequality.
- *Future Orientation*—the extent to which individuals engage in future-oriented behaviors, for example, delaying gratification, planning, and investing in the future.
- *Performance Orientation*—the degree to which a society encourages and rewards group members for performance improvement and excellence.

3.3 Sample/respondents

The study is based on a result of 17,300 middle managers from 951 organizations from the food processing, financial services, and telecommunications services industries.

3.4 Type of questions

Originally, the research was designed in the area of leadership, not with the work motivations. Nevertheless, later the questions were added about the respondents' perception of the organization or societies in which they live or work reflecting on as it really is and as it should be (desired vs. desirable).

3.5 Applications

This model could be useful besides the area of leadership also in studying aspects of intergroup and international relations (De Mooij, 2011). In marketing and advertising field, GLOBE model can serve as a framework for developing assumption concerning the perceived level of performance orientation to be expected by consumers that are exposed to advertising in the different countries (Diehl, Terlutter, Mueller, 2008). Moreover, we add that the set of identified clusters of countries based on similarity of values can be used as a factor that strongly influences the process of adapting global marketing strategies within the selected cluster, and thus reducing the costs in marketing budget and planning.

3.6 Critique

It is quite clear that the GLOBE researchers were heavily influenced by Hofstede model in their choice of variables to be measured. In fact, some of their nine dimensions share the same names as have the Hofstede's dimensions. Therefore, it is possible that some of the GLOBE scales assess some latent stereotypes rather than objective characteristics of society (McCrae, at all, 2008). Some major discrepancies have been found between the data reported by House and actual social practices in particular country (Diehl, Terlutter, Mueller, 2008).

4 SCHWARTZ MODEL

The Israeli psychologist Shalom Schwartz introduces an alternative conceptual and operational approach in obtaining cultural dimensions of values that are work-related into a study of value priorities. Originally, the research was aimed to develop a theory of a universal psychological structure of human values, later the individual-level value types were differentiated and extended to the culture level (De Mooij, 2011). From selected

56 surveyed values, 10 (individual-level) value types were derived. These 10 value types were ordered into two basic bipolar dimensions and each pole represents a higher-order value type that consists of two or more of the selected 10 culture types. De Mooij (2011) argues that Schwartz's distinctions refers rather to categories that to dimensions, because the dimensions should be statistically independent, but in case of the Swartz's value types they are overlapping. Three dimensions with different polar locations are summarized as follows:

4.1   Dimensions/categories

- Autonomy (Intellectual or Affective) vs. Embeddedness
  Autonomy values—a person is viewed as an autonomous entity who is able to pursue his or her individual, independent interests and desires, such as personal interest, self-direction, stimulation, and hedonism. A polar opposition exists of autonomy versus conservatism. Under Intellectual autonomy belong values such as individual thought, curiosity, creativity, freedom, and broadmindedness. Under Affective autonomy belong values of varied life, stimulating activity, and exciting life.
     Embeddedness—cultures that are primarily concerns with security, conformity, and tradition, with the values of devout, obedient, social order, and family security. The interests of person are not viewed very distinctive from the interests of group.
- Hierarchy vs. Egalitarianism
  Hierarchy—emphasis is placed on the legitimacy of hierarchical roles with the values such as social power, authority, influential, humble, and self-enhancement.
     Egalitarianism—emphasis is on the welfare of the other people or transcendence of selfish interests with the values such as social justice, responsible, helpful, loyal, honest, and equality.
- Mastery vs. Harmony
  Mastery—promotes active efforts to adapt one's social surroundings and get ahead of people, with the values as daring, capable, success, ambition, independence, social recognition, self-direction at the individual level.
     Harmony—emphasizes harmony with nature, with the values of social harmony, peace, social justice, helpful, and beauty.

4.2   Sample/respondents

Matched samples of students and teachers in 54 countries.

4.3   Type of questions used

Schwartz asked respondents for guiding principles in people's lives in regard to social issues. Questions were asked about social justice, humility, creativity, social order, pleasure, ambitions and so on (Schwartz, 2004).

4.4   Applications

This model serves as a guideline for researchers that choose to focus on the study of individual-level and also culture-level values. The description of Schwartz's value types are used in marketing and advertising, for example in selecting effective creative strategies including the choice of advertising appeals.

4.5   Critique

The questions chosen in the survey required respondents to evaluate the importance of abstract values, which makes the questions easier to answer for people with higher education. Since the sample consisted of only educated respondents (university students and

Table 1. Summary of the intercultural models.

Description	Hofstede model	Globe model	Schwartz model
Surveyed time period	1967–1973	1994–1997	1970–80
Respondents	Non-managers and managers	Managers	Students and teachers
Organizations surveyed	1	951	xxx
Industry	Information technology	Food processing, financial services, telecommunications services	Academics
Number of countries surveyed	66 (72)	62	54
Country of origin	The Netherlands	USA	Israel
Cultural dimensions/ Categories	Power Distance Avoiding Uncertainty Masculinity/Feminity Individualism/Collectivism Long Term Orientation Indulgence vs. Restraint	Power Distance Uncertainty Avoidance Human Orientation Collectivism I. (institutional) Collectivism II. (In-Group) Assertiveness Gender Egalitarianism Future Orientation Performance Orientation	Autonomy (Intellectual or Affective) vs. Embeddedness Hierarchy vs. Egalitarianism Mastery vs. Harmony
Areas of applications	Predicting behavior Information processing (perception and categorization) In the concepts of self, personality and identity Branding strategy in global marketing Interpersonal and mass communication Advertising (how it works)	Leadership Intergroup and international relations Performance orientations in recipients	Values (individual and culture-level) Values types and creative strategy and advertising appeals
Critique	Composition of sample Work related behavior and preferences Feminine scores not measured directly Some countries only by an estimate	Strong influence of Hofstede Discrepancies by data and actual social practice	Composition of sample Evaluation of abstract values Self-reflection vs. real behavior Different scores in different publications Unipolar scale

Source: author's adaptation.

their professors), the objective generalization of the results to an entire society can be partially doubtful (De Mooij, 2011). Moreover, the fact that the survey is based on people's self reports of the importance they attribute to values, it is critical to establish that this method of self-reflection does not need to always reflect someone's actual behavior (De Mooij, 2011). Finally, country scores are not readily available and different scores have been found in different publications (for example 1994 and 2007), which might result in confusing conclusions. Some difficulties for interpretation result also from the selected scheme of unipolar scales (i.e. opposite) that are characteristic for the Schwartz model. As the negative correlation with one does not automatically mean a positive correlation with the opposite scale (De Mooij, 2011).

5 THE SELECTION PROCESS

Based on the above introduction of the selected intercultural models, in this part of our contribution, we propose concrete set of steps in decision-making process about which intercultural model is the most appropriate in a given situation from a marketer point of view in order to adapt to local target markets effectively:

1. Specify your marketing problem and overall external and internal situation
2. Define your marketing and communication goals
3. Closely examine the advantages and disadvantages of each model
4. Choose one of the models based on the availability of the comparable data of all the countries involved (i.e. country of origin and country/countries of entry)
5. Be aware of the shortcoming of the selected models in the application process and the complexity of the culture itself
6. Revise your model's selection if necessary.

Table 1 shows a brief summary of the main information about the examined intercultural models and might serve as a basic guideline.

6 CONCLUSION

The aim of the presented paper was to examine three cultural models that are nowadays used most frequently in the area of intercultural differences with the focus on global marketing: Hofstede model, GLOBE model and Schwarz model. Specifically, we have briefly introduced each of the models, described the composition of sample and the questions used in the methodology, illustrated the desirable applications in the theory and practice and finally, concluded with the critique. All three presented models have some similarities and differences based on various conceptual and measurement approaches. Moreover, each of the models has different applications as a result of its descriptions. All three models have not been introduced primarily for the field of marketing but for the international business and management. Therefore, further research is needed for the theory and practice application to global marketing and current stage of knowledge is not satisfactory. As we can see, the major shortcomings of the introduced models are mainly in the composition of sample and therefore the generalization of the findings are to some extent questionable. On the other hand, all of the models have quantified results that allow not only to access the actual state of the surveyed country, but also to make a comparison between different countries or making a cluster of countries based on the similarities. In fact, all these factors are important for the global marketing. We believe that by comparing and contrasting all examined models, we were able to give a basic understanding on how these models work and therefore, how they can be used in decision-making process made by global marketers. Nevertheless, we understand that the issue of culture and its impact on marketing and advertising is highly complex and this paper is focused only on one special part of it.

REFERENCES

De Mooij, M. & Hofstede, G. 2010. The Hofstede Model. Application to global branding and advertising research. *International Journal of Advertising* 29 (1): 85–110.

De Mooij, M. 2010. *Global Marketing and Advertising. Understanding Cultural Paradoxes.* Thousand Oaks, CA: Sage.

De Mooij, M. 2004. *Consumer Behavior and Culture. Consequences for Global Marketing and Advertising.* Thousand Oaks, CA: Sage, 2004. 403 s. ISBN 978-1-4129-7990-0.

Diehl, S. & Terlutter, R. & Mueller B. 2008. The Influence of Culture on Responses to the Globe Dimension of Performance Orientation in Advertising message—Results from the U.S., Germany, France, Spain, and Thailand. *Advances in Consumer Research* 35: 269–275.

Hoppe, M. 2007. Culture and Leader Effectiveness: The GLOBE study. *Central European Journal of Communication.*

McCrae R.R. *et al.* 2008. Interpreting GLOBE Societal Practices Scales. *Journal of Cross-Cultural Psychology*, 39: 805–810.

Schwartz, S.H. 2004. Mapping and interpreting cultural differences around the world. In H. Vinken at all. (Eds.). *Comparing cultures: Dimensions of culture in a comparative perspective* (pp. 43–73). Leiden: Brill.

Wang, J. & Shi, X. 2011. Interpreting Hofstede Model and GLOBE Model: Which way to go for Cross-cultural research. *International Journal of Business and Management* 6, (4): 93–99.

Methodological bases of media and lifestyle

Janka Kyseľová

Faculty of Educational Sciences, University of Matej Bel Banská Bystrica, Bratislava, Slovakia

ABSTRACT: Mass communication is subject to daily changes that result in the best satisfaction of the recipient approaching him. In today's terms the media fail only informative function but it is primarily the satisfaction of beneficiaries through entertainment. The audience has become more difficult not to be content with a superficial glance, looks more and more information from all walks of life (education, science, arts, travelling). Situation arises in which one becomes slavishly dependent on specific programs through which survives even his life. It can be stated that the traditional transmission model is always changing and sharing it enriches the different models.

1 INTRODUCTION

The oldest philosophy and ethics efforts might include analysis of social phenomena and processes. The formation and the first appearance of this effort can be dated back to the period when a man began to reflect on their own lives and thinking of its nature. The actual exploration of own being its essence and factors affecting the way a person's life is not a modern phenomenon in science. The bottom line is that in the development of society especially in cultural and historical plane occurred human desire for justice in the way of life known as a variable while in theory appearing during the times of ancient Greek philosophy.

"The advantage of contracting terminology is the idea that principles of justice are taképrincípy that rational people would vote. It is on the basis of these principles can explain and justify the concept of justice … The principles of justice deal with the conflicting demands of the benefits obtained from social cooperation and concern relations between different people or groups." (Hajduk, 2013, p. 281). In the process of creation itself, both physical and spiritual resources at the same time also constitutes a form of expression and way of life as well as the very essence of people. Shaping the way of life and the nature of the process in philosophy and her attention began to emerge late sixties and early seventies when this phenomenon was given to awareness and other social sciences. At the beginning of the activity in this area mainly focused on sporadic and mostly random research activities but at present the research in this area, systematic and purposeful and we can also designate a focused, holistic and interdisciplinary. In essence, covers the entire field of observation and scientific knowledge. This condition was necessary to come as a result of social practice which highlighted the seriousness and regency of all the important issues in the development of society. *"Elementary rules of coexistence have become part of the legal norms and stabilized as a minimum unquestionable moral imperatives. Many of them adhere fully automatically as a matter of course. They are part of our culture reflect values that we accept. But for some of us for granted protruding We are not quite clear why we observe them, think about them, we question them, or even consciously accept them."* (Hajduk, 2008, p. 42).

The research is a way of life can find a large number of research papers and studies and the authors are focused on the analysis of elements, attributes and moments or lifestyle factors. In some cases, we find confusion between the concepts of life and lifestyle possibly living standards with a focus on structural components, quantitative and qualitative indicators and their definitions. For example Grulich and quite often define new way of life in terms of general sociological categories. Often however we may encounter mistakes in the characteristics of these terms, because the scope of the topic is very broad indeed.

Way of life and its definition and exploration we can only assign a sociological categories because it belongs also to philosophy, ethics, economics and political science. We can even have extend to this area practically all social sciences and their categories and requires complex solutions and approaches. Important its analysis and from several angles. However, it is necessary to take into account the unifying scientific—methodological considerations. We may assume that under comparison between expressions way of life and standard of living is just a way of life broader concept because it implies also the very essence of living. In the social sciences, it is possible to perceive the essence of the concept and way of life as aggregate category that deals with the life of man and his nature, content and structure as a whole.

2   A WAY OF LIFE

Way of life can be called those aspects which make up the ways of coexistence between people and their joint activities, also outward manifestation, that behavior and internal that subjective aspects of human life. Differentiation in this case possibly from different perspectives an example of the division in terms of social groups. In the case of the general conditions applicable can find groups which are characterized by a specific way of life that manifests specific forms of behavior, and external signs which indicate the specificity of the whole substance of this group of people. In their differences can create subcultures and cultural expression created his own group consisting of. An example of this subculture respectively specific groups of youth, but there are also groups that despite differences in the way their lives are not showing them externally. An example in this case are seniors. Standard of living which is determined primarily real income of people is part of the lifestyle.

Carries the necessary material goods while specifying the structure needs and their volume, job opportunities, housing conditions and education levels. However this includes health care, cultural life and timetable. *"Only the ethics of justice may justify the universal validity of standards that ensure every individual and every particular community equal right to the authentic realization of the good life. As far as here is, in our view, unreasonable juice authenticity realization of the good life—whether individuals or communities—against the autonomy of identity relating to his senses. Authenticity and autonomy of reason creates ideal complementary moments post-conventional identity of modern people."* (Hajduk, 2008, p. 58).

Also important is the self-realization of man and family conditions to determine the standard of living. In the case of the standard of living we talk about the category dealing with physical and economic conditions of human life. Also includes meeting the basic needs of life what is different compared with the category of lifestyle, which focuses mainly on living manifestations and forms and life processes of active and conscious controlling own living conditions. As mentioned above with terms as a way of life and standard of living is closely related to lifestyle category.

3   A WAY OF LIFESTYLE AND MEDIA

Lifestyle people manifests itself in their way of life which however does not determine the quality of this way of life. Term lifestyle or lifestyle while characterize individual; i.e. psychological, moral and social peculiarities in the way of a person's life. Lifestyle Makeover—is a way of life that a person leads a life style leaves a particular track. Lifestyle is not determined so that in the manufacturing method existed nuances. The lifestyle can say that it is affected by many factors such as geography, history and its peculiarities, psychological relationships, tradition and spiritual riches that brings this category. Consequently it is a lifestyle a way for individuals to specific behavior that stands out certain peculiarities of the case his ways, habits and inclinations. But focuses not only on individuals but also social groups.

Material and technical level of society influences the individual's lifestyle quite significantly in partnership with the results of material production. Outward forms of life of individuals or social groups clearly defined lifestyle and manifested it also organization of working and leisure time, interests, hobbies and non-work activities, family life and participation in the management of public affairs and to participate in public life as such. This means that lifestyle can be described as social, psychological and economic phenomenon. Within the lifestyle we talk about psychological and social specificities in human behavior when compared with the way of life. The lifestyle is an important and essential feature of particular manifestation of his individuality and independence which can be relative. It also performs the process self-creation own as individuals which is influenced by the idea of human spirituality, the intense, sociability and morality in life. It follows that if an individual shapes define and to develop its own personality also defines and shapes also developing their own personal lifestyle.

For individuals and groups of people is a natural formation of his own lifestyle which can't be associated solely with the effort to emphasize certain characteristics of people and groups of individuals, which in this context can be described as special. This effort should not be associated with a commitment to originality and the allocation of the company and group of people at any cost. Personal lifestyle is shaped through education and reflects the culture and moral values in society in its most general nature. "*In the process of discourse comes to justifying it receives the addressee standards that, as a rational entity must necessarily act in accordance with reason. In a situation of moral problems one wonders, will that lead to a reasonable solution. Sensible solution is consensual solution in an equal meaningful argumentation in which arguments are accepted everyone equally without considering the individual from whom they come.*" (Hajduk, 2009, p. 66).

Lifestyle can be understood as a reflection of human individuality and uniqueness which is specific to a particular individual. Lifestyle is yet formed following: a) options that are in the social system objective, b) design affecting the behavior of man, which are subjective. Lifestyle becomes an important identifying characteristic of a particular social class or group which is closely related to the social differentiation of society.

Way of life is closely linked with the perception and understanding of quality of life which is reflected in particular in science in Western Europe. Democratic capitalism in this case becomes a sort of epoch in quality of life theorists in this field. According to these experts, it is possible to realize success in the quality of life for individuals without the need for conversion of state's and monopolium foundations in capitalism. Quality of life in its essence represents a certain way of life man page which has a direct bearing on the qualitative side expresses standard satisfying material and spiritual needs of individuals and refers to individual and social life, and components that cannot be characterized quantitatively. These elements of human life expressing its qualitative aspect while you cannot even measure how it is possible for the individual elements of living. In conjunction with Western theories can be associated with quality of life implications of development that is oriented rationalism in science and technology. This concept was introduced by the American economist J.K. Galbraith in 1967 and initially the term has been used primarily for the identification of specific questions intended to environmental degradation. Quality of life is thus derived from the fundamental characteristics of Western civilization hardest focusing on all areas of social life in which they determined mainly indicators of growth. Quality of life is perceived by the German Social Democrats under continuous strengthening of private property, business functions and the market mechanism. Allocates only two basic reasons why the quality of life issue is posted at the forefront: 1. The primary material needs of the population are satisfied with the possibility for economic growth and development of science and technology. Needs of the people therefore in this respect and their growing movement to activate and update the necessary assurance and higher forms of satisfying. 2. Side effects that brings economic growth in this manner reveal and bring some form of threat to the health of human existence. Currently, together with the development of quality of life issues come under which it is necessary to solve global problems appearing in the civilization of the 21st century. Gradually emerging evidence that despite developmental differences countries worldwide problems in these countries affect the existence of mankind as a whole and therefore each country directly. This means that it is necessary to identify solutions to resolve this situation or suggest options to solve problems. More than forty years ago appeared the first signs of common starting points of the problems in the form of founding the Club of Rome. Its establishment was initiated by the different world capacities and personalities of the theory and Sciences, and his goal was a dialogue aimed at solving a rational approach to problems and their definition. The establishment of this institution dates back precisely to 1968 and its main founder was an Italian industrialist A. Peccei who simultaneously became its first president. Club of Rome was mainly informal and non-governmental association which oversaw scientists from different fields of science and from different countries.

The objective to which the association focus was the creation of a global forum that would allow the best discussion of politicians, statesmen and scholars on topics dangers of global crises. The essence should also be finding solutions and ways to prevent these problems or prevent them. In our conditions, the club and its activities perceived in terms of addressing environmental damage but its activities and objectives were clearly aimed at a wider area. Between the present and the past especially the ancient world there is a huge difference in the perception of nature. While in the history of nature was perceived at the highest point scale of values the company at present it as much as we try to evaluate. Anthropocentrism that period of history however the procedure has changed and people have started to realize the need to protect their life surroundings. Establishment of the club preceded the occurrence of crisis phenomena which discussed economic and scientific—technical development areas of the country in Western societies. The development of the countries however had a very significant negative side effect that was the danger of ecological crisis. This danger steadily risen in the foreground there were concerns about whether humanity is capable of solving various problems of an economic, social, political, moral, cultural and religious or ecological character at the beginning of the new millennium. The increase in these problems which at that time dial the fact could change the global catastrophe. *"Moral standards are created by man and values that become through their discourse mediated argued content corresponding skills and human performance. Functionality moral standards depends on the ability of humans to understand their meaning, identify with them so that they become part of his own moral attitudes."* (Hajduk, 2009, p. 67).

Club of Rome began to expand its activities and in 1972 began with the issue of reporting cycles called "warnings to mankind." These reports were individually examined by scientists around the world and their attitudes were different. In Western countries, theorists focused on the club notice as to ways and means of modeling the future of humanity, in the Eastern Europe countries theorists looked at these reports with criticism, mainly focus on the content itself the Club of Rome. Criticism against the club however was mainly ideological. We select several authors such as I.T. Frolov Perspectives of man (1983) or V. M. Lejbin Club of Rome and his ideas (1985).

4   CONCLUSIONS

Mass communication is subject to daily changes that result in the best satisfaction of the recipient approaching him. In today's terms the media fail only informative function but it is primarily the satisfaction of beneficiaries through entertainment. The audience has become more difficult not to be content with a superficial glance, looks more and more information from all walks of life (education, science, arts, travelling). Situation arises in which one becomes slavishly dependent on specific programs through which survives even his life. It can be stated that the traditional transmission model is always changing and sharing it enriches the following models

- "Ritual—expressive model" (management processes, sharing, common expressions of commonly accepted values and Forging community cohesion)
- "Promotional model" (assumed as the main aspect of the media presentation and demonstration of claiming attention) and
- "Income model" (based on abstract and general semiological analysis of cultural significance, encoding, decoding etc.).

Recent years have encountered in mass media with the presentation of advertisements. They are everywhere you look whether on television in newspapers and even on the radio. Television advertising includes all mentioned accents because it is not only information about the product or brand but also promote and manipulate the recipient to adopt new guidelines to consumerism. The point is to keep the individual in just the right layer to be "in". The study of mass media wants to watch all the structures that are generated during communication processes from which these processes generate. Structures means all departments and regularity arising from the technological situation and practice, whether tangible or intangible files (data, rules, systems) and layer (part of), institutions, implementing units, the nature of the contractual relationship and i. From the broadest perspective can be distinguished approaches and perspectives according to which the widest possible contexts advocate its hearing on the media:

a. economic considerations (whether economic—industrial)
b. The political aspect (or politico—economic)
c. technological aspects. The public interest (mixed aspect)—somehow sums up the pathos of other normative concept, including criticism of capitalism and the protection of the individual (a "semantic power" of the individual) against the interests of national and international political and economic institutions.

Mass communication is broken down by area, sector (newspapers, film, television, etc.), distribution (national level, international or regional), according to various media (station), or specific individual products (song book), or by the establishment or providers. The audience is the main objective of media or mass media or mass communication, that recipients who receive products such forms of communication. Information which is conveyed same group are carefully mediated.

And the main core group is its structure, interests, size and behavior. Marketing potable content is based on the focus of media research what are the requirements of the recipient—what is the consumerist content.

REFERENCES

Eco, U. 2000. *Mysl a smysl*. Praha: Vize 97.

Gadamer, H.G. 1970. *The Power of Reason*. In Man and World, vol. 3. p. 5–15.

Hajduk, Ľ. 2008. *Liberalizmus a komunitarizmus*. (Liberalism and communitarism).Bratislava: Štátny pedagogický ústav.

Hajduk, Ľ. 2009. *Filozofia spravodlivosti* (A philosophy of justice). Bratislava: Štátny pedagogický ústav.

Hajduk, Ľ. 2013. Utilitaristický alebo deontologický concept sociálnej spravodlivosti. In Kyuchukov, H. et al. 2013. *The educational and social sciences in the 21 century*. Bratislava: St. Elisabeth University, pp. 278–288.

Le Bon, G. 1994. *Psychologie davu*. Praha: KRA.

McQuail, D. 2000. *McQuail's Mass Communication Theory*. 4th Edition by Sage Publications Ltd.

McQuail, D. 2002. *Úvod do teorie masové komunikace*. Praha: Portál.

Meadows, D.L. 1972. *The Limits to Growth*. New York: Universe Book.

Nový, I. 1997. *Sociológie pro ekonomy*. Praha: Grada Publishing.

Ortega y Gasset, J. 1993. *Vzpoura davů*. Praha: Naše vojsko.

Creative personality in relation to the Big Five personality model

Richard Keklak
Faculty of Mass Media, Institute of Media Studies, Pan-European University in Bratislava, Bratislava, Slovakia

ABSTRACT: While thinking about the relationship of individual's creativity and personality, it may arise a question how the creativity influences individual's personality and on the other hand to what extent is the creativity influenced by personal characteristics of the individual. The study describes to what extent act personality traits (Neuroticism, Extraversion, Openness, Agreeableness, and Conscientiousness) as pre-requisite for creativity and to what extent are the part of it.

Creative personality in this study is outlined in the means of personal traits taxonomy grouped into five factors, so-called Big Five. The aim of this study is to verify the existence of intrapsychic relations between creativity and personal dimensions of the Big Five personality model.

Achieved results point out that individuals with higher creative potential are more self-confident, active, talkative, cheerful and optimistic than individuals with lower creative potential. Creative individuals are also more sympathetic and trustful and they prefer mutual cooperation. Higher level of mental unstableness, imbalance and lower level of resistance to mental exhaustion is typical for the individuals with lower creative potential. We also suppose that the individuals with lower creative potential are commonly characterized by these emotions: uncertainty, anxiety, fear, worry and sadness.

1 INTRODUCTION

The most significant representatives of psychology of personality has tried to explain to what extent act personality traits as pre-requisite for creativity and to what extent are the part of it. This was done on the basis of studies of personalities.

Specialists as M. Csikszentmihalyi (1996), T. Amabile (1983) or H. Eysenck (in M.A. Runco, 1996) and M. Jurčová (2003, 2009), M. Zelina (1995), V. Dočkal, M. Matejík (1996) and others within the Slovak Republic were dealing with the researches concerning the personality traits that are defined as a cause of creativity or as a primary part of it.

Personality has been conceptualized from a variety of theoretical perspectives, and at various levels of abstraction or breadth. Many personality researchers had hoped that they might devise the structure that would specify the personality in a common language. (P. Halama, 2006). According to O.P John and S. Srivastava (1999), such integration was not to be achieved by any one researcher or by any one theoretical perspective. After decades of research, the field is approaching consensus on a descriptive model, general taxonomy of personality traits, the "Big Five" personality dimensions. P. Halama and col. (2006) claim that these dimensions do not represent a particular theoretical perspective and do not proceed from any specific theoretical school but were derived from analyses of the natural-language terms people use to describe themselves and others. In other words, natural language is the origin of the scientific taxonomy. Since 80' the so-called model of big five factors is markedly used within the personality research. Even in 90' we can see the strong position of this model in theoretical and research field (e.g. J.M. Digman 1990; R.R. McCrae and P.T. Costa 1991; M. Hřebíčková and I. Čermák 1996; I. Ruisel 1998). Authors of five factors theory of personality, R.R. McCrae and P.T. Costa (1991), consider the human nature to be knowable, rational, variable and active. "Knowable" means that the human being can be an object of some scientific research. Rationality expresses that people can understand the others and themselves as well, and that they act according to their wishes and ideas. People are intelligent observers of their own thoughts, emotions and behaviour. Source of information about personality is important to gain not only through self-knowing but also through the way in which the others can see us.

After researching the life and works of creative people that lasts for 30 years and on the basis of the answers of examined individuals came M. Csikszentmihalyi (1996) to a following conclusion or finding: each of us was born with two contradictory sets of instructions, self-conservative tendency, made of instincts for self-preservation and an expanding tendency made up of exploring, for enjoying novelty, risk, etc. The first one require little encouragement or support from outside to motivate the behavior and the second can wilt, if it is not cultivated. The real creativity never appears as a result of a sudden outburst but as a result of a long-term hard work. Result of a creativity future contributes to the life enrichment and complexity; creative development appears mostly in above-average quality conditions, which is usually associated with high standard of well-being; creativity often develops in a crossroads of cultures, where different beliefs, ways of life and information from different traditions are exchanged and synthesized and the individual can more easily understand the new combination of ideas in such environment; creative individuals are not tend to be so-called suffering geniuses, they love their work and their motivation results from their work, and from the joy of work and not from the desire for fame and money; creative individuals really know their work is good and they can differentiate bad ideas from the good ones. Creative individuals can also feedback themselves.

Each of us have a certain amount of creative potential and development of such a potential is influenced by individual's physical traits, and individual's needs, interests, motives and volitional characters, which tend the individual to be resistant against psychological stress, work-related stress and stress related to live difficulties. Criteria for creative personality identification are not united, but personality traits typical for creative individuals were defined by psychologists. This was done on the basis of characterization of significant individuals carrying out real creative activities. Such personality traits facilitate the development of creativity and consequently assist while facing the life difficulties and burnout. These traits are: persistence, self-confidence, courage, curiosity, intuition, tirelessness, ability to work intensively, ability to put the work to an end, independence and self-reliance, ability to take a risk together with the ability to accept the defeat, willingness to overcome the obstacles and difficulties and to cope the traits defined as conflicting.

Numerous specifications of creativity are dependent on theoretical concepts. We can define creativity in general as ability, attitude and process. It is an ability to imagine or invent something new—we do not speak about the process of creation of something from nothing. Furthermore, it is an ability to create ideas, solutions, thoughts and piece of work by using combinations, changes and replications of existing ideas. The attitude of the individual is characterized by agreement, acceptance of change and something new, willingness to cope with ideas and thoughts and by flexibility. Creative process is characterized by hard work, continuous process of thoughts while thinking about the solutions, improvisation and clarity.

Proceeding from the 4P interaction (four component model) M. Jurčová (2009) defines creativity as a wide-ranging phenomenon that is determined by numerous factors and variety of manifestation in all spheres of human activities. Creativity is based on the product that is new and convenient as well. It arises as a result of the creative process. Personal traits are considered to be the potential (prerequisite) for creative activities of the human. Creative product is a result of the author's cooperation with socio-cultural and historical conditions.

Ability to use the knowledge about creativity in practice is expressed mostly by 4P interaction of key areas in psychology of creativity (product, process, personality, environment), which becomes the framework for structuring the knowledge about creativity. This framework became also the base for the first systematic analysis for creativity factors in Slovak

Republic. M. Zelina (1990) defines creativity as an interaction between subject and object in which the subject is changing the world and creating a new, useful subject or reference group/population of significant value.

According to K.K. Urban (in T. Kováč, 2002) the creativity appears in new and surprising product and that's why the creativity needs to be understood as an ability to create a new surprising product as a solution of sensitively perceived issue while taking into account the broadest circumstances concerning the given information. This can be also done through analysis and flexible processing focused on solution, through unusual associations, through reconstructing and combination of introduced information together with data of individual's own experience and imagination. Another part in a process is also synthesis, structuring and forming such information, components and structures with the aim to elaborate new solutions. Result of the process is to create a product that is finally understood by others (through their communication) as useful.

R.R. McCrae (1987) was dealing with relation between several creativity levels and personality trait of Openness to experience. He discovered positive relations of six measures of divergent thinking (verbal, mental, association fluency) and all sub-scales of Openness factor. C. Martindale (1989) says in his theoretical study (referring to R.R. McCrae) that discriminant validity of this personal trait is questionable and he thinks that openness to experience and creativity are synonyms, i.e. openness do not bring into creativity conception nothing new.

Study of L.A. King and col. (1996) is dealing with the creative personality in relation to the five factors personality model. Positive relation between extraversion and openness and verbal creativity was detected in this study. Negative correlation relation between agreeableness and creative action was found out as well. Comparing the results of R.R. McCrae (1987) the study of L.A. King also stated that the most significant positive relation appears to be the one between the factor of openness to experience and verbal creativity. Openness correlated with creative actions even significantly. Results demonstrated that individuals with higher creativity level are able to produce creative actions in case they are open to new experiences and are ready to explore them.

L.A. King (1996) claims that openness may play a role of certain catalyst used for expressing someone's creative abilities, it may increase the inner motivation of the individual and it helps individuals to evaluate the given situation as potentially creative. As to the creativity development interventions, the group of individuals with higher degree of openness but lower level of creativity seems to be very promising. Already mentioned study provides also information about relations between the levels of creative abilities, creative actions and personality trait of conscientiousness. Conscientiousness is in a positive relation with creative actions, but only in case when creative abilities are of the lower level. In the middle and higher level of creativity the negative relation between creative action and conscientiousness was detected. This discovery is surprising because (on the contrary with the common myths concerning the creativity) persistence, self-discipline or hard working (associated with conscientiousness) are inevitable for achieving the excellent creative actions. This may be a result of a higher standard (more strictly given criteria) for assessing their own creative actions (as to the more conscientious creative individuals), which would be the reason of the fewer amount of such individuals in contrast to those that are less conscientious.

2   THE PRESENT RESEARCH

2.1   *Research question and hypotheses*

The aim of this research is to analyses the relation of examined personality traits variables and creativity. Specifically, we are expecting the existence of positive relation of openness and extraversion concerning the creative thinking. We assume that the higher degree of openness to experience and extraversion is the prerequisite for the individual's higher creative potential. The bases for this research problem are the studies of R.R. McCrae (1987) and L.A. King

(1996). Results demonstrated that individuals with higher creativity level are able to produce creative actions in case they are open to new experiences and are ready to explore them. We are also interested in the quality of relation between creative abilities level and other examined personality dimensions. We assume the existence of higher degree of neuroticism in negative relation towards creative thinking.

1. We are expecting the existence of positive relation between personality dimensions of openness and extraversion towards creative thinking.
2. Higher degree of neuroticism will be in negative relation with creative potential.

2.2 *Participants*

623 participants were involved in this research.
Basic demographical characteristic of the sample:

Age range: 19–50 years (M = 31.28)
Gender: men 24.6% (N = 153), women 75.4% (N = 470)
Marital status: single 44.8% (N = 279), married 52.2% (N = 344)
Education: secondary school with LE 53.5% (N = 333), University 46.5% (N = 290).

2.3 *Methods*

NEO Five Factor Inventory of P.T. Costa and R.R. McCrae; Slovak translation by Ruisel, Halama, 2007).

It is a multi-dimensional questionnaire working on the basis of factors-analysis approaches and five factors personality model. The authors of this questionnaire are P.T. Costa and R.R. McCrae (1991). Each of these five factors (Neuroticism, Extraversion, and Openness to Experience, Agreeableness and Conscientiousness) is further-specified by another six sub-scales.

Urban test of creative thinking (T. Kováč, 2002) was used to determine creativity; creative individual.

It is a tool through which the individual's creative potential can be examined. Test helps to identify the identifications of the high level of creative capacities and it also discovers the individuals with creative capacities characterized as below-average. This test is a kind of a non-verbal examination, in which the picture is shown to an examined person. Picture consists of five figures situated within the square and the last figure, the sixth one, is situated outside the square. The task of the examined person is to make the drawing by finishing these figures according to its own fantasy.

3 RESULTS

Results of relational analysis between creative potential and personality traits of the whole studied sample (see Table 1) indicate that our first hypothesis was not significantly validated. What is significant, is the negative relation between creative potential and neuroticism ($r = -0.096*$).

Table 1. Correlation analysis between examined personality traits variables and creative potential.

Variables	Neuroticism	Extraversion	Openness	Agreeableness	Conscientiousness
Creative potential					
r	−0.096*	0.068	0.002	0.050	−0.074
Sig.	0.016	0.088	0.951	0.212	0.065

* $p < 0.05$.
r – Pearson' correlation coefficient.

On the basis of the achieved creative potential score examined people were divided into the groups of those with higher and lower creative potential. Through the comparative analysis we were looking for some probable differences within achieved score in a specific examined personality factors. Results of this comparative analysis are confirming the relations in correlation analysis.

Statistically significant differences were detected in two cases (see Table 2). Individuals with higher creative potential are presented as those with higher level of extraversion ($t = 3.466$***) and agreeableness ($t = 1.850$*). This specific result confirms our stated assumption of positive relation between creative thinking and extraversion.

Due to the heterogeneity of the whole sample, we decided to divide it into four groups: individuals that are in a close interaction with others, individuals working in professions that do not require the interaction with others, group of unemployed people and students (see Table 3).

In the group with higher level of interaction, two significant differences were detected, more specifically, in personality traits variables of Extraversion ($t = 1.907$*) and Agreeableness ($t = 1.951$*). Another differences (statistically not so significant) indicate that more creative individuals in this group have the tendency to be open to experience ($t = 0,843$), conscientiousness ($t = 0.857$) and they are more resistant to psychical exhaustion ($t = -0.421$).

While comparing the groups of individuals with higher and lower creative potential working in professions that require lower level of interaction with others, statistically significant differences were detected in personality traits variable of Extraversion ($t = 2.040$*) and Openness to experience ($t = 1.978$*). These results are confirming the hypothesis, in which we have assumed a positive relation between creative thinking and personality dimensions of Openness and extraversion.

Comparison of groups of individuals with higher and lower level of creative thinking in the group of unemployed people is showing the significant differences in personality traits variable of Neuroticism ($z = -2.245$*) and Extraversion ($z = -2.598$**).

Table 2. Differences between the groups with higher and lower creative potential in examined personality traits variables.

Personality traits	Higher creative potential (N = 289)		Lower creative potential (N = 334)		
	Mean	SD	Mean	SD	t
Neuroticism	21.84	5.572	22.27	5.665	−0.935
Extraversion	29.10	4.062	28.01	3.796	3.466***
Openness	25.43	3.822	24.94	3.986	1.556
Agreeableness	26.31	3.988	25.72	4.032	1.850*
Conscientiousness	28.80	3.539	29.02	3.911	−0.747

* $p < 0.05$.
*** $p < 0.001$.
t – Student's t-test for two independent selections.

Table 3. Frequency of research sample.

	Frequency	Percentage	Cumulative percentage
Helping professions	202	32.4	32.4
Non-helping professions	218	35.0	67.4
Unemployed	48	7.7	75.1
Students	155	24.9	100.0
Total	623	100.0	

4 CONCLUSIONS

It is a well-known fact, that R.R. McCrae (1987) was dealing with the connection of creativity and personality traits of extraversion and openness to experience, which are two polar factors of the big five model. Our next assumption was determined on the basis of these results. L.A. King, L.M. Walker and S.J. Broylers (1996) discovered the negative relation between the higher level of creativity and conscientiousness. Above mentioned authors explain that creative individuals have higher standard (more strictly given criteria) for assessing their own creative actions.

According to the results, the higher level of mental unstableness, imbalance and the lower level of resistance to mental exhaustion is typical for the individuals with lower creative potential. And on the contrary, individuals with lower creative potential are commonly characterized by emotions such is uncertainty, anxiety, fear, worry and sadness.

Achieved results point out that the individuals with higher creative potential are more self-confident, active, talkative, cheerful and optimistic than individuals with lower creative potential. Creative individuals are also more sympathetic and trustful and they prefer mutual cooperation.

Another differences (although not so significant) show that individuals with higher creative potential are more resistant to mental exhaustion, they behave in unconventional way and prefer changes in their lives, but on the other way they are non-systematic, unstable and unreliable. These achieved results are supported by the study of L.A. King and col. (1996).

What is important is, that individuals working in helping professions and with the assumption of higher interaction and higher creative potential show the higher degree of extraversion and agreeableness. Extraversion ensures the higher quality and quantity of interpersonal interactions, level of activation and the need of stimulation, which are crucial for professions requiring the high level of interaction. Agreeableness ensures the quality of interpersonal orientation as well.

Higher level of creative thinking concerning the individuals working in professions that do not require direct orientation on helping and communication with others ensures that these individuals are much more warm-hearted and oriented to others. Higher creative potential also ensures that individuals are looking for new experience, that they are tolerant towards the unknown and open to discover it.

Within the group of unemployed persons we came to a conclusion that even though these individuals are creative, they have to cope with negative emotions such is fear, depression and indecision (in comparison with the group of individuals with the lower creative potential). Less creative individuals with rigid thinking do not see other possibilities of the reality in their lives and they just accept their current life as unchangeable reality. On the contrary, more creative individuals are thinking about the varieties that life can offer and unreality, indisposition or inability to use these unreachable varieties of life cause their psychical imbalance. Individuals with higher creative potential in this group (as well as in all groups of professions) are more sociable, they prefer situations when they feel themselves pleasantly and they are looking forward the interaction with others. Differences identified in other personality variables describe the group of unemployed people with higher creative potential as individuals which are able to think about several ways to improve things, they are open to new ideas and ready to make experiments. They are sympathetic toward the others, they are willing to help the others and believe the others will help them as well. Disability to change the current reality of their lives that was already mentioned decreases the motivation and persistence to achieve the goal of a focused behavior, despite their high level of creative thinking.

REFERENCES

Amabile, T. 1983. *The Social Psychology of Creativity.* New York: Springer Verlag, 1983. 234 p. ISBN: 0-38-790830-7.
Basham, A.L. 2008. *The wonder that was India.* Calcutta: Fontana Books.
Berry, J.W. 2008. Globalization and acculturation. *International Journal of Intercultural Relations,* 32: 328–336.
Bortz, J. & Döring, N. 2005. *Forschungsmethoden und Evaluation* [Research methods and evaluation]. Berlin (Germany): Springer.

Csikszentmihalyi, M. 1996. *Creativity, Flow and the Psychology of Discovery and Invention.* New York: Harper Perennial, 1996. 464 p. ISBN: 0–06–092820–4.

Digman, J.M. 1990. Personality structure: Emergence of the Five Factor model. In *Annual Review Psychology,* 1990, vol. 41, p. 417–440.

Dočkal, V. & Matejík, M. 1996. Zámerné rozvíjanie detskej tvorivosti: korelácie s anxietou. In *Československá psychologie,*1996, roč. 40, č. 1, s. 1–13.

Epstein, S., Pacini, R., DenesRaj, V., & Heier, H. 1996: Individual differences in intuitive-experiential and analytical-rational thinking styles. *Journal of Personality and Social Psychology,* 71: 390–405.

Forsyth, D.R. 1980. A taxonomy of ethical ideologies. *Journal of Personality and Social Psychology,* 39: 175–184.

Gorsuch, R.L. & McPherson, S.E. 1989.: Intrinsic/extrinsic measurement: I/E revised and single-item scales. *Journal for the Scientific Study of Religion, 28,* 348–354.

Graham, J., Nosek, B.A., Haidt, J., Iyer, R., Koleva, S., & Ditto, P.H. 2011. Mapping the moral domain. *Journal of Personality and Social Psychology,* 101: 366–385.

Haidt, J. & Kesebir, S. 2010. Morality. In S.T. Fiske, D.T. Gilbert, & G. Lindzey (eds), *Handbook of social psychology* (5th Ed.): 797–832. Hoboken, NJ: Wiley.

Halama, P. a kol. 2006. *Religiozita, spiritualita a osobnosť. Vybrané kapitoly z psychológie náboženstva.* Bratislava: Ústav experimentálnej psychológie SAV, 2006, 156 s. ISBN: 80–88910–22–6.

Hřebíčková, M. & Čermak, I. 1996. Vnitřni konzistence české verze dotazniku NEO-FFI. In *Československá psychológie,* 1996, roč. 40, č. 3, s. 208–216.

Jensen, L.A. 2011. Navigating local and global worlds: Opportunities and risks for adolescent cultural identity development. *Psychological Studies,* 56: 62–70.

John, O.P. & Srivastava, S. 1999. The Big Five Trait Taxonomy: History, Measurement, and Theoretical Perspectives. In Pervin, L.A. - John, O.P. (Eds) Handbook of Personality. *Theory and Research.* New York: The Guilford Press, 1999.

Jurčová, M. 2003. Tvorivosť—jej koncepčný rámec a výskumný potenciál v CEVIT. In *Človek a spoločnosť,* 2003, roč. 6, č. 3.

Jurčová, M. 2009. *Tvorivosť v každodennom živote a vo výskume.* Banská Bystrica: IRIS, 2009. 268 s. ISBN: 978-80-89256-42-6.

Khare, A. 2011. Impact of Indian cultural values and lifestyles on meaning of branded products: Study on university students in India. *Journal of International Consumer Marketing,* 23: 365–379.

King, L.A., Walker, L.M., Broylers, S.J. 1996. Creativity and Five Factor model. In *Journal of Reseach in Personality,* 1996, vol. 30, no. 2, p. 189–203.

Kirmayer, L.J. & Minas, H. 2000. The future of cultural psychiatry: An international perspective. *Canadian Journal of Psychiatry,* 45: 438–446.

Kováč, T. 2002. *Urbanov test tvorivosti.* Príručka. Bratislava: Psychodiagnostika, 2002, 57 s.

Martindale, C. 1989. Personality, situation and creativity. In J.A. Glover et al. (Eds.) *Handbook of creativity (Perspectives on individual differences).* New York: Plenum Press, 1989, s. 211–232.

McCrae, R.R. 1987. Creativity, divergent thinking and openness to experience. In *Journal of Personality and Social Psychology,* 1987, vol. 52, no. 6, p. 1258–1265.

McCrae, R.R. & Costa, P.T. 1991. The NEO Personality Inventory: Using the Five Factor model in counseling. In *Journal of Counseling and Development,* vol. 69, no. 3–4, p. 367–376.

Paulhus, D.L. 1991. Measurement and control of response biases. In J.P. Robinson et al. (eds*)* *Measures of personality and social psychological attitudes:* 17–59. San Diego: Academic Press, 1991. ISBN: 978-0-12-590241-0.

Redfern, K. & Crawford, J. 2010. Regional differences in business ethics in the People's Republic of China: A multi-dimensional approach to the effects of modernisation. *Asia Pacific Journal of Management,* 27: 215–235.

Ruisel, I. 1998. Päť veľkých faktorov ako nová paradigma výskumu osobnosti. In Sarmány Schuller I. - Košč, M. – Jaššová, E. (Eds.) *Človek na počiatku nového tisícročia: zborník z konferencie.* Bratislava: Slovenská psychologická spoločnosť pri SAV, 1998, s. 41–44.

Runco, M.A. 1996. Personal creativity: Definition and developmental issues. In *New Directions for Child Development,* 1996, 72, p. 3–30.

Shweder, R.A. 2008. The cultural psychology of suffering: The many meanings of health in Orissa, India (and elsewhere). *Ethos,* 36, 60–77, 2008.

Zelina, M. & Zelinová, M. 1990. *Rozvoj tvorivosti detí a mládeže.* Bratislava: SPN, 1990. 136 s. ISBN: 80-08-00442-8.

Zelina, M. 1995. *Výchova tvorivej osobnosti.* Bratislava: Univerzita Komenského, 1995, 156 s. ISBN: 80-223-0713-0.

Current Issues of Science and Research in the Global World – Kunova & Dolinsky (Eds)
© *2015 Taylor & Francis Group, London, ISBN 978-1-138-02739-8*

Virtual media reality—second and third life

Yvonne Vavrová
Pan-European University, Bratislava, Slovakia

ABSTRACT: Virtual reality is used in many fields: in architecture, medicine, sports, art and mainly in the military. VR has become a normal part of our lives. We can ask ourselves 'What is really "real" and what is really virtual? Should we be afraid of further development of virtual reality in the media? Are there any boundaries to where it is still safe and healthy? Who will be the next Avatar?'

1 INTRODUCTION

We all probably agree on one point: We only live one life. At least here, in this world, in a material body, in the material world. With the exception of some religions that are based on the principle of reincarnation like Buddhism, all religious and philosophical opinion unites in the fact that we have only one life here on earth. So everybody wants to really enjoy this one life.

Everybody has own unique path. Some party their life away, others devote their life to their family, others study or educate themselves throughout their whole life, others devote it to charity, and yet others waste their life. And then there is a large group of mostly young people who have decided that one life is not enough and escape into the second and third lives, which are delivered to us by modern media.

But let's go back a few years. Entitled Second Life, developed by Linden Lab in 2003, is a very popular computer game that allows players on a computer to enter a virtual world and survive their so called second life. Life in a different body, with a different character, in a different flat or house with different family members. You can put away your own physical body somewhere at home, on a hanger, and jump into a young muscular male or curvy female body, which are healthy, full of energy and freshness. Who would not want to have that? You can become someone you have dreamed all your life of being, you can have a job which in reality you did not get perhaps because in the real world you were either not bold enough, or perhaps not qualified to get, or maybe simply somebody snatched the job right in front of you.

In 2008 more than twelve million people in the Czech Republic were registered on Second Life and in the real economy they spent more than 30 million crowns. The program allows you to move to different levels, which depend on the budget, as well as on age of the player. Immediately after the payment of an entry fee a new island resident or visitor is allowed to use a limited space on the island and is separated from the rest of the rest of the Second Life world. A new resident can only communicate with other residents through their avatars, can chat with others, sell each other things, travel or educate himself. Just like in the real world. *(1)*

This is nothing special. After all, playing is fun and also educational. Comenius teaches us that learning through play is the most effective way to learn. But history did not stop at Second Life and computer games designers did not stop innovating and hoping for great profits they carried on inventing more and more games, where players are not limited to just one extra life, but also a third and fourth and so on.

It would be unusual for producers of media products, such as commercial and public broadcasters, film studios, even educational products not to follow this trend.

But more on that later.

We can say that virtual reality can be experienced when reading a detective story or a romantic novel. Our sense of hearing is increased, maybe even our blood pressure, our hands sweat when we put ourselves in the position of an escaping victims, our leg muscles twitch when they shoot us or our character. Virtual reality has been part of our lives for a very long time even long before the invention of the printing press by Gutenberg. For example stories of people talking by the fire place, when they fried mammoth or some other wild animal meant they experienced dramatic moments in their heads and their veins. Our ancestors and forefathers and foremothers did not know that life will be enriched by a virtual reality, but which, unlike the current one, is projected only in their heads ...

The brain has an amazing ability: It can be in the here and now, but can also imagine being in a totally different space. After all, the principles of quantum physics also apply to our brain. The same particle can be at the same time in several places at the same time. Our mind can be physically here, but also elsewhere. Our body can react to stimuli and experience things that directly affect our physical body, but can also react to experiences occurring only in our minds. I can read thoughts that someone is sending thousands of kilometers away. Brilliant scientific ideas emerge at the same moment in different parts of the world. This is quantum physics. While our material body rests on an office chair in front of a monitor, the second quantum particle goes to another virtual body where we experience strong emotions as well as to the material body. Schizophrenia? No... just the miraculous ability of our body and our brain.

Many areas of activity would not survive without virtual reality today. Medicine, architecture or the military all depend on it. It would be difficult for pilots to learn how to fly without simulators, difficult for surgeons to operate and difficult for architects to build bridges. Virtual reality allows them to move in the space without victims, where the hard laws of physics do not apply. This is a huge benefit of virtual reality.

It seems that there is nothing to worry about. Virtual reality helps wherever the fantasy tries to connect with the reality. Is there a boundary beyond which virtual reality becomes dangerous and unhealthy?

'Avatar'. The name was not invented by James Cameron, the creator of a hugely successful sci-fi movie of the same name. The origin of the word Avatar comes from a Hindu concept and means an earthly incarnation of spiritual beings. It is primarily a visual representation of the user in the virtual reality world. It can exist as a three-and two-dimensional object, or just as a character. An Avatar occurs as the interpreted consciousness of a particular person, e.g. even though his/her body may have died, his/her avatar is able to continue to communicate. Online games are infested with avatars and virtual space is overwhelmed by them.

Everywhere is teeming with avatars. And not only in computer games. Differences start to blur—'who am I and who is my Avatar'. While I'm weak and poor, my avatar is powerful and rich. It may dispose of people, who are standing in my way, and who prevent me from becoming infinitely powerful. And while in real life we suffer in bad interpersonal relationships, our media avatars have different rules of behaviour. If you have money, you have a strong avatar. Do you have money? Then you are healthy super macho or a super sexy blonde. I, the virtual avatar, do not need to burden myself with relationships, family interactions, do not need to touch anyone, I do not feel the reality of a slap on my face.

Let's put ourselves in the bodies of handicapped people, people who have never felt what it's like to have legs, how it is run, what it's like to hug someone. Virtual reality of the third millennium has offered just such an opportunity. When stimulating certain nerve endings in the brain, it mobilizes a centre that allows physically handicapped people to literally experience real walking, running and flying. What better outcome could virtual reality offer?

I remember 15 years back in Canada, I visited one of the first 3D virtual cinema in the world. The film itself only lasted a few minutes, but I will never forget my beating heart, the increased heart rate, the dizziness and my sweaty hands as I was leaving the movie theatre. The film was screened in the cinema, which in addition to 3D film offered also moving and shaking chairs, smell and dust and sound-pressure waves that shook the theatre. I survived that 10 minutes of my life. I flew into the air, bumped into things, experienced free fall from the 50th floor, drowned in water, dissolved in hot lava volcano. If anyone wanted to knock me over, it would have been no problem during those ten minutes. I realized that our brain is a diabolical instrument. Or maybe angelic...?

Neuromarketing as a new field of activity began to spread not only in order to better explain the functioning of our brain, but also to help sell various products and services. Advertising companies will happily buy different neuromarketing studies, so they can tune the advertisement to make customers intuitively and unconsciously long for their products using tomography to monitor brain activity while watching the ads.

"The investigation of magnetic resonance imaging is used, which measures changes in certain brain regions. Furthermore, electroencephalography (EEG) is used, which measures the specific activity spectra of the brain or changes in physiological state (galvanic skin response, heart rate, respiratory rate).

Subliminal effect affects human without realizing it..."

However, it is questionable whether it is ethical and whether the average customer is able to resist it.

Virtual reality media will never be erased from our lives. Whether we want it or not we will have to learn to live with it. We adults have that much easier in some way. Our virtual avatars lie somewhere hidden deep in our minds as our parents read bedtime stories. Avatars are presented to today's generation and no one asks them whether they invited or not. They are in fact aliens against which many people are defenceles.

However, modern-day avatars can be very personalized and they can be very appealing. Players have endless possibilities to give their avatar a very individual form, including various options for action. Physical characteristics can be customized in all sorts of way—change their height and weight, skin colour, posture, colour and shape of hair, clothes and accessories. The Avatar Marketplace system—something closely resembling our shopping centre Aupark or Eurovea, is somewhere (virtual) players can buy for their avatars clothing and accessories as well as products from famous international brands.

Recently, a movie by American director Spike Jonze titled Her appeared in our cinemas. There's a story of very near future, where for a fee, we will be able to buy an OS partner—an operating system in the form of a small ear gadget. The OS partner can have a very individualized form, can be male or female, can be old or young. We can purchase any character traits. This operating system accompanies us from morning to night, to work, in bed, and can virtually even have sex with us. Its great advantage is that we can disable it. It was interesting to investigate, how lonely people became dependent on these OS partners, to extent that they replaced real people. I would like the idea of delivering such OS partners for abandoned and old people, who would be able to project their deceased spouses or partners, or even children they had lost. Whether that's too big a step for science fiction I could not say. But life shows us that it probably is not. Today we communicate in the virtual environment of digital media without physical contact with the world and with hundreds, or thousands, or millions of people at simultaneously all around the world. We can even experience different emotional states, when connecting with object avatars. Even services such as online sex is today real in a computer environment. What was stunning in this movie was that it showed us that avatars can quickly move closer to our soul, can be empathetic, sensitive, fun, joyful and understanding. Sometimes even more than an ordinary person. It showed also that the loss of your avatar could result in experiencing strong emotions of sadness just like a loss of a person or the nearest family member.

One very interesting study examined to what extent do looks in the virtual world, changes people's behaviour in the real world and the behaviour of other people. Perhaps not surprisingly, in experiments using various Avatars, it was found that people who used very attractive, more beautiful avatars are acting in the on-line games more confidently and far more willing to establish contacts with others. Much more interesting is the finding that similar effects can also spread into the real world in a limited way, and that the people in the on-line games using attractive avatars may show a higher degree of confidence after a certain period of time even in the real world. It is difficult to determine precisely the effects of such findings but it appears that the process of creating an identity and "identify" in the "Second Life" are part of a sense of personal identity and the future of the human identity can be partly based on the "dialogue" between the different ways of how people relate to it and diverse forms, which constitute the different version of "me".

Many studies carried out by the renowned German neuro-psychiatrist Manfred Spitzer showed that while the middle and older generation is immune to virtual reality generated by all the different media, the younger generation is much more delicate and vulnerable. In his book Digital Dementia, he states that computers and in a way the current digital media makes them insensitive, even more primitive than humans than we might expect. People tend to talk about digital intelligence, but excessive use of digital media leads on the contrary to mental and physical dementia and social isolation.

Paying attention to the environment of virtual worlds raises concerns some of which is related to neglect in real life. There are warning stories about the "war widows" that can leave her husband because of the game World of Warcraft, or a married couple whose baby died without any care, while the couple was devoted to playing online games with virtual avatars. The most important feature of virtual worlds is that they are bringing some hope—they are highly interactive and more participants are engaged—as opposed to the so-called linear media such as television. People who are active in a virtual world people do not inevitably become isolated and it is not simply a matter of "unreality but can also be a way to conduct and a new level of engagement in world events."

2 CONCLUSIONS

Virtual reality refers mainly to the media. It is only a matter of time before virtual reality penetrates into other scientific fields. Given that the media has become a mass culture today we have to recognize the pluses and minuses carefully and with great understanding. We boarded the train, which can no longer be stopped. However, no one can see around corners.

REFERENCES

Chatfield, T. 2013. *Digitální svět 50 myšlenek, které musíte znát*: Nakladatelství Slovart: 209–210.
Spitzer, M. 2014. *Digitální demence*. Brno: Host, 2014, 341.

Problem of an intellectual and his position in the society

Mária Macková
Faculty of Mass Media, Paneuropean University, Bratislava, Slovakia

ABSTRACT: The study focuses on current problems of a personality in relation to which there arises the problem of an intellectual and his/her position in the society in the past and at present. In the topic's context it deals also with other aspects—nature of the age, values of life, moral qualities of a personality, an objective view of current scientific, artistic, political, economic and spiritual life. It gives a brief analysis of the transformation of intellectual's identity in the past. It is an appeal for the dialogue of professionals from various scientific and cultural fields and for trust in creativity and rational knowledge.

Great spirits have always encountered opposition from mediocre minds. The mediocre mind is incapable of understanding the man who refuses to bow blindly to conventional prejudices and chooses instead to express his opinions courageously and honestly.

Albert Einstein

An integral view of a personality is not possible without respecting the social and historical aspect. Moreover, psychological and philosophical-social conceptions of a personality are closely related and the both views are in dialectic unity. It is due to man's mission and position. Presently, there is a trend to understand and explain man, personality and his/her creative activity in more complex terms. Based on the teachings of materialistic nature of reality (materialistic monism) which holds that matter is the fundamental substance in nature considered to be primary in relation to consciousness or anything spiritual and in the sense of ethical relativism—a principle according to which moral semantic formations (moral notions, concepts etc.) have a relative and conditional nature—the concept of an individual comes to the fore.

Man—a personality is born with inner dispositions into a certain environment. Extraneous influences interfere in development and formation of the personality, either through education or through various social stimuli and relations. On the other hand, these relations influence psychological structure of a person. A deeper knowledge of psychical relations of a person gives a better precondition and basis in order to create an integral view of the personality in the social and historical environment. In brief, if you know person's psyche you will be able to assess more responsibly his/her quality, value, but particularly the motives of his/her behaviour and the action at all. Equally, the issue of position of a personality in the cultural space, which has become conspicuous recently, however, which might have existed for a long time, or have developed since the beginnings of culture, is essentially related to man.

The paradox of the previous century is, on the one hand, an astounding advancement of human knowledge, revolutionary discoveries in science, technology, production or other areas of social life, and on the other hand, global problems of humanity which need to be solved acceleratingly. Also the issues of cultural transformation, education and development of human potential belong among these problems, even if they seem to be less important in comparison with ecological, health, demographic problems etc.

The question why is the position of a contemporary man in the society a relentless concern is answered most appositely using the quote of the French philosopher, theologist, anthropologist Pierre Teilhard de Chardin: "Man's place in nature... we may well wonder why, as science progresses, this question becomes continually more important and fascinating to us.

In the first place, no doubt, it is for the eternal, entirely subjective, and therefore somewhat suspect, reason that the problem touches what is very close to us—our own selves. Even more, however (and here there is no suggestion of anthropomorphic weakness) it is because we are beginning to realize in our minds—and this as a direct function of the most recent advances in our knowledge—that man occupies a key position in the World, a position as the principal axis, a polar position. To have understood the universe it would be sufficient for us to understand man..." (T. de Chardin, 1990, 26.)

Based on similar considerations, but in the first place on facts, you need to take into regard that liberation of the society and cultures from ideologies of the previous decades which tried to control man—man's inner life, have not brought the new meaningful reflection.

At the same time with the personality issue there arises another problem—the problem of an intellectual and his/her position in the society—in the past and at present. Also some related questions arise—about nature of the age, values of life, moral qualities of a personality, an objective view of our scientific, artistic, political, economic and spiritual life.

When translated, the concept intellectual (French l´intellectuel; Latin intellegere) means to understand, comprehend. If you search for the meaning of the word in the contemporary Vocabulary of Slovak Language (Short Dictionary Slovak language) you will learn that an intellectual is man with higher educational level, standing at high spiritual level, usually working mentally, simply a scholar. (htttp://slovniky.juls.savba.sk) The Wikipedia monolingual dictionary reads that "An intellectual is a person for whom ideas, science, art and culture are so important that these determine not only the goals of his or her everyday life, but also the basis of his or her political thinking and action". (http://sk.wikipedia.org/vykladovy slovnik).

From history you learn that people, who are considered to be intellectuals nowadays, have emerged since the 18th century as independent writers without any fixed institutional relations, who liked to give their opinion about public issues.

The word intellectual started to be used at the end of the 19th century as a derogatory name of the critics of Dreyfus affair, and they were blamed by politicians (Georges Clémenceau, 1892) for elitism, insufficient patriotism and loyalty. However, its typology goes back to Greeks, to Plato who defines a philosopher as a thinking person who includes in his knowledge an ideological and social unit (polis) and becomes also a politician.

Let us spare a thought for several opinions of research workers who dealt with the issue of an intellectual. The curious thing is that each of them had a different motive.

Jan Bakoš, professor of history and art, gave a cultural and historical analysis of two seemingly remote topics: the transformation of intellectual's identity and the metamorphosis of the idea of historical monument in his book Intelektuál & Pamiatka (Intellectual & Historical Monument). The knowledgeable scientist, renowned in Slovakia and abroad, appeals in this book for a dialogue of professionals from the various fields of science and culture "to pay a bit more attention to each other than up to now, not to be shy to show interest in the work of another also if it seems that the spheres of their interests are unbridgeably remote. Not to trivialize what connects them: the faith in genuine creativity and rational knowledge. So that the words about the need of inter-, multi- or post-disciplinary research would not remain empty phrases." (Bakoš, 2004, s. 11)

In the historical analysis of an intellectual, completed by the author in the 19th century, you can find many interesting and curious thoughts. We selected one from the part Critique of a scholar from the end of the 18th century. The Treatise on the Health of Men of Letters by a Swiss physician Tissot reads: "Most scholars senselessly waste their time and health. Some of them compile what has been known for a long time. The second ones repeat what has been said a hundred times. The third ones engage in absolutely useless researches. They work with unrevealing documents and dead boring works, and no one realizes what damages they induce and what little worth it has for the audience." (Chartier, 2004, s. 162) It seems that the thought might be relevant also today.

In spite of that it is necessary to emphasize that science is based on research personalities who are respected by the scientific community in their own country, but also in the context of an international scientific communication. They become founders of scientific schools, set the long-term direction in their field and represent the scientific community. "The authority of science in the society has become unquestionable and irreplaceable, and most of world religions have actually co-opted it into their belief systems. But this unusual consolidation, irreplaceability of science for the contemporary society modified also the status of an intellectual: the intellectual became a worker of science, a force in the factory of science, a scientific worker, a small and anonymous wheel in the huge train of wheels of science industry." (Bakoš, 2004, 15.)

In considerations about the phenomenon of Slovak intellectual we need to go back into the past, too. Also Slovakia has had its position in the arena of world science. As soon as in the joint multinational Hungary, when most of the country's industry was concentrated in our territory, we had significant personalities in various fields of science. In spite of that, since the times of the Saints Cyril and Methodius Slovakia has been more a science receiver rather than an initiator. There were several causes to this, but it seems that one of them has been the fact that in our country we have not learned to esteem the performance of wise men, we rather, except for some cases, despised them.

Ivan Hulin says: "The pursuit of broader generalization brings an intellectual in the arms of philosophy where he brings phenomena into relation with the fundamental questions of human existence. The assessment of phenomena from the point of view of fundamental questions of philosophy will necessarily lead to the consideration of moral and value aspects. ... The mission of an intellectual is the interpretation of the world in the name of movement, in the name of development, and change. An intellectual is necessarily in permanent opposition to the existing state of the society, as he is obliged to think through the consequences of the chosen solutions." (Hulin, et al., 2005, 307)

An urgent discourse on the topic of intellectual was started in Slovakia by a philosopher, professor Vincent Šabík in the book Intellectual as a homo politicus, with consciousness and an appeal—not only to remind, but insistently draw attention—to restoration of an adequate position of intellectuals in the society. At many places in the book he emphasizes that it is not correct if most of the population of today's Slovak society know intellectual personalities less than the actors of mass entertainment, e.g. actors, singers, presenters, comics etc.

In his considerations about intellectual he also goes back into history of European and world culture to find answers to the questions—who is actually an intellectual, how we should understand him, how he defines himself, what is the difference of an intelligent person. He might not give a definition, however, his ideas, for example about the need of self-reflection of intellectuals or "the loss of historical identity and vision capability" are inspiring in many aspects.

In this context it is appropriate to recall the basic attributes of intellectualism which are the following: "sensibility, perception, ability to react sensitively to impulses and changes, dispositions to formulate clearly, precisely and decidedly hot issues of the age, to apply the knowledge of the philosophy of history, to generalize and re-interpret experiences, to anticipate problems and consequences. An interesting opinion is the exaltation of "socratism in intellectuals' thinking, the ability of critical reflection, questioning the given. Even if intellectuals succumb to Hegel's, slyness of reason' the intellectuality will give them an autonomy of position if it is auto critical also in relation to power, politics, which otherwise defends itself against public critique (authoritative politics also in repressive ways, democratic rather by ignoring)." (Compare: Šabík, 2007, 135) According to the quoted author, today the intellectuals defend themselves against the unifying trends of modern societal systems. Particularly, postmodernists refuse a unified opinion and pluralize intellectual discourse.

It follows from the above that definition of personality has to be based on two aspects of the perception of man. On the one hand, as a member of a certain "mass", and on the other hand, as an individual—an intellectual. Unfortunately, the society (politicians, economists, organizers of show business) is interested mainly in the "mass" people.

However, history convinces us that qualities of creative personalities from the fields of science, art and spiritual life must stand the test of time and in most cases they got the credit only after their death. Therefore the considerations or questions what is the real position of intellectuals in social, cultural and spiritual processes are mostly topical and worth looking for an answer.

That is to say, an intellectual is neither a genius loci nor a myth sized citizen—a scientist or artist, nor an accidental heroes, but a creative, single-minded, active man with distinctive abilities, who is unique and unreproducible, even if with many human imperfections and faults. It is a real human being conscious of his or her lot in life which he or she accepts at the given time and space and follows it meaningfully, at all times in relation to the others, to the society, nation and humanity.

We will finish the considerations on the topic of "intellectual" with the thoughts of a literary scientist, professor Peter Zajac who stated that the concept of an intellectual went through very contradictory developments in the 20th century, "... it adumbrated the role of an intellectual as an incorruptible truth seeker and uttered, even at the expense that he will stand up with his critical attitude against the majority opinion of the nation, class, race, religion. In this sense an intellectual fulfilled the cautionary Cassandra's function or the role of public conscience." (Zajac, 2007, 78)

Finally, a brief look in the past will remind us of the roots of creative personality in the cultural space. However, we are not going to follow in detail the historical transformations of man who lived on craft. A brief overview will be started by skipping to a scholar who became a citizen, thus losing the purely spiritual nature (unlike middle-ages or renaissance man). Also the period of developing capitalism should be mentioned, as the technical progress, democratization of political life and secularization of culture of that time should have become a proof of the winning of modern man believing in the outspoken enlightenment-positivistic rationalism.

The optimism of the positivistic faith in science and knowledge was disturbed for the first time by the pioneer of cognition Immanuel Kant by his noetics according to which man, apart from his consciousness, cannot in fact know any outer fact. Transferring the truth into human mind, he removed the frontiers of matter and spirit. In spite of that, romanticism born out of Kant's noetics and the poetic sight of Jean-Jacques Rousseau turned inward human soul. And after that, at the end of the 19th century, scepticism about the impossibility to know the objective truth broke out at full blast. The most frequent topics of the contemporary artists were the theories of "trendy" philosophers such as "the will to hold power" of Arthur Schopenhauer, the phantasm of "Ubermensch" of Friedrich Nietsze, "intuitivism" of Henrich Bergson according to whom the only place of reality and experience of life is consciousness. An extreme subjectivism with an effort to release man from responsibility for his actions was topped at the turn of the centuries by Sigmund Freud with his psychoanalysis. (Compare: Marek, 1996, 222)

In Slovakia, the concept of an intellectual—was born by Art Déco. According to historians, the Slovak intellectual was recruited mainly from the representatives of the after-revolution literary generation which, together with the incoming generations, not only impressed a modern national nature on Slovakia, but also co-created its overall cultural environment. (Jakšičová, 2012, 38)

The ways of Slovak intellectual have always been tortuous. The Slovak intellectual of the 19th century lived in the middle of two problems. One of them was the problem of the existence of national community and the other its organization. In this relation the fundamental question, on which the intellectual had to focus, was the issue of political, economic and legal unity. In two key historical moments—1849 and the period of 1861–1875—he did not achieve his goal: the full independence of Slovak nation and experienced a big disappointment, the way out of which was to bring politics into culture, particularly into literature, but later also the vision of a fusion of the Slovak nation with another nation. "very stifling was the contradiction between intellectual—and individual—necessity of a critical attitude towards national movements and the requirements of national unity, which tried to eliminate, or at least to restrict, criticality into their own ranks. This led to the situation that the critics of own national imperfections became national, 'sinners' and 'outsiders' (today we might call them 'dissidents')." (Zajac, 2007, 78)

REFERENCES

Bakoš, J. 2004. *Intelektuál & Pamiatka.* Bratislava: Kalligram.

Balcar, K. 1983. *Úvod do psychologie osobnosti.* Praha: SPN.

Berďajev, N.A. 1997. *O otroctví a slobodě člověka.* Praha: OIKOYMENÉ.

Chartier, R. 1998. Die Gelehrte, In: Vovelle, Michel (Hg.), *Der Mensch der Aufklärung,* Frankfurt a. M.

Hulín, I., et al. 2005. *Úvod do vedeckého bádania. Dialógy, úvahy a zamyslenia.* 2. Bratislava: SAP.

Jakšičová, V. 2012. *Kultúra v dejinách—dejiny v kultúre. Moderná slovenská kultúra a slovenský intelektuál v siločiarach prvej polovice 20. storočia.* Bratislava: Veda.

Jung, C., G. 1998. *Duše moderního člověka.* Praha: Academie.

Krátky slovník slovenského jazyka, 4. doplnené a upravené vydanie, Bratislava: Veda, 2003.

Marek, J. 1996. *Česká moderná kultúra.* Praha: Mladá fronta.

Pervin, L., A. 1993. *Personality. Theory and research.* New York.

Rogers, C., R. 1999. O osobnej moci. Bratislava: Iris.

Ruisel, I. 2008. *Osobnosť a poznávanie.* Bratislava: Ikar.

Slovenská otázka dnes. Bratislava: Kalligram, 2007.

Šabík, V. 2007. *Intelektuál ako homo politicus.* Bratislava: PROCOM, s. r. o.

Teilhard de Chardin, P. 1990. *Vesmír a lidstvo.* Praha: Vyšehrad.

*Faculty of informatics, section new trends
in information technology applications*

Current Issues of Science and Research in the Global World – Kunova & Dolinsky (Eds)
© 2015 Taylor & Francis Group, London, ISBN 978-1-138-02739-8

Crowd simulation using potential fields and markers

J. Běhal Dadová
Faculty of Mathematics, Physics and Informatics, Comenius University, Bratislava, Slovakia

ABSTRACT: We are mostly concerned with path finding algorithms and decision making in the crowd. We present crowd simulation approach for virtual museum. Visitors in the museum are simulated as agents that are influenced by attractors and markers. Markers are used for collision avoidance and for attractors we use potential fields. Potential fields are widely used in robotics and real time games. They are important for crowd simulations and agent based systems. We present path planning algorithm that is based on decision making extracted from the observation of the real crowds. Our algorithm uses traditional potential fields approach with additional directional vectors. Agents decision process is based on personality and behavior of a real crowd. Agents need to choose next step based on their goal, intentions, potential field force in the neighborhood and collision avoidance strategy. Our approach combines well known techniques, but variables are added to create a realistic result.

1 INTRODUCTION

Adding crowds to the computer games and motion picture is more and more possible thanks to the available computational power. Crowds are interesting not only visually, but they create more believable, life-like scenes. In interactive computer games and virtual museum tours crowds influence user personal emotions of the environments. Moreover analyzing human behavior in selected environments can be used for safety purposes, to identify best evacuation routes and exits to avoid traffic jams during evacuation. Even positions for exhibits in museum can be properly set if the virtual crowd in museum shows some problems with them.

2 OBSERVATION OF A REAL CROWD

Studying the behavior of a real crowd is important for the believable simulation result. According to Bates (1994), believability of an agent is based on the right timing of an action and is made possible by the expressing emotions at the right moment. Therefore not only type of behavioral patterns is important but also their right timing and the analysis of the behavior is important. There are behavioral patterns in each location type that are not common in other types. We discuss in this section these specific patterns for the visitors in a small museum and for the people visiting city squares.

Firstly, we tracked positions of the visitors in the city during their visit using GPS (Global Positioning System) tracking system. For this tracking to be possible, tracked visitor used GPS tracking application for the mobile devices. Mostly, these applications are used to track sport workouts (running, walking, bicycling). We have tried two most popular sport tracking applications for the Android OS, as this operating system is nowadays more common for the users of smartphones and other mobile devices. These sport tracking applications are SportsTracker and Endomondo. Both have versions that can be downloaded from the Google Play Market for free. After tests, where we tested accuracy of the tracked path and available exporting options, we decided to use Endomondo, because Sports Tracker seems to have worse accuracy in some cases.

We tracked visitors in two different, historically interesting locations in Bratislava: Hviezdoslavovo námestie a Hlavné námestie. For the export of the tracked path we decided to use exported path to online version of Google maps, because their use is available, accurate and free for our purposes.

We gained knowledge about the general path of the visitors, because in the urban environment, GPS chip and location itself independent on the application used, has limited accuracy. In smartphones used for our research, GPS was enhanced with assisted GPS, that has better accuracy, but is still limited. It must meet the U.S. Federal Communications Commission (FCC) requirements for location information to support E911 services. The FCC requires that handset-based systems locate the caller to within 50 m for 67% of calls and to 150 m for 95% of calls (Zandbergen 2009). Which means that assisted GPS if accurate to 150 m for 95% of cases, which is not suitable for our purpose. But we tried the tracking applications and results were still suitable to gain data from those tracked positions. We also evaluated the data by actually observing the tracked visitors and manually discarding data that were not accurate.

Another problem with GPS tracked path as behavior evaluation is that final result was only visitor's path with additional information about the length of the path, duration, speed. We needed also information about length of stops during the path and places where visitor slowed down. These data can be extracted by using GPS positioning, but not using sport tracking applications.

Therefore another option that we used to observe behavior of a visitors was video capture from various locations. It is also possible to use video capture for indoor spaces, which is more suitable for out problem. Because we wanted to observe also behavior of museum visitors, not only city visitors and those are mostly indoor spaces. We captured various videos in different days during the year and week. There are two different outdoor locations—Michalská ulica a Hlavné námestie, for the both videos are captured from 20 m above the ground on the View positions available to the public (Michalská brána and Stará radnica). We also captured videos from indoor locations of the Malokarpatské múzeum Pezinok (The Small Carpathian Museum in Pezinok). This institution is being shaped as a regional museum with a strong accent on viticulture, grape growing and notably wine from the Small Carpathian viticulture region. There are displayed various small artifacts, text informations and audio guides available. However, visitors during our capturing process did not have audio guides available.

2.1 Case study—Malokarpatské múzeum Pezinok (museum)

This indoor environment consists of 10 small rooms, where are displayed small artifacts and texts. We captured only videos inside of the rooms, without GPS location. We calculated average speed of the visitors during their visit in the rooms, but their speed varied according to the density of artifacts, or text information. If there were texts, visitors stopped to read it. Therefore these stops lowered average speed. According to the captured data, visitor stopped for 8 seconds to read the texts (0:42–0:50). Average speed for all visitors, that we observed was 0.33 km/h and slightly increased to 35 km/h as the fatigue level was higher and their visit longer.

2.2 Case study—Michalská ulica (street)

Michalská ulica is a narrow street with various souvenir shops, cafés, five consequent street that are exits and entrances for the recorded area. Only two are visible on the captured videos, because three are under the building from which the video was taken (Michalská brána). We have only captured videos from this location, without GPS tracking. Mostly people were only trespassing, creating straight line patterns. 10% of people were entering or exiting shops on the street. Moreover this video was captured during the lunch time and we assume that most of the people were going to/from the lunch, therefore they were not visitors.

2.3 Case study—Hviezdoslavovo námestie (square)

Hviezdoslavovo námestie is a 350 m long square with various interesting statues along the walk, fountains and surrounding buildings. Moreover on one side of the square there are souvenir stands where visitors can buy traditional Slovak gifts. We used only GPS tracked path, because this square too long to fit in camera field of view and there are lots of trees, that occlude view from the higher positions. According to the captured data, visitor walked through whole square and stopped to buy gifts or just to observe them in one of the stands. Also visible pattern is approaching interesting objects closer, as visitor did in front of the theatre and near Hans Christian Andersen statue. On the other hand, there is no visible change in the path near Pavol Országh Hviezdoslav statue. Either the reason is that visitor was not interested in the object, or observing the statue is good enough on the straight path through the square. The statue is positioned in the central part of the square and visitor need to come round it to avoid collision. Average speed of all the tracked visitors was 3.6 km/h, depending how low they bought gifts or took pictures.

2.4 Case study—Hlavné námestie (square)

Main square in Bratislava is 55 m long and 40 m wide and creates almost a shape of a rectangle. As there are only few low trees, it is well suitable for capturing the video. It is even possible to capture video from the top view of the Bratislava City Museum in the Old Town Hall. Moreover we also tracked visitors in this location. There are three different chosen paths, that show various entrance and exit combinations. Also behavior of the tourist were different, because there are three interesting objects and at least one interesting building and each tourist was interesting in another objects according to the tracked path. It is also possible, that some places were too crowded, therefore tourist chose to avoid the object instead of waiting ans approaching it. That is another problem of the GPS path tracking, that it provides information only about one person and all crowd information such as collision avoidance, crowd density are lost.

From the captured videos, there are also behavioral patterns that tourists showed. We captured videos before the lunch time, therefore there are more tourists than was in the Michalská ulica case study. Crowd in this square is not dense, because larger groups that visit this location, usually stay during the commentary in the free space, not near the interesting objects. The most popular object is Maximiliánova fontána (fountain), where most of the tourists took pictures and Napoleonov vojak (statue), because it is on the bench, where visitors can sit and take picture. Space around these objects is often crowded and interested visitor must wait.

Almost half of the people that were on the square during our capturing time were not visitors, just regular citizens, who only crossed the square with the straightest line possible. Their average speed is close to the normal walking speed 5 km/h, but differs depending on the collision avoidance and disability of the person. Another example is person with faster average speed, because of the bicycle used. This location is in the center of the town, where no cars are allowed, therefore bicycle is fast option for the movement. Average occurrence of such a person with bicycle is one in 10 minutes. Their speed is higher, therefore we did not considered them in our pedestrian average speed calculation. Another example are tourists that stopped their walk to observe or take picture of an interesting object. Their average speed is significantly lower to the normal walking speed, which is expected. The last example is the person who takes picture of someone else near the interesting object. The person walks back to fit object and another person in the camera, therefore this patterns is meaningful only when smaller groups, or at least two people are available. This behavior is very common for the statues, and fountains, which are smaller comparing to the buildings and can fit in the camera. Also this behavior is mostly visible for the outdoor scenes, because of the light conditions and most of the museums does not have taking pictures available.

3 BEHAVIORAL PATTERNS

In the previous sections we discussed observed data and behavioral patterns that are visible in the examples. In this section we would like to discuss how we can incorporate these patterns

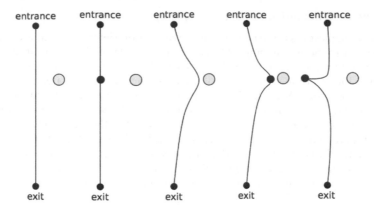

Figure 1. Behavior of the people on the square. (a) citizen, that only crosses the square (b) visitors, that stops to take picture, or observe object (c) visitor, that comes closer to the object for observation, or wants to be on the picture (d) visitor, that comes closed to the object and stops to read information (e) visitor, that walks back to take picture of an object.

in the simulation process. There are two main categories that will enhance simulation of visitors behavior. These include different zones of influence and different path patterns for each behavior. Zones of influence define circle around the object where people show predefined behavior, different for each zone. We observed four different zones: read zone (when some text is on the interesting object visitors stops to read it), observe zone (visitor stops or slow down to observe the object), take picture zone (visitor stops to take picture of an object, or walk back to this zone to take a picture), be on the picture zone (visitor come closer to fit him/her with the object in the camera held by another person). These zones are different for the small indoor museum and larger outdoor environment. Also they are different for the smaller and larger objects. Because visitor need to be further away to fit object in the camera.

Another data categories are different paths that are created by the visitors. Not all of them can be incorporated in our current solution. These paths are shown on Figure 1. Currently our solution does not allow to create smaller groups, not even two people as a separate group, therefore last path, where person walks back to take picture of another person is not possible with our current solution. However, the same path is created when person discovers that object would not fit camera field of view and need to walk back. Patterns that we discovered for the tourists in the city and visitors in the museum should significantly enhance our simulation when incorporated in it.

4 POTENTIAL FIELDS AS ATTRACTORS

Potential fields are widely used in robotics and real time games. They are also important for crowd simulations and agent based system. In this work we present a path planning algorithm for virtual museum visitors. Our algorithm uses traditional potential fields approach with additional directional vectors. Moreover these fields are not used as repulsion from the obstacles, but as attraction force to the interesting objects (attractors), such as displayed art objects. Potential fields only influence motion of agents, but they are not main force, that direct their movement, which is controlled by agent's decision process. This enhancement allows agents or museum visitors to calculate their preferred path according to their personal preference.

4.1 *Potential fields calculation*

Our environment is divided in the regular rectangular cell grid for easier movement calculation, where cells are twice as large as radius of an agent representation to avoid unnatural movement. There are different types of cells (see Fig. 2), entrance (blue), exit (red), interesting art object (yellow, attractor with different attraction value) and walls (green). We calculate potential field for each interesting object separately using the following equation:

$$y = I - 2^{\frac{x.c}{I}},$$

where x is euclidean distance from the interesting object, I is attraction at the interesting object, y is resulting attraction at current cell and c is constant global value that defines attractor influence. Example of this function is shown on the graph (Fig. 2).

Moreover, mentioned equation calculates only attraction values for each cell that is in range of the attractor. We add also directional vector for these cells. With this addition, agent can be navigated to the attractor from each cell that are included in the influence range. Attraction values and corresponding attractor vectors for one interesting point are shown on Figure 3. When two or more attractors influence the same cell, higher influence value and also all vectors to the attractors are saved, see Figure 2.

4.2 Path calculation

The final path for an agent is calculated by decision process in each step. Agent starts at entrance point (blue cell) and path is goal directed. Currently goal for an agent is exit (red cell). In the advancement we set goals as subset of all interesting point as simulation of a museum guide, which navigates users to the most interesting points, similarly to the behavior in exhibition hall.

According to the psychological study of museum visitors (Robinson et al. 1928), people tend to approach interesting art objects during their stay in the museum until fatigue level is higher

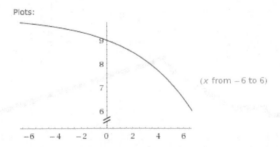

Figure 2.   Example of the function we use for potential field calculation. Attraction value is here set to 10. As x parameter we use Euclidean distance from the attractor.

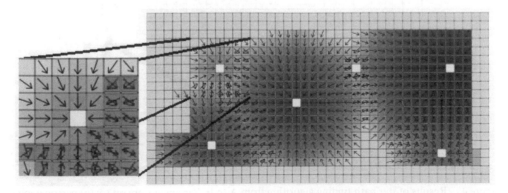

Figure 3.   Results of potential fields calculation with direction to the attractors. Potential fields for all attractors. More than one direction vector could be associated to a single cell (red cells).

than the specified value. This value depends on personal preference. When fatigue level is high enough, visitor aims directly to the exit. We use this information for calculating final decision.

Decision process uses global and local values, most important local value is distracted direction, which is direction to the attractor according to the occupied cell. This direction is normalized vector which is calculated according to the potential field calculation (above). When there are given more directions from the cell, then sum of vectors is set as distracted direction. If agent already visits some attractor, potential field to this attractor will be avoided from the future calculations. Another local value is personal preference of distraction, which depends on each individual and corresponds to the fatigue. Final decision, if the agent moves in each step towards the goal or towards interesting point we evaluate as in the following equation:

$$b = 1 - \frac{|X_t X_d|}{|X_d X_0|},$$
$$d = \frac{I_t}{I_p},$$
$$I = (1 - \alpha) + b + c + d,$$

where α is small (defined by the designer of the system, currently it is 30° angle between direction to the goal, and distracted direction in current position, X_t current position, X_d position of a final goal, X_0 position in $t = 0$, I_t attraction at current cell, I_p attraction at interesting point, c personal preference of distraction, I is final attraction. When final attraction is lower than

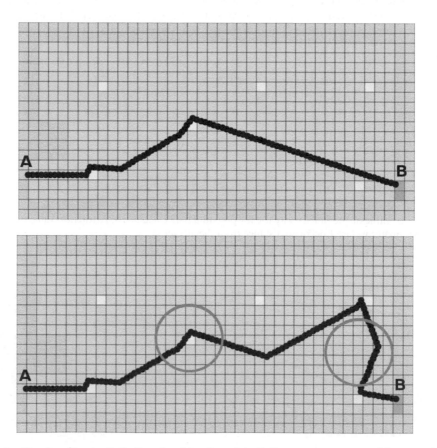

Figure 4.   Results of our path finding algorithm from A to B. Parameters are set to: (a) personal interest = 0.5, global threshold = 1.0 (b) personal interest = 0.2, global threshold = 2.0, red circles show abrupt changes during the movement.

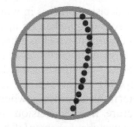

Figure 5. Results of a smoothed path. (a) pre-calculated path (b) smoothed path (black dots as representation of an agent have smaller radius to make changes in the path more visible) (c) both paths together.

global threshold, final direction is towards goal, when it is higher, final direction is towards interesting point. Different global thresholds simulate attraction of the whole museum to the visitors. Examples of final paths, with different parameters are shown on Figure 4. Red circles on the figure indicate problems with the abrupt changes in the direction during the movement. Abrupt change is when the angle between previous step and next step is larger than 30°.

We solve this problem by pre-calculating next few steps (in our case 4) and smoothing path created with these steps. For the smoothing we used standard cubic Bèzier curve (Farin 2002), where control points are pre-calculated positions in time. When the curve is calculated we find smoothed positions on the curve with the respect to the distance between positions in the pre-calculated curve. Positions of smoothed points and pre-calculated positions will be in the convex hull of pre-calculated points, from the definition of the Bèzier curve, because pre-calculated points are the same as control points, see Figure 5.

Moreover regular movement without change in speed is realistic only in unconstrained environment. Our environment is different, we study exhibition hall and museum with interesting objects (art objects, exhibits and others). From the study of museum visitors results, that visitors tend to slow down and look at the interesting objects, when they approach them. We also incorporated this behavioral pattern in our solution and shorten steps when agent approached interesting objects Shortening the steps makes agent movement slower. Also agents approach interesting object from the position of best view. More about best views is discussed by Varhanikova (2013).

5 RESULTS

For the implementation purpose we used architecture that is proposed by Pelechano et al. (2007). We have used various environments with various exits and entrance points to show our solution. However, all of them are simplification of room in a small museum, or exhibition hall (rectangular shape, entrance points, exit points, with walls and interesting objects) or a simplification of a city square (rectangular shape, entrance points, exit points, surrounding buildings and interesting objects). We used the same rectangular shape and interesting points to have better comparison of the behavioral patterns created. Color coding of the cells in the examples is following: green cells are walls, blue cells are entrances, red cells are exits, yellow cells are interesting objects and dots are in most cases agents. Black dots are either current positions of more agents, when whole crowd is displayed, or position of one agent in various time steps. Blue dots are positions of one selected agent in various time steps when the crowd is displayed.

After study of the human navigation behavior, we modified our method to include behavioral patterns according to this study. It is possible to set each interesting object as large object, such as fountain, or building, which are more common in outdoor environment. It is also possible to set each interesting object as a text object, where it is possible to read some information, which is more common in indoor environment, such as museums. These settings allow us to create scalable behavior, suitable for the museums and also for the city squares.

Moreover, it was necessary to create decision process in which visitor choose behavioral pattern that is suitable for him and surrounding, when he approaches interesting objects. The suitable behavioral pattern is agents intention, which is a high-level goal that an agent is committed to achieve and agent has exactly one intention at any given time (Thalmann & Musse, 2013). This allows us to use Finite State Machines for the decision process.

With this decision process, it depends on the visitors actual preference if picture of the object will be taken. It is not a deterministic process, where, if visitors is near and has the camera then he/she will take the picture. In each decision step, where question is asked, visitors decides with personal preference in current moment if he/she is willing to take the picture. Personal preference for each question from the graph is set at the beginning of the simulation for each agent and it is a percentage value that will tell how much is agent willing to do something. Example of the path is that visitor took picture of the first two objects and in both cases needed to step back (objects were large), but near the third object only slowed down to observe the object. Also from the crowd study, when visitor took picture of the fountain on the square, it did not mean, that he took also pictures of the statues.

Moreover, we also introduced a leaflet to our solution. A leaflet serves as a guide for suitable next goal decision, when visitor is willing to observe the environment, not only one selected object. Without the leaflet, final path was unrealistic, because visitor walked two times around the same point and whole path was too long, violating the Least Effort Law. The guide includes ordered list of goals (order is predefined depending on the entrance point and is same for each visitor, as the guide in real museum).

The guide contains only preferred list of the objects, that is offered to each agent, but agent decides to choose, which interesting objects are suitable for him from the guide. This selected objects are then set as goals for the agent in the same order as they are in the guide, but not all of them are selected. If agents visits some of the goals along the path to the previous goal, then this object as a goal is deleted from the list. It serves as a memory for an agent. If some object was already visited, agent will remember it and would not approach the object again. The path calculated using a guide is shown on the Figure 6. In this example, visitors chose from the guide following goals: no. 1, no. 2, no. 3, no. 4 and no. 6. Agent visited also object no. 5, because potential field around this object is strong enough and agent's personal preference is high enough to be guided also by the potential field.

The solution process described above allows one visitor to follow behavior that is well suitable for the museum, exhibition hall and city square and is according to the study of a real human behavior. On the other hand, there is usually more than one visitor in such environments. Therefore also some collision avoidance technique is needed. We decided to

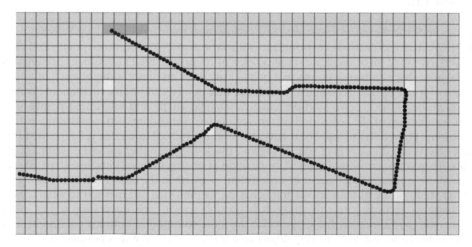

Figure 6. Example of the final path with potential fields and a guide with following selected goals: no. 1, no. 2, no. 3, no. 4, no. 6.

choose method, that uses markers (de Lima Bicho 2009). It is scalable enough for use also with special points, as are in our case interesting objects, where agents can get stuck easier, because they are attracted to these points. Therefore we distributed markers more sparsely around these interesting objects, so agents will be repelled from interesting objects by the markers. Because markers influence final movement only partially, they avoid collisions with other agents, but are still attracted to the interesting points.

When we added markers to the solution, final path of an agent was not correctly calculated because it violated Least Effort Law. The main problem was, that markers are suitable solution for avoiding collision, but if there are no possible collisions between agents (for example agent is alone in the environment) then path should be straight. Therefore we first check if there are other agents in neighborhood cells, if there are no agents, and collision is not possible (step is smaller that the cell size) then markers do not influence agent at all. Then the final path with the same goals is straight.

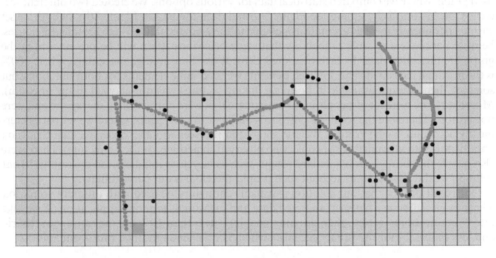

Figure 7. Example of the final path influenced by collision avoidance. Black dots are representations of agents, blue dots are positions of one agent during the simulation.

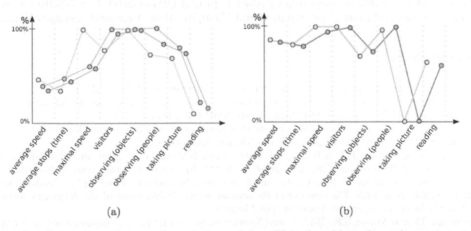

(a)   (b)

Figure 8. Comparison of our results and observed data. Observing objects shows the percentage of objects that were observed by some people/agents, and observed people shows the percentage of people comparing to all analyzed people/agents. (a) yellow—observed data from the historical city square, blue—observed data, where only observing visitors are taken into account, green—data from our application (b) yellow—observed data from the museum, green—data from our application.

Afterward we let more than one agent enter our environment using entrance with the frequency similar to the data from the behavior study. Also average speed and behavioral patterns that are suitable for the environment are set according to this study. The agents in our simulation enter empty environment, then before closing, we closed entrances and let agents walk through the environment until they leave using exits. This behavior is according to the observation from the museums, but entering and exiting can be also set to infinite where agents enter and exit until user stops the simulation which is behavior that is observed in city square (they usually do not have opening hours). The final result is shown in the Figure 7, where agents are represented as black dots and path of a selected agent during simulation is shown with blue dots.

6 CONCLUSION

For comparison of results from our method and data from the observation of a real humans we used graph, where we compared statistical data for various options. We created two different scenarios, the historical city square, see Figure 8(a) and the museum, see Figure 8(b). In both we set various parameters to be as accurate as possible to the analyzed data. There are still differences, because parameters set only probability that some behavior pattern will be present and in the graph are measured data. In each measured element we show average percentage, where 100% is the maximal measured value for each element. We used same elements for both situations, but some of them are not available in both (e.g. taking pictures was not allowed in the museum). Moreover there were lot of people in the city square that were only crossing the square and were not visitors. Therefore the resulting polyline is different to our results, where we set parameters mostly according to the visitors. We also showed in this scenario also observed data that took only visitors into account (those people, who observed something, or took a picture). There were also people that brought non-standard results, such as bicyclists and distorted the results. In our solution we did not take into account such people, therefore is our curve different.

Our result proofs, that our solution is scalable enough to be used with different scenarios. The average difference between analyzed data and measured from our application (if we take into account more visitors, than crossing people) is less than 10% for the museum scenario and less than 15% for the square scenario, which is small enough to be proof our solution.

ACKNOWLEDGEMENT

This project was partially supported by ASFEU project OPVaV-2011/4.2/07-SORO: Comeniana, methods and tools of presentation and 3D digitization of cultural heritage objects.

REFERENCES

Bates, J. 1994. The role of emotion in believable agents. *Commun. ACM*, 37(7):122–125.
de Lima Bicho, A. 2009. *Da modelagem de plantas a dinamica de multidoes: um modelo de animacao comportamental bio-inspirado*. PhD thesis, UNICAMP, Campinas.
Farin, G.E. 2002. *Curves and surfaces for CAGD: a practical guide*. The Morgan Kaufmann series in computer graphics and geometric modeling. Morgan Kaufmann.
Pelechano, N. & Allbeck, J.M. & Badler, N.I. 2007. Controlling individual agents in high-density crowd simulation. *In Proceedings of the 2007 ACM SIGGRAPH/Eurographics symposium on Computer animation*, SCA '07, pages 99–108, Aire-la-Ville, Switzerland, Switzerland. Eurographics Association.
Robinson, S.E. et al. 1928. *The behavior of the museum visitor*. Publications of the American Association of Museums. American Association of Museums.
Thalmann, D. and Musse, S.R. 2013. *Crowd Simulation*, Second Edition. Computer science. Springer, 2nd edition.
Varhaníková, I. 2013. *Optimalization of scene geometry and animation curves for virtual environments*. PhD thesis, Faculty of mathematics, physics and informatics, Comenius University, Bratislava, Slovakia.
Zandbergen, P.A. 2009. *Accuracy of iphone locations: A comparison of assisted gps, wifi and cellular positioning*. Transactions in GIS, 13:5–25.

Current Issues of Science and Research in the Global World – Kunova & Dolinsky (Eds)
© *2015 Taylor & Francis Group, London, ISBN 978-1-138-02739-8*

Improving the search algorithm for new binary Error-Control Codes

Ján Doboš
Faculty of Informatics, Institute of Applied Informatics, Pan-European University, Bratislava, Slovakia

Martin Rakús
Faculty of Electrical Engineering and Information Technology, Department of Transmission System, Institute of Telecommunications, Bratislava, Slovakia

ABSTRACT: In digital transmission and storage systems, Error-Control Codes (ECC) provide a security mechanism allowing the detection and correction of a certain amount of errors, which can be caused by the presence of noise in these systems. ECC are denoted as $[n, k, d_{min}]$, where "n" denotes the codeword length, "k" number of information symbols and "d_{min}" is the minimum code distance. The minimum code distance determines the error correction/detection capabilities of a given code. Theoretically computed bounds of minimum code distance for many ECC have been found but to find generator matrices for many codes reaching theoretical bounds still remains a research problem. The presented paper describes an improved version of a proposed algorithm used for searching for the generator matrices of binary ECC. The described algorithm is based on an alteration of generator matrices of already found ECC. The presented algorithm reduces the total amount of tests required by the basic algorithm in order to reduce the execution time of the whole search process. The final part of this paper compares the execution time and memory requirements of the basic and improved version of the search algorithm.

1 INTRODUCTION

In state of the art digital storage and transmission systems, the presence of noise can cause formation of errors. In order to detect and correct a certain amount of errors, a security mechanism is provided by Error-Control Codes (*ECC*).

The detection and correction capabilities of these codes are determined by the minimum distance of the used *ECC* code. Computed lower and upper bounds of minimum code distances d_{min} of linear $[n, k, d_{min}]$-codes over $GF(q)$ maintained by Markus Grassl can be found online (Grassl, 2008). Most recent improvements on the upper bound of binary codes with the proof or reference to the proof are visualized in a table form online (University of Bayreuth, 2014).

A large number of linear codes with prescribed minimum code distance have been constructed based on a method first described in series of papers by Braun, Kohnert and Wassermann (Braun et al., 2005), (Braun et al., 2005). Generator matrices have been constructed by solving the corresponding Diophantine linear systems by a special program developed by Zwanger (Zwanger, 2008). A database of found generator matrices of codes over $GF(2)$ and the dimension k lower than 18 can be found online (University of Bayreuth, 2014). However, generator matrices of large number of codes reaching the theoretically computed upper bound still have not been found and represent a complex research problem.

In this paper a different approach to the search for generator matrices of codes reaching optimal bounds is analyzed. The described approach is based on modification of already found generator matrices. The paper is organized as follows. In part 2 the basic preliminaries from linear algebra and coding theory required to understand the proposed approach are presented. Part 3 focuses on the description of the selected modification algorithm and techniques leading to its improvement (i.e. reduction of total execution time). Part 4 provides a comparison of the basic and improved version of the algorithm. In the last part some concluding remarks are made.

2 THEORETICAL PRELIMINARIES

For understanding the method described in this article, a basic understanding of linear algebra and coding theory is required. This part of the document contains the minimum knowledge base taken from (Hall, 2010) and (Rosen, 2006) with emphasis on binary linear block codes. A more complex view reader can find in (Kuttler, 2012), (Cherney et al., 2013) or (Roth, 2006).

2.1 Mathematical preliminaries

It is important to introduce the concept of an echelon matrix and linear independency, which will be used in the following parts of this document. Echelon matrices can appear in two forms: the row echelon form (abbreviated ref) and the reduced row echelon matrix (abbreviated rref). A matrix is in row echelon form when the following conditions are satisfied:

- the leading entry (pivot; the first non-zero element of each row) is 1,
- each leading entry is in a column to the right of the leading entry in the previous row,
- rows with all elements equal to 0 are below the rows containing a non-zero element.

The reduced row echelon form must satisfy an additional condition:
- the leading entry of each row is the only non-zero element in its column

It is possible, by using a series of elementary row operations, to transform any matrix into its rref form. The transformation algorithm for binary matrices can be described by 2 steps:

1. Pivoting the matrix
   - find the pivot in the first column of the matrix,
   - by interchanging the row, move the pivot row to the first row,
   - XOR (Exclusive OR) each of the lower rows where the pivot column contains a 1 with the pivot row, so every element in the pivot column of the lower rows is equal to 0.
2. Getting the matrix in row echelon form, repeating the pivoting
   - repeat the algorithm from step 1 for a sub-matrix where the previous pivot rows and columns are ignored,
   - continue the processing until there are no more pivots to be processed.

If a vector equals to the sum of scalar multiplies of other vectors, it is said to be a linear combination of these vectors. A set of vectors is linearly independent, if no vector of the set is a scalar multiple of another member of the set or the linear combination of the vectors in this set. We can think of a k-by-n matrix as a set of k (n) row (column) vectors having n (k) elements. The maximum number of linearly independent row (column) vectors of the matrix is denoted as the rank of this matrix. Based on the characteristics of row echelon forms, it is quite simple to find the rank of any matrix: the number of linearly independent vectors in a matrix is equal to the total number of non-zero rows in its row echelon form.

2.2 Coding theory preliminaries

Let $B = \{b_0, b_1\} = \{0, 1\}$ be an alphabet. Then values b_0, b_1 are called the symbols of the alphabet B. A block code C of length $n \in N$ constructed over B is a non-empty subset of B^n. A vector $c \in C$ is called a codeword. The number of vectors in C, denoted by $|C|$, is called the size of the code. A code of length n and size k is called an $[n, k]$-code.

In order to talk about the number of errors, their detection and correction it is important to introduce the Hamming distance and the minimum distance of a code. Let $x = x_1, \ldots, x_n$

$\in B^n$ and $y = y_1, \ldots, y_n \in B^n$ be codewords. Then Hamming distance of codewords x and y is defined by equation (1).

$$d(x,y) = \sum_{i=1}^{n} (x_i \neq y_i) \tag{1}$$

Also a simpler definition of Hamming distance can be provided: Hamming distance is the number of elements in which the analyzed codewords x and y differ from each other.

Let $C \subset B^n$ be a code. The minimum distance of the code, denoted d_{min}, is defined by equation (2).

$$d_{min} = d(C) = \min\{d(x, y)|x, y \in C : x \neq y)\} \tag{2}$$

An $[n, k]$-code of distance d_{min} is called an $[n, k, d_{min}]$-code. The values n, k, d_{min} are called the parameters of the code. The aim of coding theory is the construction of codes with a short n, large k and d_{min}. The code's capability of detecting and correcting errors is directly connected with the value of d_{min} (Hall, 2010).

Linear codes are a special subset of block codes, whose codewords are strongly related to each other. This relation allows a shorter and more compact definition of these codes. A set $C \subset B^n$, which is a linear subspace of B^n, is called a linear code.

Now we need to show the connection between the weight and the distance of the code. To define the weight of a code, first, we need to define the Hamming weight. Let $x = x_1, \ldots, x_n \in B^n$. The Hamming weight of x, denoted $hwt(x)$ is the number of its non-zero coordinates (3).

$$hwt(x) = \sum_{i=1}^{n} (x_i \neq 0) = d(x,0) \tag{3}$$

Let C be a code (not necessarily linear). The weight of C, denoted $hwt(C)$, is defined by (4).

$$hwt(C) = \min\{hwt(x)|x \in C : x \neq 0)\} \tag{4}$$

Furthermore, if C is a linear code, its minimum distance d_{min} equals to its weight:

$$hwt(C) = d(C) = d_{min} \tag{5}$$

To finish the coding-theory preliminaries, we need to define the generator matrix and its parity sub-matrix. A generator matrix G_C of a linear code C is a matrix whose rows form the basis of code C. If C is a linear $[n, k]$-code then G is a k-by-n matrix. The rows of a generator matrix are linearly independent (code basis). A generator matrix of a linear block code C is said to be in standard (systematic) form if it is in form $G_C = (I_k \ \vdots \ P)$, where I_k denotes a k-by-k identity matrix and a k-by-$(n-k)$ matrix P denotes the parity sub-matrix.

Two $[n, k]$-codes are equivalent if one can be derived from the other by a permutation of its coordinate entries. Different forms of equivalencies are described in (Hall, 2010).

Based on preliminaries the following advantages of linear codes can be highlighted:
- a linear code can be described using its basis (can be found via Gaussian elimination of matrix comprised of the codewords as rows)
- the distance of the code is equal to its weight.

3 SEARCH ALGORITHM

3.1 *General code modification algorithms, formulation of key search properties*

If in some sense, an existing code has good detection and correction capabilities, it could be possible to find new codes (reaching an upper bound for d_{min}) based on the good one.

According to (Hall, 2010), an [n, k]-code can be modified by six basic techniques: augmentation (fix n, increase k); expurgation (fix n, decrease k); extension (fix k, increase n); puncture (fix k, decrease n); lengthening (increase n and k) and shortening (decrease n and k).

By combining these basic techniques, more complex techniques can be created. Based on the selected modification method, not only the length and/or the dimension of the code change, but the minimum code distance may change as well.

Based on the parameters of the source code the dimension of the input generator matrix and its parity sub-matrix can be specified (property P1).

Based on the theoretical preliminaries from part 2, d_{min} of a linear block code is determined by the lowest Hamming weight of its non-zero codewords. Therefore, the input parameter d_{min} directly determines the Hamming weight of the vectors forming the rows of the code's generator matrix. The Hamming weight of these rows and their linear combinations must be equal or greater d_{min}. Moreover, the Hamming weight of each row of I_k is equal to 1. This implies, that the Hamming weight of vectors forming the rows of the parity sub-matrix P must be equal or greater than $d_{min} - 1$ (property P2).

The Hamming weight of a result of linear combination of l rows must be equal or greater than $(d_{min} - 1)$ (property P3).

Moreover, the test of Hamming weight, bound by property P3 satisfies the following condition: from a certain value of l (the number of linearly combined rows), the test bound is equal or lower than zero. This statement can be denoted mathematically as (6).

$$1 \geq d_{min} \Rightarrow (d_{min} - 1) \leq 0 \tag{6}$$

Hamming weight of each row of the parity sub-matrix is equal or greater than 0. This implies, that l-sized combinations, whose Hamming weight test bound of property P3 would be lower or equal to zero, are not needed to be tested (property P4).

3.2 Basic search algorithm

The original version of the search algorithm that authors seek to improve is based on the combination of code puncture (gen. matrix column removal) and code augmentation (gen. matrix row addition). The goal of this modification is to achieve the following code transformation described by (7).

$$C:[n,k,d_{min}] \rightarrow C^*:[n^*,k^*,d_{min}^*]:n^* = (n-1);k^* = (k+1);d_{min}^* \leq d_{min} \tag{7}$$

Input parameters provided at the beginning of the search process are the length of codewords n, dimension k, minimum code distance d_{min} and the generator matrix G_C of the selected source code. The value of requested d_{min}^* is specified during input as well. The result of the search is the generator matrix G_C^* (it does not have to be in the standard systematic form) of the requested code C^* with parameters satisfying the transformation.

In the following parts of this document, this simple method of code modification will be denoted as the UPRIGHT search algorithm/method.

The UPRIGHT search algorithm consists of the following steps:

1. take a generator matrix G_C of an existing [n, k, d_{min}]-code C (generator matrix may be in non-systematic form)
2. remove a selected column (this operation leads to $n^* = (n-1)$)
3. add a new row with Hamming weight from closed interval $\langle d_{min}^*, n^* \rangle$
4. test if the minimum code distance of the code with generator matrix G_C modified by steps 2 and 3 is equal to chosen d_{min}^*
5. if the test in step 4 fails iterate through steps 3 and 4 until all possible vectors are generated and tested. If step 4 is successful, save the found generator matrix for future purposes of analysis and deployment.

3.3 *Improving the UPRIGHT search algorithm*

The first approach to improve the described method is based on the properties described at part 3.1 and migration of the whole search process to the parity sub-matrix with dimensions satisfying property P1. The migration to the parity sub-matrix lowers the total number of combinations required to test P3. The total number of tests (worst case scenario when all of the combinations are analyzed) can be calculated based on equation provided in Table 2 at part 4 of this document. Exact values obtained by experimental run of the software are also located in the evaluation part.

The first improvement of the UPRIGHT algorithm consists of the following steps:

1. take a generator matrix G_C of an existing $[n, k, d_{min}]$-code C (generator matrix may be in non-systematic form)
2. remove a selected column (this operation leads to $n^* = (n-1)$)
3. add a new row with Hamming weight from closed interval $\langle d^*_{min}, n^* \rangle$
4. transform the modified generator matrix into standard systematic form
5. test if the minimum code distance of the code with generator matrix G_C modified by steps 2 to 4 is equal to chosen d^*_{min}
6. if the test in step 5 fails iterate through steps 3 to 5 until all possible vectors are generated and tested. If the test is successful, save the modified generator matrix from step 4 for future purposes of analysis and deployment.

However, focusing the search on parity sub-matrix extends the basic algorithm by step 4 required to transform the modified generator matrix into its standard systematic form. This requires a transformation of the matrix into its rref form. The pseudo-code and example software implementations of the transformation using different programming languages can be found online (Rosetta Code, n.d.). If the matrix in rref form is still in a non-systematic form, further execution of column permutation is required (creation of an equivalent code). The operation of matrix transformation into systematic form introduces additional processing power requirements. Moreover, the transformation must be repeated for each generated vector we add.

The final improvement minimizes the number of required rref transformation processes to 1 and further lowers the total number of generated vectors. The main idea of the reduction is based on linear algebra and the understanding of this improvement is supported by shape analysis of matrices provided in Table 1.

Consider an input generator matrix G_C of an existing $[n, k, d_{min}]$ – code. Modify it by removing a selected column and transform the matrix to its standard systematic form G^m_C. The modified k-by-$(n-1)$ generator matrix consists of a k-by-k identity matrix I_m and a k-by-$(n-1-k)$ parity sub-matrix P_m as illustrated in Table 1 (A).

Next, the step of row addition is executed. If the target generator matrix G^{tr}_C has to be in standard systematic form (see Table 1 (B)), the shape of the row that needs to be generated and added to G^m_C can be defined as (8). The beginning part of the vector (row) consists of k zero elements followed by a single element 1. The rest of the vector consisting of elements X is then generated.

$$row\_vector = (0 \quad 0 \quad \cdots \quad 1 \quad X \quad X \quad X \quad X) \tag{8}$$

Table 1. (A) Shape of the modified generator matrix.

Modified generator matrix

$$G^m_{\ C} = (I^m \mid P^m) = \begin{pmatrix} 1 & 0 & \dots & 0 & P^m_{1,1} & \dots & P^m_{1,n-1-k} \\ 0 & 1 & \dots & 0 & P^m_{2,1} & \dots & P^m_{2,n-1-k} \\ \vdots & \vdots & \ddots & \vdots & \vdots & \ddots & \vdots \\ 0 & 0 & \dots & 1 & P^m_{k,1} & \dots & P^m_{k,n-1-k} \end{pmatrix}$$

Table 1. (B) Shape of the target generator matrix.

Target generator matrix

$$G^{tr}_C = (I^{tr} \mid P^{tr}) = \begin{pmatrix} 1 & 0 & \dots & 0 & 0 & P^{tr}_{1,1} & \dots & P^{tr}_{1,n^*-k^*} \\ 0 & 1 & \dots & 0 & 0 & P^{tr}_{2,1} & \dots & P^{tr}_{2,n^*-k^*} \\ \vdots & \vdots & \ddots & \vdots & \vdots & \vdots & \ddots & \vdots \\ 0 & 0 & \dots & 1 & 0 & P^{tr}_{k^*-1,1} & \dots & P^{tr}_{k^*-1,n^*-k^*} \\ 0 & 0 & \dots & 0 & 1 & P^{tr}_{k^*,1} & \dots & P^{tr}_{k^*,n^*-k^*} \end{pmatrix}$$

Table 2. Total number of tests—equations (worst case scenario).

Algorithm	Basic	Final
# of vectors	$TV = \sum_{i=d^*_{min}}^{n-1} \binom{n-1}{i}$	$TV = \sum_{i=d^*_{min}-1}^{n^*-k^*} \binom{n^*-k^*}{i}$
# of combinations	$TC = \sum_{i=1}^{k} \binom{k}{i}$	$TC = \sum_{i=1}^{d^*_{min}-2} \binom{k}{i}$
# of tests	$TT = TC + TV * TC$	$TT = TC + TV * TC$

Regarding the required shape of the row vector, only vectors (elements X) with length equal to $(n^* - k^*)$ with Hamming weight from closed interval $\langle d^*_{min} - 1, n^* - k^* \rangle$ are needed to be generated. The reduction of the total number of generated vectors gained by this shaping is analyzed in part 4 of this document.

In order to achieve the systematic form G^{tr}_C, Gaussian elimination is utilized on the parity sub-matrix P_m. The process of Gaussian elimination depends on the values stored in elements $P^m_{i,1} : i \in \langle 1, k \rangle$. Since we are dealing with binary codes, elements can contain only values 0 and 1. If the value in $P^m_{i,1}$ is equal to zero, the rest of the i-th row will not be affected by Gaussian elimination. Taking properties P1 to P4 into account, in order to test the minimum distance of the code with transformed matrix G^{tr}_C we need to test property P3 on the combinations of its k^* elements with size from the closed interval $\langle 2, d^*_{min} - 1 \rangle$. The total number of generated combinations can be expressed by equations provided by Table 2 in part 4 of this paper.

Moreover, based on the values of $P^m_{i,1} : i \in \langle 1, k \rangle$, the whole spectrum of the generated combinations can be divided into two groups. This division supports the creation of an effective two phase algorithm used for testing of property P3. The classification of each generated combination is based on XOR operation of the values $P^m_{i,1}$ of i-th rows corresponding the analyzed combination. If the result is equal to zero, the combination must satisfy the following conditions:

– the Hamming weight of the result of XOR operation of the all rows of parity sub-matrix P^m_i (excluding element $P^m_{i,1}$) corresponding the combination must be equal or greater than $(d^*_{min} - j)$, where j denotes the size of the combination (in the following text these combinations are referred as native)

– the Hamming weight of the result of XOR operation of the all rows of parity sub-matrix P^m_i (excluding element $P^m_{i,1}$) corresponding the combination subsequently XOR-ed with the added row must be equal or greater than $[d^*_{min} - (j+1)]$, where j denotes the size of the combination (in the following text these combinations are referred as standard).

The conditions for the second group (XOR result of the corresponding values of $P^m_{i,1}$ equals to 1) can be formulated analogically:

– the Hamming weight of the result of XOR operation of the all rows of parity sub-matrix P^m_i (excluding element $P^m_{i,1}$) corresponding the combination must be equal or greater than $[d^*_{min} - (j+1)]$, where j denotes the size of the combination (native combinations)

- the Hamming weight of the result of XOR operation of the all rows of parity sub-matrix P_i^m (excluding element $P_{i,1}^m$) corresponding the combination subsequently XOR-ed with the added row must be equal or greater than $(d_{min}^* - j)$, where j denotes the size of the combination (standard combinations).

If any of the added rows satisfies all properties being tested, it is the correct one. In order to get the generator matrix G_C^{tr} of the code we are searching for, we need to XOR the rows whose $P_{i,1}^m$ is equal to 1 with the property satisfying row (to finish the Gaussian elimination process).

To further verify the correctness of the results, the weight spectrum of the found code is calculated using the MATLAB script the authors have developed. The script utilizes GPU to speed up the calculations.

4 EVALUATION OF THE ALGORITHMS

This part of the document compares the basic (B) and final improved (F) version of the UPRIGHT search algorithm. The comparison focuses on 3 main indicators: the total number of executed tests, memory requirements and total execution time.

The evaluation of the algorithms required specific changes of the software implementation, which led to the creation of means for search scaling and partitioning by prompting the user to specify the minimum and maximum Hamming weight of the words to be generated and added. This useful feature allows not only partial distribution of the search process to a larger amount of workers/computers, but also to focus on a certain group of generated vectors (for example: based on the statistical analysis of the code, the target Hamming weight of generated words with the highest probability to produce a correct result can be selected). Collection of exact amount of executed tests required an implementation of a test counter.

The analysis and comparison of the total number of executed tests can be based on the worst case scenario calculated by equation provided in Table 2 and on values collected by experiments. The experimental results are aggregated in Table 3.

It is important to note, that values TC and TV from Table 2 are considered as the Worst Case Scenario (WCS) values, when all combinations (see property P2) are used during the test of d_{min}^* for each added row. However, if the test fails for example on the combination of 3 elements, the rest of the combinations are skipped (4 to k^*-sized) and the whole test process starts over for the following added row. The largest number of tests is executed for the group with Hamming weight equal to the half of the length of the generated vectors.

As we can see (Table 3), the original UPRIGHT search algorithm is so time consuming, that in order to get some experimental data, we needed to reduce the Hamming weight interval of the generated rows. Even when this restriction was utilized, the total number of the tests was still bigger than the total number of the tests required by the improved version of the algorithm operating on all possible Hamming weights.

In order to speed up the search process, certain elements and partial results (for example the result of row XOR operation described in part 3.3) are stored in memory during the run of the algorithm. Measured memory requirements are aggregated in Table 4.

In case of memory requirements gathered and aggregated in Table 4, the advantage of the improved version over the basic one of the UPRIGHT search algorithm was confirmed.

Table 3. Total number of tests—experimental results.

Transformation	[37,16,10] to [36,17,9]		[47,15,16] to [46,16,15]	
Analyzed *hwt*	30–36	8–19	41–46	14–30
Algorithm	B	F	B	F
# vector	2 391 494	135 666	5 894 556	134 666
# tests (EXP)	3.07e+009	7.96e+006	2.05e+008	5.08e+009

Table 4. Memory requirements—comparison of realized algorithms.

Transformation	[37,16,10] to [36,17,9]		[47,15,16] to [46,16,15]	
Analysed *hwt*	30–36	8–19	41–46	14–30
Algorithm	B	F	B	F
Memory req.	327 kB	196 kB	196 kB	131 kB

Table 5. Execution time—comparison of realized algorithms.

Parameters	[37,16,10] to [36,17,9]		[47,15,16] to [46,16,15]	
Analysed *hwt*	30–36	8–19	41–46	14–30
Algorithm	B	F	B	F
Exec time	503 s	1 s	90 s	746 s

The last, but also the most important indicator is the runtime of the algorithm. Time is the key parameter, since we want to obtain results in a "real-time" interval (days, weeks or a couple of months). Results, even a correct generator matrix of a code, obtained after a larger period of time, for example in a matter of years, can already be "obsolete". The total dominance of the improved version over the basic one was confirmed by focusing on the execution time (see Table 5). In most of the cases, the total execution time of the basic version was so high, we had to restrict the Hamming weight interval of the generated words.

Moreover, energy is also consumed during the runtime of the algorithm. In ideal case we want to use the energy at our disposal to obtain relevant results. Based on the theory and experimental data, the comparison indicators shows strong dependency from the input search parameters. The most important values are the difference $(n^* - k^*)$ and the minimum code distance d_{min}^*. To minimize the execution time and the total number of tests, value $(n^* - k^*)$ should be minimized and value d_{min}^* should be maximized. This knowledge provides an important mechanism, allowing us to focus on codes, where the search process would provide results in a "reasonable" time interval.

The most limiting factor of the experimental comparison are the execution time of the basic version and the public availability of source codes of other state of the art code-modification search algorithms. The time required to execute the full spectrum of the required test is measured in the interval of weeks and months. This led to the described reduction of the test spectrum, which may "obscure" the algorithm comparison. The lack of publicly available source codes of state of the art search algorithm based on similar generator matrix transformation also limits the inter-algorithm evaluation of our improved method. Evaluation against algorithms based on different approaches would provide data in a slightly different concept. More complex comparison scenario and experimental data collection will be evaluated in our future work.

5 CONCLUSION

The search of new error-control codes is an interesting and complex research area providing a wide spectrum of different approaches. In this paper, an improvement of a search method based on modification of a generator matrix of an existing code by combination of code puncture and code augmentation was addressed.

The modification and wider specification of the search target led to the collection of theoretical properties formulation enabling the speed-up of the selected search algorithm. The process of its improvement and implementation required a creation of a memory efficient method to store binary vector and matrix data. A generator of vectors of a given length and Hamming weight, generator of *l*-sized combinations of *m*-elements and a function for transformation of generator matrix into its standard systematic form has been created. Due to complexity of the software implementation, the pseudo and source codes were omitted. In case of interest, the codes can be obtained by contacting the authors.

The basic and the improved version of the UPRIGHT search method have been analyzed and compared based on theoretical calculations and experiments. The improved version succeeded to minimize all the three monitored parameters (number of tests, memory and time requirements). The experiment also led to a formulation of an efficient mechanism for input and target code selection from online available tables (Grassl, 2008) or (University of Bayreuth, 2014).

Search methods based on modification of an existing code are the most suitable alternative from the aspect of time and energy requirements. However, the success of these methods is directly determined by luck or the quality of the statistical analysis used to select the input generator matrix. Another great problem presents the availability of generator matrices of already found codes.

Different approaches to the search based on utilization of massive parallelism on GPUs, further improvement of the software implementation of the algorithms or alternative algorithms based on a step-by-step generator matrix construction provide further fields for future research.

ACKNOWLEDGEMENT

This work was supported by Visegrad Fund and National Scientific Council of Taiwan under IVF–NSC, Taiwan Joint Research Projects Program no. 21280013 "The Smoke in the Chimney—An Intelligent Sensor—based TeleCare Solution for Homes".

Also, we would like to thank to professor Ing. Peter Farkaš, DrSc. whose successes in code search inspired our research.

REFERENCES

Braun, M., Kohnert, A. & Wassermann, A., 2005. Construction of (sometimes) optimal codes. *Proceedings of ALCOMA conference.*

Braun, M., Kohnert, A. & Wassermann, A., 2005. Optimal linear codes from matrix groups. *IEEE Transactions on Information Theory.*

Cherney, D., Denton, T. & Waldron, A., 2013. *Linear algebra.*

Grassl, M., 2008. *Code Tables: Bounds on the parameters of various types of codes.* [Online] Available at: HYPERLINK "http://codetables.de" http://codetables.de.

Hall, J.I., 2010. *Notes on coding theory.* [Online] Available at: HYPERLINK "http://www.mth.msu.edu/~jhall/classes/codenotes/coding-notes.html" http://www.mth.msu.edu/~jhall/classes/codenotes/coding-notes.html.

Kuttler, K., 2012. *Elementary linear algebra.*

Rosen, K.H., 2006. *Discrete mathematics and its applications.*

Rosetta Code, n.d. *Reduced row echelon form.* [Online] Available at: HYPERLINK "http://rosettacode.org/wiki/Reduced_row_echelon_form" http://rosettacode.org/wiki/Reduced_row_echelon_form.

Roth, R., 2006. *Introduction to coding theory.*

University of Bayreuth, 2014. *Best linear codes.* [Online] Available at: HYPERLINK "http://www.algorithm.uni-bayreuth.de/en/research/Coding_Theory/Linear_Codes_BKW/index.html" http://www.algorithm.uni-bayreuth.de/en/research/Coding_Theory/Linear_Codes_BKW/index.html.

University of Bayreuth, 2014. *Bounds on the minimum distance of a linear code over GF(2).* [Online] Available at: HYPERLINK "http://www.algorithm.uni-bayreuth.de/en/research/Coding_Theory/Bounds/q2.html" http://www.algorithm.uni-bayreuth.de/en/research/Coding_Theory/Bounds/q2.html.

Zwanger, J., 2008. A heuristic algorithm for the construction of good linear codes. *IEEE Transactions on Information Theory.*

Current Issues of Science and Research in the Global World – Kunova & Dolinsky (Eds)
© *2015 Taylor & Francis Group, London, ISBN 978-1-138-02739-8*

Snore detection with android OS mobile terminals

Tomáš Čechvala
Institute of Telecommunications, Slovak University of Technology in Bratislava, Bratislava, Slovakia

Peter Farkaš
Faculty of Informatics, Institute of Applied Informatics, Pan-European University,
Bratislava, Slovakia
Faculty of Electrical Engineering and Information Technology, Institute of Telecommunications,
Slovak University of Technology in Bratislava, Bratislava, Slovakia

Eugen Ružický
Faculty of Informatics, Institute of Applied Informatics, Pan-European University,
Bratislava, Slovakia

Milan Šimek
Brno University of Technology, Brno, Czech Republic

Attila Vidács & Lórant Vajda
Budapest University of Technology and Economics, Budapest, Hungary

ABSTRACT: Recently Principal Component Analysis (PCA) and Linear Discriminant Analysis (LDA) were studied and incorporated into implementation of snore detection system in mobile phone environment. In this manuscript modified methods for automatic snore detection, inspired by existing scientific articles, are presented. The main goal of these modifications is to increase the robustness of snore/non-snore signal classification. First the weak spots of each solution are analyzed. For both PCA and LDA analysis, features were computed from output of Short Time Fourier Transform (STFT)—a square sum of amplitudes in given equally sized sub-band (15×500 Hz) of the frequency range 0–7500 Hz at sampling frequency equal to 16 kHz. PCA and LDA based detection methods were simulated and compared. 15-dimensional features are calculated and projected into two-dimensional classification sub-space. The results of the comparison showed that due to different behavior of PCA and LDA, utilization of different classification methods is required. Following the knowledge obtained from the analysis one implementation is proposed and described.

1 INTRODUCTION

Recently sleep monitoring and snoring detection, important tools of medical diagnostics, are becoming parts of different e-Health systems. Tissues of the human body are relaxed during sleep. Relaxation may cause constrictions along the upper airway, and breathing triggers mechanical oscillations of the tissues such as soft palate or tongue around the constriction. Snoring is the result of the oscillatory motion of these tissues.

In the last 15 years, the snoring problem has entered the realm of clinical medicine. It is a prevalent symptom, and about 50% of the adult population snore frequently (Lugaresi et al 1980, Norton and Dunn 1985). It has been reported as a risk factor for the development of diseases such as ischemic brain infraction, systemic hyper arterial hypertension, coronary artery disease and sleep disturbance (Wilson et al 1999). Several studies have also shown the relationship between snoring and Obstructive Sleep Apnoea Syndrome (OSAS), which is usually associated with loud and heavy snoring (Lucas et al 1988, Wilkin 1985).

It is a common clinical practice to examine patient's sleep characteristics via whole night polysomnography records, which requires the individual to spend a full-night in a somnolab facility (Azarbarzin & Moussavi, 2010). However, some patients feel discomfort and are looking for alternative solutions. Therefore, the popularity of snoring detection systems in mobile device environment is significantly rising.

The paper is organized as follows. In Chapter 2 a brief introduction and explanation of state of the art classification method is presented. Chapter 3 covers he mathematical background of PCA and LDA methods. Classification algorithms are described in Chapter 4. Implementation details are discussed in Chapter 5. In the last part some concluding remarks are made based on our experiment results and possible improvements and future research are suggested.

2   STATE OF THE ART

Known snore detection algorithms are based on dimensionality reduction. Our work was inspired by (Azarbarzin & Moussavi, 2010), where 15 dimensional feature space was reduced into two dimensional. To find this 2D orthogonal feature subspace, utilization of PCA or LDA analysis is required. There is a significant difference between these two procedures.

Imagine a cloud of data points (dots in space). PCA returns an orthogonal set of vectors in a direction of the highest variance of the points. These vectors are called principal components. The first principal component is located in such a direction in which data are spread the most. The second component is orthogonal to the found first principal component and oriented in the direction of highest possible variance. Other orthogonal components can be found in the same way. Principal components are computed out of snoring sound.

However, LDA analysis is much different. It also returns a set of orthogonal vectors but not in the direction with the highest variance. Let us think of more than one data cloud. Each of them represents specific type of data. LDA returns a set of vectors in such a direction in which the separability of these data classes is maximized.

3   PCA AND LDA—THE MATHEMATICAL BACKGROUND

For effective determination of inter- and intra-patient sound intensity variation by PCA, the energy of each 500 Hz sub-band is normalized by the total energy of the episode. For the k-th episode consisting of N_k sub-frames containing 1600 samples, the i-th element ξ_i^k of its feature vector ξ^k can be computed by Equation (1).

$$\xi_i^k = \frac{\sum_{j=1}^{N_k} \sum_{f=500(i-1)}^{500i} |y(j,f)|^2}{\sum_{j=1}^{N_k} \sum_{f=0}^{7500} |y(j,f)|^2} \qquad i = 1, 2, ..., 15 \qquad (1)$$

where $y(j,f)$ is the STFT (using the Hanning window) of the j-th frame of the episode.

The dimensionality of snoring sound feature vectors is then studied via Principal component Analysis. To find the principal components, first, the covariance matrix C of all snoring sound feature vectors ξ^k in the training database needs to be computed based on Equation (2).

$$C = \frac{1}{K} \sum_k (\xi^k - \bar{\xi})(\xi^k - \bar{\xi})^T \qquad (2)$$

where $\bar{\xi}$ is the mean of snoring feature vectors obtained from the training data set and K is the total number of snoring feature vectors.

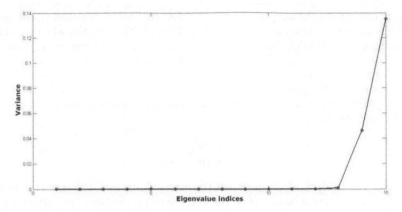

Figure 1. Spectrogram of a sample recording.

The eigenvectors corresponding to the largest eigenvalues of the covariance matrix are the basis vectors of the subspace. These eigenvectors span the new classification space.

According to Figure 1, it can be seen, that two eigenvalues have much higher values than the others. This indicates, that projection into two dimensional space is accurate. Equation (3) describes feature vector projection onto 2D subspace required to compute the new features.

$$\hat{\xi}^k = \begin{bmatrix} x_k \\ y_k \end{bmatrix} = W^T \xi^k \tag{3}$$

where the columns of W are the two eigenvectors corresponding to the largest two eigenvalues of the covariance matrix (Azarbarzin & Moussavi, 2010).

Unlike PCA method, the LDA method tries to find the subspace that best discriminates different data classes. The within-class scatter matrix represents variations in appearance of the same individual while the between-class scatter matrix represents variations in appearance due to difference in identity. The between-class scatter matrix S_b and the within-class scatter matrix S_w are defined as Equations (4) and (5).

$$S_w = \sum_{j=1}^{C} \sum_{i=1}^{N_j} \left(\Gamma_i^j - \mu_j \right)\left(\Gamma_i^j - \mu_j \right)^T \tag{4}$$

$$S_b = \sum_{j=1}^{C} (\mu_j - \mu)(\mu_j - \mu)^T \tag{5}$$

where Γ_i^j is the i-th sample of class j, μ_j is the mean of class j, C is the number of classes, N_j is the number of samples in class j and μ represents the mean of all classes.

The goal of LDA is to maximize S_b while minimizing S_w, in other words, to maximize the ratio $det|S_b|/det|S_w|$. It is maximized when the column vectors of the projection matrix (6) are the eigenvectors of $S_w^{-1} S_b$.

$$W_{opt} = [w_1 \ w_2 \ ... \ w_m] \tag{6}$$

where $\{w_i: i = 1, 2, ... m\}$ is the set of generalized eigenvectors of S_b and S_w corresponding to set of decreasing generalized eigenvalues $\{\lambda_i: i = 1, 2, ... m\}$ (Cayusoglu et al., 2007).

4 PCA AND LDA—CLASSIFICATION ALGORITHM

To classify snore/non-snore signals, authors in (Azarbarzin & Moussavi, 2010) utilized an iterated weighted robust regression algorithm. The reason why the algorithm is weighted is to make the regression resistant to outlying points (those too far are assigned with low weight).

However, in case of the chosen parameters (15×500 Hz and PCA analysis), the projections of snore features are forming a shape of a line. This problem occurs when the number of outliers is significant. It might happen when too many non-snore segments come into computation. As a result, the regression line might deviate from the correct position.

To overcome this situation, we have raised the robustness of the regression algorithm by incorporation of the Hough transformation. It allows us to detect a line on a 2D plot. Hough transformation returns an angle and a distance (7) from certain position, in our case from the upper-left corner of image.

$$y = -\frac{\cos\theta}{\sin\theta}x + \frac{r}{\sin\theta} \tag{7}$$

This allowed us to algebraically interpret the detected line and determine the distances of each point of the feature subspace from the line. Based on the values of distance, a decision whether a feature is snore or non-snore can be made.

In case of using LDA, different classification algorithm needs to be utilized. The main reason is that the snore features are not projected into a line as in the case of PCA. Instead, they form a shape of a cluster. While the snore features are concentrated into one cluster, the non-snore are spread elsewhere (see Fig. 3). To compute the projected feature space, we considered snoring as a separate data class. Normal breathing process was considered the second data class. The separation into two classes is the reason why snore features are concentrated into a shape of a cluster.

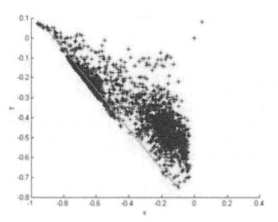

Figure 2.   PCA projection.

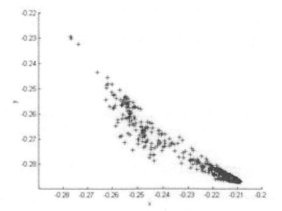

Figure 3.   LDA projection.

Once the feature vectors are formed, the *K-means* clustering algorithm according to (Cayusoglu et al., 2007) was deployed to label each segment as snore, breath, or noise, in which both breath and noise clusters were considered as parts of the non-snore class. K-means clustering algorithm aims at minimizing the objective function, in our case a squared error function defined by Equation (8).

$$J = \sum\nolimits_{i=1}^{C} \sum\nolimits_{j=1}^{N} \| \hat{\xi}^j - C_i \|^2 \tag{8}$$

where C_i is the two-dimensional center of the cluster i, C is the number of clusters (here $C = 3$).

The algorithm, which implementation is described in Chapter 5 of this document, is composed of the following steps:

1. Select and place the initial points in the feature space representing the initial cluster centroids.
2. Assign each 2D feature (or data point) vector to the cluster with closest centroid to that data point.
3. Recalculate the new positions of centroids (by calculation of average value of the data points in each cluster).
4. Repeat step 2 and 3 until the position of the centroids is stabilized.

The most concentrated cluster is classified to be the one with snore features.

5 SOFTWARE IMPLEMENTATION IN MOBILE PHONE ENVIRONMENT

First, the recorded signal is divided into segments with 50 ms shift and 100 ms length (the audio data segments of length 100 ms have a 50% overlap). In case when nothing but noise is captured, the signal contains silent episodes. We expect the segments corresponding to these episodes to have very low energy. Moreover, we expect that there will be at least one quarter of silent episodes. Otherwise the recording would be intentionally too much disturbed. Another property of segments carrying silent signal is that their energy vary very little from each other.

Taking all these expectations into consideration, the energy threshold for non-silent episodes can be computed as follows:

1. Sort the segments by energy (lowest first)
2. Consider sorted energies as statistical data. In this data find the value of lower quartile $Q1$.
3. $Q1$ is trusted to belong to silent segment. However we do not know total number of silent segments. By knowing they do not vary much from each other, we say that every segment with energy lower than $2*Q1$ is considered to be silent one.

In worst case scenario, when a person is not only snoring but breathing loudly, the threshold will probably cut out segments which belong to breathing. This expectation has been confirmed by the experiment we conducted.

After we rule out low energy segments, there might be very short discontinuities in remains of the fragments. For example a low energy sound event lost most of its segments in previous step. We consider an event as non-snoring and rule it out too if only a few segments remain. In our case, a sequence of segments shorter than 300 ms is considered to be non-snore. At this moment, we call the remaining segments active. This reduced the total number of frames for which STFT needs to be computed.

As mentioned earlier snoring is classified after the monitoring phase is finished. Computing all the features at once would be extremely time and memory consuming. The best way is to compute features from active frames and store them in memory for classification. The principals of algorithm are depicted on Figure 4.

First, we record a 40 seconds of signal. The recorded signal is passed for processing and another 40 seconds start to be recorded. Meanwhile we need to find active segments

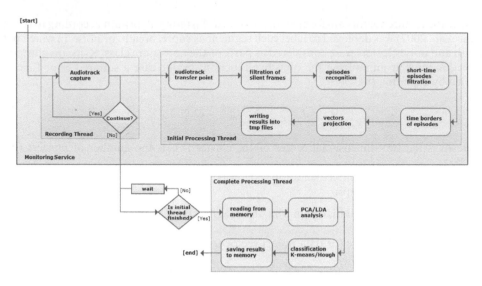

Figure 4. Algorithm designed for android OS.

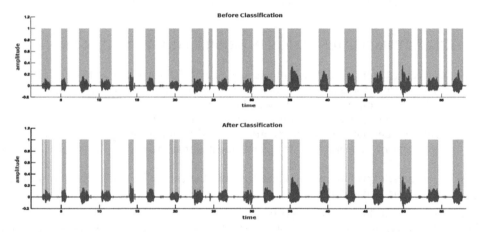

Figure 5. Experiment result: PCA analysis and Hough transform.

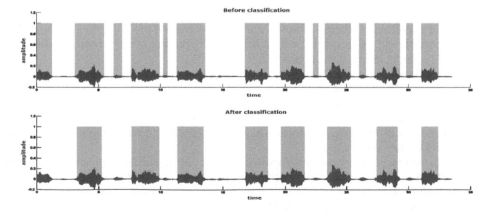

Figure 6. Experiment result: LDA analysis and K-means clustering.

in recorded signal and compute features from them. Android device Servis class is used to handle the monitoring phase. The main advantage of this approach is that our application is multi-thread ready. This means that an instance of thread class is created every time a short term operation is required to be done. Computed features are projected and saved into memory. After the finish of the monitoring phase, we recover 2D features from the memory and classify them. In case the projection is made by PCA analysis, we utilize Hough transformation (see Fig. 5); in case of LDA usage we use K-means algorithm (see Fig. 6).

6 CONCLUSION

Both sleep monitoring and snoring detection present an interesting and complex area providing a wide spectrum of different approaches to the classification process. In this paper, we analyzed mobile environment implementation suitability of existing approached by means of their simulation.

We focused on performance comparison of PCA and LDA analysis. In initial testing LDA analysis turned out to be producing more accurate results in distinguishing between snore and non-snore data. PCA has also performed well, but the probability of decision (classification) error turned out to be higher. Moreover, in PCA accurate classification border is required to be determined more properly.

To simulate PCA analysis we followed article (Azarbarzin & Moussavi, 2010). In case of LDA, we used the same computation of feature vectors as in PCA analysis. When projecting data with PCA, we introduced classification with Hough transform algorithm. When using LDA analysis, the classification was done with k-means clustering.

The same parameters were used as in article (Azarbarzin & Moussavi, 2010)—sampling frequency 16 kHz, scope frequencies up to 7500 Hz, and 15 sub-bands in size of 500 Hz. To make classification more robust, we need to improve input data quality. It means that we tried to minimize number of potential non-snore segments to be active. It was also crucial to decide whether snoring even occurred or not during monitoring. First we tried to detect snoring with Zero-Crossing Rate (ZCR) of signal in each segment. Later we ruled this idea out because ZCR values vary too much for both, snoring and breathing signal. At the moment, another indication is being tested—standard deviation (STD). Consider feature vector. Standard deviation of its element could indicate, whether snoring occurred. The values for breathing are mostly up to 0.15. Deviation of snore features varies from 0.16 to 0.25. This idea is currently tested. Using energies to determine potential snoring episodes would not be efficient. The distance from microphone to user might vary.

If PCA analysis is used, classification should be done using Hough transformation. If LDA analysis is used, classification is done using k-means clustering.

We tried to alter and improve their performance to better fit the environment of mobile devices. Our research resulted in a software implementation of the designed algorithm for a mobile device operating on Android OS platform.

Future research areas are presented by analysis of other modern classification methods or further testing and development of our software product.

ACKNOWLEDGEMENT

This work was supported by, Slovak Research and Development Agency under contracts SK-AT-0020-12 and SK-PT-0014-12, by Scientific Grant Agency of Ministry of Education of Slovak Republic and Slovak Academy of Sciences under contract VEGA 1/0518/13, by EU RTD Framework Programme under ICT COST Action IC 1104 and by Visegrad Fund and National Scientific Council of Taiwan under IVF–NSC, Taiwan Joint Research Projects Program no. 21280013 "The Smoke in the Chimney—An Intelligent Sensor—based TeleCare Solution for Homes".

REFERENCES

Azarbarzin, A. & Moussavi, Z., 2010. Unsupervised classification of respiratory sound signal into snore/no-snore classes. *32nd Annual International Conference of the IEEE EMBS.*

Azarbarzin, A. & Moussavi, Z., 2011. Automatic and unsupervised snore sound extraction. *IEEE Transactions on Biomedical Engineering*, 5(58).

Borade, S.N. & Adgaonkar, R.P., 2011. Comparative analysis of PCA and LDA. *Business, engineering and industrial applications.*

Cayusoglu, M. et al., 2007. An efficient method for snore/nonsnore classification of sleep sounds: Physiol. Meas. 25.

Emoto, T. et al., 2007. Feature extraction for snore sound via neural network processing. *Conference of the IEEE EMBS.*

Jané, R., Solá-Soler, J., Fiz, J.A. & Morera, J., 2000. Automatic detection of snoring signals: validation with simple snorers and OSAS patients. *Engineering in medicine and biology society.*

Mathworks, n.d. *Mathworks documentation center: least-square fitting.* [Online] Available at: "http://www.mathworks.com/help/curvefit/least-squares-fitting.html" http://www.mathworks.com/help/curvefit/least-squares-fitting.html.

Yuzhe, J. & Bhaskar, D.R., 2010. Algorithms for robust linear regression by exploating the connection to sparse signal recovery. *Acoustics speech and signal processing.*

The research activities of the Faculty of Informatics PEU and possibilities of their applications

Eugen Ružický & Peter Farkaš
Pan-European University, Bratislava, Slovakia

ABSTRACT: The Faculty of Informatics of Pan-European University (PEU) participates in five international research and development projects. From these the most important is the TeleCalmPlus system, which develops a field of wireless sensor applications for remote medical care of disabled and elderly people at their homes. We build a comprehensive Usability laboratory for testing the final product of running projects as well as for further development in the future. Some investigations are approved as internal grant by PEU. These are mainly focused on the issues of Business Informatics and Visualization. Along this line our effort is to apply knowledge from all projects at the faculty into one joint research project. We are looking for possibilities of linking research with other faculties of PEU as well.

1   INTRODUCTION

Recently, due to the extensive development of wireless communication systems, it has become a real challenge for the amount of data transferred to use cloud computing. One of the most important issue of European ICT research COST is Action IC1104 "Random Network Coding and Designs". Two main research directions in coding are: construction of optimal network codes and efficient encoding and decoding schemes for a given network code. We participate in this programme IC1104 with prestige European universities.

Another our research work focus on solving problems with Austrian and Portuguese Universities for parallel algorithms (Doboš et al. 2014, Schindler & Páleník 2014). This scheme is supported by Slovak Research and Development Agency.

We are working on international development programme not only with European universities but also from the Far East such as the University in Taiwan. From 2013 we participate in international project TeleCalmPlus granted by Visegrad Fund and National Science Council Taiwan. Our team cooperate with most respected international institutions from Hungary, Czech Republic and Taiwan which gives us new progress opportunities. The aim of this project is to develop a system for remote medical care for seniors and disabled people in their home environment.

Another European project "Quality Education in Pan-European University with International Cooperation" prepares curriculum program Applied Informatics in English using ESF Operational Programme. The main our outcome of the project will be new innovative textbooks, lecture notes and presentations for our study program in collaboration with foreign universities. All results from mentioned projects, we use in process of teaching at our faculty. In addition, there is also research and development that takes place outside of approved projects mainly in Business informatics and Application of Computer Graphics (Lvovich 2013, Řepa 2014, Voříšek 2012, Novotný et al. 2014, and Ružický 2012).

2   RESEARCH OF CODING AND PARALLEL ALGORITHMS

Our activity in the area of "Random Network Coding and Designs" is to solve the problems of data coding for different applications. (Farkaš & Ružický 2013). This problem is important so that detailed studies are addressed from different viewpoints for data transfer encoding and options for error-control codes (Doboš & Rakús 2014). Two bilateral projects are currently in progress in the area of parallel algorithms. The first one is Software and algorithm parallelization for simulation of 4G systems, prepared in cooperation with the University of Aveiro in Portugal. The second bilateral project focuses on solving the problem of parallel algorithms and software for programmable Graphic Processing Units (GPU) and Wireless Sensor Networks with the Vienna University. This project presents a method for structuring existing decoding algorithms for massively parallel decoding of linear block codes. These technologies are widely used in the most recent communication standards such as LTE-A. Our design is quantified by presenting a comparison to the original acceleration achieved using GPU.

3 APPLICATION IN HEALTHCARE

The aim of the TeleCalmPlus system is to develop system for remote medical care for seniors and disabled people in their home environment (Vidács et al. 2014). Using home sensor equipment for measuring blood pressure, temperature, body motoric and other, the characteristic patterns of behaviour will be monitored continuously and the data will be sent to the database in Cloud computing. Information about the person will be evaluated by powerful analysis algorithms that will be run in the cloud. This complex information will be processed to form a simple user-friendly indicator for family members who will be updated about the condition of their elders through application on their smartphones or tablets. Similarly, even more detailed information can be accessed by staff of medical care. Figure 1 shows a system for processing information taken from the TeleCalm project (Vidács et al. 2014).

The main system components are: Intelligent Home Environment (sensors), Server Backend, Cloud Computing, and User smartphones or tablets. Home environment sensors are responsible for the collection of the measurement results from the various data sources. The hardware devices are installed in the home environment and lightweight software has to run

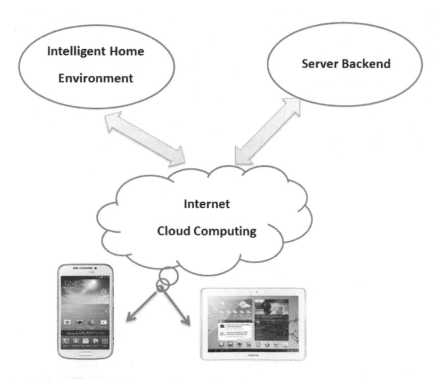

Figure 1.   Draft architecture of proposed system.

on the personal computer of the relative. We use different equipment from global manufacturers such as Japanese company Omron or American company e-Health and Taiwanese special cube of Chiao-Tung University. These devices provide information about the blood pressure, electrical activity of the heart (ECG) and physical activity, which are transmitted to the server backend through Bluetooth or Wi-Fi. In addition, we use other sensors that monitor information about the sleep phases of a person.

Measurement results are periodically updated to a central database where simple predictions are made in order to determine the health status of elderly and disabled people. The measurement and prediction results are made available for query through web services. Smartphones and tablets are used by the relatives of the elderly or disabled people. The simplest patient result is presented on the main screen of the mobile devices as an icon with three possible colours: green—everything is all right, yellow—the relative should pay attention as something is not in conformance with the normal behaviour, and red—the relative should check on the monitored individual. We are looking for various suitable sensors for measuring parameters of human health, as well as programs for the suitable user interface. Goal of different tasks is to find practical sensors and solutions to be used in homes by the monitored persons. The levels of sensors, localizations and data compression methods are considered. Our task in this project will be not only to coordinate mobility and dissemination and thus publishing of the results, but also in cooperation with National Chiao-Tung University (NCTU—the best academic workplace in field of Informatics in Taiwan) to develop compression algorithms for communication in sensor network. We will participate in the demonstration and testing of the final system. We have prepared a new project with leading European and Japanese universities for Horizon 2020 (Farkaš & Ružický 2014).

4 USABILITY LAB

We build up a comprehensive "Usability Lab" for approved research projects as well as for the projects in the future. In this way we apply knowledge from several examined projects at the faculty. Our Usability Lab is divided in two rooms with one-way mirror. First room is for usability testing of programme or device with the all necessary equipment such as video-audio recording devices or eye motion tracker. Second (observer) room allows the designers, developers, and other parties involved in the project to observe and realize that some things which they had found to be intuitively good within developer team, seem more complex during testing.

We would like to test TeleCalmPlus prototype in our Usability lab. Choosing participants for testing involves consideration by doctors, psychologists, sociologists or similar experts. Seniors and disabled people will test all the systems in the laboratory like at home. Thus in advance we will define whether all the technical equipment and software will be suitable for them. Inseparable part of the laboratory utilisation shall be not only the research projects, but also student's projects. They will be able to control their own created system such as software or hardware application to find errors or irregularities, efficiency and overall look and feel of their products. In addition, they will be expected to propose the ways to eliminate all found deficiencies.

5 CONCLUSIONS

We try to use all the results from the above mentioned projects in the teaching process "Applied Informatics" and to continue in the line of its practical orientation of concrete problems from real life experiences, see (Černáková 2014, Farkaš 2014, Fogel 2014, Jurišová 2014, Kultán 2014, Lacko & Ružický 2014, Páleník 2014, Řepa 2014, Palko 2014, Šperka & Lvovich 2014, Šperka 2014, Schindler 2014, Schindler 2014, Voříšek, 2014). The Faculty of Informatics of PEU offers education on legal, economic, informational and psychological aspects of "Applied Informatics" and prepares its students in accordance with the current demands in the area of information technologies. Finally we are looking for possibilities of linking research with other faculties of PEU as well as with the faculties of STU.

ACKNOWLEDGEMENT

Papers and books mentioned in the conclusion have been supported, by ESF Operational Programme Education under contract NFP 26140230012, by Slovak Research and Development Agency under contracts SK-AT-0020-12 and SK-PT-0014-12, by EU RTD Framework Programme under ICT COST Action IC 1104 and by Visegrad Fund and National Scientific Council of Taiwan under IVF–NSC, Program application no. 21280013.

REFERENCES

Černáková, I., 2014. *Business Information Systems.* Nitra: Forpress.

Doboš, J., Páleník, T., Rakús, M. & Ralbovský A. 2014. Intelligent Interferer in LTE and LTE-A. In *Current Issues of Science and Research in the Global World*, Vienna Conference. London: CRC press, Taylor & Francis group.

Doboš, J. & Rakús, M. 2014. Improving the Search Algorithm for New Binary Error-Control Codes, In *Current Issues of Science and Research in the Global World*, Vienna Conference. London: CRC press, Taylor & Francis group.

Farkaš, P. & Ružický, E. 2013. Physical layer network coding with feedback for one simple body area wireless sensor network. In *The International Conference on Health Informatics.* Vilamoura, Portugal. Cham: Springer, 2014. p. 152–154.

Farkaš, P., 2014. *Information and Communication Systems.* Praha: Wolters Kluwer.

Farkaš, P. & Ružický, E. 2014. Multi-Cloud Hosting IOT Based Big Data Service Platform Issues and One Heuristic Proposal How To Possibly Approach Some of them. In *Current Issues of Science and Research in the Global World*, Vienna Conference. London: CRC press, Taylor & Francis group.

Fogel, J., 2014. *Operating Systems.* Nitra: Forpress.

Jurišová, E., 2014. *English for IT Students.* Nitra: Forpress.

Kultán, J., 2014. *Introduction into Databases.* Nitra: Forpress.

Lacko, J. & Ružický, E., 2014. *Web Technologies and Design.* Nitra: Forpress.

Lvovich, I. & Kostrova, V. 2013. Enterprise activity management rationalization on the basis of dynamic network structures and decision making support system modeling. Voronezh: Science book.

Novotný, M., Lacko, J. & Samuelčík, M. 2013. Applications of multi-touch augmented reality in education and presentation of virtual heritage In *Procedia Computer Science*, Vol. 25, London: Elsevier B.V. p. 231–235.

Páleník, T., 2014. *Computer Networks.* Nitra: Forpress.

Palko, V., 2014. *Discrete Mathematics.* Nitra: Forpress.

Řepa, V., 2014. *Process Management and Modelling.* Nitra: Forpress.

Řepa, V. & Voříšek, J. 2014. Role of the Process-Driven Management in Informatics and Vice Versa. In *Current Issues of Science and Research in the Global World*, Vienna Conference. London: CRC press, Taylor & Francis group.

Ružický, E., 2012. Visualization of Economic Data for Interactive Analysis. In Information Technology Applications, Bratislava: PEU and Eurokodex, No.1, p. 36–41.

Schindler, F. & Páleník, T. 2014. Massively parallel decoding of tanner-graph defined error correcting codes. In: Trudy mezhdunarodnogo lektoriya: posvyashchennogo 30-letiyu kafedry "Sistemy avtomatizirovannogo proyektirovaniya i informacionnyye sistemy" Voronezh: Voronesh Institute of High Technologies. p. 72–81.

Schindler, F., 2014. *Data Structures and Algorithms.* Praha: Wolters Kluwer.

Schindler, F., 2014. *Essentials of Programming.* Nitra: Forpress.

Šperka, M. & Lvovich, I., 2014. *Introduction to Information Technology.* Nitra: Forpress.

Šperka, M., 2014. *Graphical User Interface in Java.* Praha: Wolters Kluwer.

Vidács A., Vajda L., Šimek M., Tseng Y., Ren Y., Farkaš P., Páleník T. & Doboš J. 2014. The Smoke in the Chimney—International Cooperation at Pan-European University. In *Current Issues of Science and Research in the Global World*, Vienna Conference. London: CRC press, Taylor & Francis group.

Voříšek, J., Bruckner, T., Buchalcevová, A., Stanovská, I., Chlapek, D. & Řepa, V. 2012. *Tvorba informačních systémů: principy, metodiky, architektury.* Praha: Grada.

Voříšek, J., 2014. *Principles of Development and Operation of Business Information Systems.* Praha: Wolters Kluwer.

Current Issues of Science and Research in the Global World – Kunova & Dolinsky (Eds)
© 2015 Taylor & Francis Group, London, ISBN 978-1-138-02739-8

Linked List Shadow Mapping

Michal Ferko

Faculty of Mathematics, Physics and Informatics, Comenius University, Bratislava, Slovakia

ABSTRACT: We present a novel method for real-time display of alias-free hard shadows called Linked List Shadow Mapping (LLSM). Our new method is derived from shadow mapping methods that produce alias-free hard shadows, but are computationally too complex to be real-time. We avoid standard rasterization during shadow map generation and instead create linked lists of view samples per shadow map texel. Such a data structure is generated directly on the GPU and then used to generate alias-free hard shadows through GPU-accelerated irregular rasterization.

1 INTRODUCTION

Shadows provide important visual cues to the observer regarding position and size of objects in the scene. When displaying realistic 3D images, shadows must be included. However, highly dynamic environments in modern 3D games make it impossible to pre-compute shadows. Therefore, the need for real-time shadowing algorithms has grown rapidly in the last few years.

Viable shadowing algorithms for real-time are derived either from shadow volumes (Crow, 1977) or from shadow mapping (Williams, 1978). Shadow volumes generate large amounts of geometry but provide alias-free shadows. Shadow mapping, on the other hand, leverages standard Z-buffer rasterization and is thus faster to compute. However, it suffers from aliasing artifacts due to its image-based nature.

We focus on extending standard shadow mapping to avoid aliasing. Our method guarantees alias-free hard shadows and is fully GPU-accelerated thanks to recent GPU features—geometry shaders, atomic counters and shader image atomics.

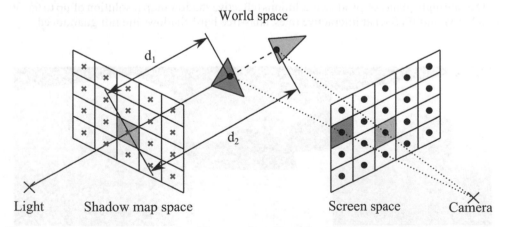

Figure 1. Standard shadow mapping. The scene is rendered from the light, storing depths at shadow map texel centers (blue crosses). The red triangle is the closest surface for the highlighted shadow map texel. It has red triangle's depth d_1 stored in the texel. The point on the green triangle is in shadow ($d_2 > d_1$), but the point on the red triangle is lit ($d_1 = d_1$).

Shadow mapping rasterizes the scene from the light's point of view and stores the computed Z-buffer into a texture called a shadow map. During rasterization from the camera's point of view, the shadow map texel which corresponds to the current camera pixel (called a view sample) is queried. Its depth in shadow map space is computed and compared with the value stored in the shadow map. If the stored value is smaller, the surface is in shadow. This is shown in Figure 1.

2 RELATED WORK

Despite its simple nature and easy implementation on the GPU, shadow mapping suffers from aliasing errors. Our method aims to overcome them. There are several modifications that produce alias-free hard shadows, but they are computationally expensive.

To generate alias-free hard shadows, we need to have for each view sample a texel (shadow map sample) that is exactly aligned to the view sample position after projection into shadow map space. However, if we project view samples into shadow map space, the required shadow map sample positions are irregularly distributed.

Even simple geometry can produce highly irregular patterns, as shown in Figure 2. Standard shadow mapping would in this case compute large portions of the shadow map which are irrelevant for subsequent computation of shadows. In other parts, the resolution of the shadow map is insufficient. There are several classes of approaches that use irregular sampling which alleviates these problems.

2.1 *Adaptive partitioning*

Adaptive partitioning methods analyze view samples and subdivide shadow map space (light clip space) into smaller rectangles of arbitrary sizes. For each rectangle, a shadow map of the required resolution is rasterized. We divide these approaches into two groups:

1. *Iterative refinement*. These methods progressively improve image quality where the output image would suffer from aliasing. These include Adaptive Shadow Maps (ASM) (Fernando et al., 2001) and Queried Virtual Shadow Maps (QVSM) (Giegl & Wimmer, 2007b).
2. *Direct hierarchy evaluation*. These methods first identify all shadow maps for every tile that needs to be rasterized. Actual shadow map rasterization is performed afterwards. These include Fitted Virtual Shadow Maps (FVSM) (Giegl & Wimmer, 2007a) and Resolution-Matched Shadow Maps (RMSM) (Lefohn et al., 2007).

Despite high quality of produced solutions (effective shadow map resolution of up to 65536^2 in RMSM and FVSM) at interactive rates, alias-free hard shadows are not guaranteed.

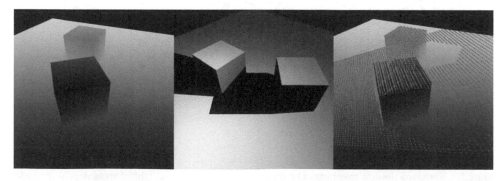

Figure 2.   View samples projected into shadow map space. Shadow map with depth coded as a gray-scale value (left), camera view determining the view samples (middle) and view samples projected into shadow map space (right).

2.2 *View-sample mapping*

View-sample mapping algorithms create a 2D acceleration data structure (kD-tree or grid of lists) in shadow map space. View samples are projected into shadow map space and inserted into the spatial structure. When rendering the shadow map, the exact positions of view samples in shadow map space are used as sample positions. Since Z-buffer rasterization relies on samples aligned in a grid, other rasterization approaches are used.

Alias-Free Shadow Maps (AFSM) (Aila & Laine, 2004) create a 2D kD-tree from view samples. The shadow map rendering step then projects all scene triangles into shadow map space and then identifies inside which kD-tree nodes the triangle is. All points stored in the hit nodes are then tested with a 2D point-inside-triangle test. Points that are actually inside the triangle have some Z value stored, and a new value is computed from the triangle in 3D and the point's position. The result is an irregular rasterization while storing Z values in the shadow map space for view samples. Afterwards, each view sample has exactly one corresponding shadow map sample, and the depth comparison works as usual.

The irregular Z-buffer (IrrZ) (Johnson et al., 2004, 2005) is equivalent to AFSM, but uses a grid-of-lists instead of a kD-tree as the supporting data structure.

3 LINKED LIST SHADOW MAPPING

Our proposed method, *Linked List Shadow Mapping* (LLSM), is a GPU implementation of a view-sample mapping algorithm. We use a grid-of-lists acceleration structure (much like IrrZ), and perform a special rasterization step that identifies the correct view sample candidates. LLSM is fully GPU-accelerated.

We built the algorithm to fit into a deferred shading pipeline (Deering et al., 1988; Saito & Takahashi, 1990). The first pass of deferred shading, usually referred to as G-buffer creation, gives us the Z-buffer from the camera's point of view in a texture. This texture can be used to reconstruct 3D world space positions of all view samples. The algorithm together with deferred shading can be divided into the following steps:

1. Create G-buffer by rendering the scene from the camera and storing the standard G-buffer data—albedo, view-space normal and depth. Our algorithm requires only G-buffer depth to function.
2. Create grid-of-lists by projecting view samples into shadow map space and constructing linked lists in parallel. The resulting data structure will look like the one in Figure 3.

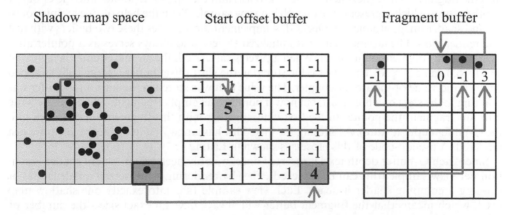

Figure 3. Data structures used in LLSM. We split shadow map space into tiles based on start offset buffer resolution. View samples are projected into shadow map space (black circles) and a linked list of view samples is created for each shadow map texel. View sample positions are irregular, causing variable length of the linked lists.

255

3. Render scene from the light using conservative rasterization, which identifies all grid tiles that a triangle hit. Simulate Z-buffer behavior for overlapping view samples.
4. Perform depth comparison for each view sample in shadow map space and store the result into a screen-space binary shadow texture.
5. Read from the binary shadow texture during evaluation of the shading term.

3.1 Creating linked lists

Our algorithm abandons the standard shadow map texture, where each texel contains one depth value corresponding to a sample placed in the center of the texel. Instead, each shadow map texel contains a linked list of samples that are inside the texel's cell (see Fig. 3). For each sample, we have one depth value—the distance of the closest surface to the light along the ray represented by the sample. The 2D sample positions are determined by projecting view samples (which we have as depth values in the G-buffer depth texture) into shadow map space.

To create such a data structure, we borrow ideas from (Thibieroz & Gruen, 2010), where authors describe a method for rendering transparent objects with alpha blending. Contrary to similar methods, there is no need to sort geometry back-to-front before rendering (or render scene geometry multiple times). This is referred to as *order independent transparency*. The algorithm simply renders transparent objects, as seen by the camera. However, instead of storing the closest surface (standard Z-buffering) for a pixel, all transparent surfaces are stored in a linked list. To produce correct alpha-blended geometry, the linked list is then sorted according to depth of samples.

Our approach is similar in implementation, but instead of storing different layers of depth per pixel, we store samples with different positions in a shadow map texel. The implementation in (Thibieroz & Gruen, 2010) requires OpenGL 4 hardware and support for shader atomic counters, shader image load and store and shader image atomics.

Our implementation uses two buffers (2D textures): the start offset buffer and the fragment buffer. The start offset buffer stores for each shadow map texel the start of the linked list. The actual linked list items (view samples) are then stored in the fragment buffer (see Fig. 3).

The start offset buffer's resolution determines the actual size of texel cells in shadow map space and is an important parameter of the whole LLSM algorithm (smaller texel cells mean shorter linked lists). The start offset buffer stores in each texel just one integer value, which is interpreted as a pointer to the fragment buffer (or NULL if the value is −1). It points to the first item in the linked list of view samples for the corresponding shadow map texel.

The fragment buffer has resolution equal to screen resolution, since every view sample will be stored in it exactly once. One texel of a fragment buffer stores two integers. One is a pointer into the fragment buffer (next item in the linked list) and the other is a pointer into the G-buffer depth texture, which represents the underlying view sample. With the help of one atomic counter and shader image atomics, an OpenGL 4 implementation creates these two buffers with just one render pass. The atomic counter is initialized to zero and always serves as a pointer after the last filled fragment buffer texel. When a pixel is rasterized, the atomic counter is increased and its old value is read. The offset value stored in the start offset buffer is atomically swapped with the old value of the atomic counter. The value that was stored in the start offset buffer was pointing to the start of a linked list, but now it points to an empty fragment. Therefore, we write into this fragment the pointer to the first item, together with this fragment's data (index into the G-buffer depth texture). As a result, the linked list has a new first item, the fragment we just rendered. A pseudo-code of this process is shown in Listing 1.

Since each G-buffer depth texel is inserted once, we only need one fragment shader execution per view sample (this can be achieved either by rendering a full-screen quadrilateral or running a compute shader instead). Each view sample falls into exactly one shadow map texel, which means that the fragment buffer will always have an exact size—the number of view samples.

Traversal of linked lists is relatively simple. We read the initial offset from the start offset buffer and then read values with new offset values from the fragment buffer until we read an offset with value −1.

```
// Used variables in global GPU memory
Texture2D gBufferDepth;
AtomicCounterUnsignedInt fragmentBufferSize;
Texture2D startOffsetBuffer;
Texture2D fragmentBuffer;

void processGBufferTexel()
{
    // Project view sample into 2D shadow map space
    float depth = readTexture(gBufferDepth, fragmentPosition);
    Point2D shadowMapPosition = projectToShadowMapSpace(fragmentPosition, depth);

    // Atomic read and consecutive increment of the atomic counter
    // We reserve one texel in the fragment buffer
    uint fragmentIndex = atomicCounterIncrement(fragmentBufferSize);

    // fragmentIndex now points to a reserved texel for the current fragment

    // Exchange stored index pointing to the list start with the index of this
    // fragment in the start offset buffer
    uint linkedListStart = imageAtomicExchange(startOffset, pixelPos, fragIndex);

    // The start offset buffer now points to the reserved texel in the
    // fragment buffer

    // Store this fragment's data into the empty texel
    uint ptrNextItem = linkedListStart;
    uint ptrGBuffer = fragmentPosition;
    imageStore(fragmentBuffer, fragmentIndex, pack(ptrNextItem, ptrGBuffer));
}
```

Listing 1. Linked list generation pseudo-code using one execution per view sample. The fragment buffer stores only two unsigned integer values.

3.2 *Irregular rasterization*

Once we have created the linked lists for each shadow map texel, we need to generate a depth value for each of the view samples. This depth value will represent the distance of the view sample from the light source. We refer to this step as shadow map rendering, however, it is an irregular shadow map.

In fact, we cannot use standard Z-buffer rasterization to compute these depth values. We send all scene triangles to the rasterization pipeline as usual, rendering them from the light's point of view. Each triangle hits several shadow map texels after rasterization. We need to compute the new depth value of only those view samples that are occluded by the triangle. In other words, each view sample projected into shadow map space that lies inside the 2D projection of the triangle should be considered. For these samples, actual depth values for the triangle are computed and a simulated Z-buffer less-equal test determines if the stored depth value should be overwritten.

3.3 *Conservative rasterization*

When rasterizing scene triangles, only shadow map texels whose center lies inside the triangle are considered. However, since view sample positions after projection are at arbitrary positions in the shadow map texel, we must take into account all shadow map texels that the triangle could hit. Therefore, we must use a modified rasterization approach that executes a fragment shader for each pixel hit by the triangle, not only pixels whose center lies inside the triangle. This special rasterization is referred to as *overestimated conservative rasterization* (Hasselgren et al., 2005; Akenine-Möller & Aila, 2005).

To execute a fragment shader for each pixel cell overlapping a triangle, we encapsulate the triangle into a convex bounding polygon. This can be thought of as the image-processing

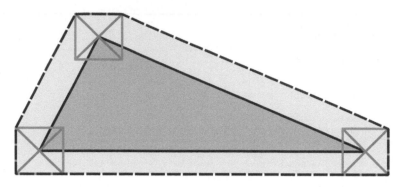

Figure 4. Conservative rasterization. The original triangle has its vertices extended based on pixel cell size and we render the bounding polygon as a triangle strip instead. Each vertex can be replaced with up to 3 vertices, therefore, the bounding polygon will have at most 9 vertices.

operation dilation, with radius equal to pixel cell size. This convex polygon is created by extending the triangle edges, so that centers of pixels that should be taken into account are inside the convex polygon. An example of the algorithm is shown in Figure 4.

We perform this step of conservative rasterization in a geometry shader, which receives one triangle and outputs its bounding polygon with respect to the current resolution. The algorithm guarantees that the bounding polygon will not have more than 9 vertices, and we output the bounding polygon from the geometry shader as a single triangle strip. This algorithm is described in detail in (Hasselgren et al., 2005).

3.4 Fitting the light viewing volume

There are several optimizations we can apply to the LLSM algorithm. To minimize the number of empty shadow map texels, we take into account only portions of the scene that are visible by the camera. The light viewing volume is reduced to encapsulate all view samples tightly (this is referred to as *shadow map fitting* (Eisemann et al., 2011)).

We calculate this tight fit using a GPU-accelerated approach that computes positions of view samples in shadow map space and then identifies an axis-aligned bounding rectangle of those view sample positions in shadow map space. The light viewing volume is then shrunk to cover only the computed bounding rectangle in shadow map space. The algorithm thus reduces the light viewing volume only to those parts where there are view samples, resulting in fewer empty shadow map texels. This approach was adopted from (Ferko, 2013).

Note that even without fitting, if the start offset buffer has the same resolution as the screen, the average length of a linked list is 1. By decreasing the number of unused texels, we achieve a better distribution of linked list items. Optimally, each texel would have a linked list containing just one item, which would result in perfectly balanced parallel execution.

4 EXPERIMENTAL EVALUATION

We have implemented Linked List Shadow Mapping (LLSM) in OpenGL 4 with shaders written in GLSL. Due to the nature of the algorithm, the alias-free quality of output hard shadows is guaranteed (our computation actually produces ray-traced hard shadows). We tested the algorithm on a PC with an Intel Core i7-3770K processor, 16 GB of RAM and a NVIDIA GeForce GTX 680 graphics card.

Table 1 shows the performance characteristics of different steps of our algorithm during a pre-defined camera fly-through. The separate steps of our algorithm are very fast (an average below 0.5 ms), except for irregular rasterization. In this step, conservative rasterization identifies which shadow map texels a triangle possibly hits and then simulates Z-buffering on view samples inside each texel.

Table 1. Execution times (in milliseconds) of individual stages of the LLSM algorithm at 1280×720 screen resolution and 1280×720 start offset buffer resolution. Showing minimum (MIN), average (AVG) and maximum runtime of the algorithm for a pre-defined walkthrough. Recorded on the Dragon scene.

	Fitting	Clear buffers	Generate linked lists	Irregular rasterization	Depth comparison
MIN	0.28	0.07	0.07	7.28	0.12
AVG	0.38	0.12	0.26	23.28	0.31
MAX	6.65	1.21	3.85	70.69	0.97

Figure 5. Used test scenes. The Dragon scene (left), the Hairball scene (middle), and the Sponza scene (right).

Table 2. Showing performance of linked list shadow mapping of recorded walkthroughs through several scenes. Screen resolution was 1280×720. Execution times are shown in milliseconds and show minimum (MIN), average (AVG) and maximum (MAX) runtime of the algorithm.

Scene	Triangle count		Start offset buffer resolution				
			640×360	1280×720	1920×1080	2560×1440	4096×4096
Dragon	864 732	MIN	15.53	**7.95**	8.15	8.35	9.97
		AVG	49.14	24.34	18.85	**17.08**	20.25
		MAX	175.49	71.78	44.28	**33.55**	34.78
Sponza	279 163	MIN	**3.19**	3.21	3.51	3.86	8.64
		AVG	29.80	21.74	20.25	**20.11**	29.70
		MAX	488.10	198.74	125.92	91.85	**65.08**
Hairball	2 850 012	MIN	155.28	87.98	33.21	**31.69**	42.72
		AVG	306.17	160.84	120.76	106.58	**105.84**
		MAX	921.54	481.45	312.15	251.20	**170.81**

We have tested the algorithm at 1280×720 resolution on several different scenes (see Fig. 5). The performance results are shown in Table 2.

The only input parameter of the current algorithm is the resolution of the start offset buffer (shadow map resolution), which determines how large individual shadow map texels will be. The resolution also affects the average length of linked lists. If the resolution is equal to screen resolution, each linked list will have one item on average. Higher resolution means shorter lists and decreases the chance that a large number of view samples will fall into a single shadow map texel. Lower resolution results in longer lists, but smaller memory consumption.

In Table 2, results show that smaller start offset buffer resolution has proven less effective than using a larger buffer. The average and maximum durations of individual steps decrease as we increase the start offset buffer resolution.

However, by increasing the start offset buffer resolution, we increase the number of rasterized pixels and required memory. Increasing the buffer resolution too much will have a negative impact on performance (rasterization will take too long) and we will use a large, but mostly empty, start offset buffer. In Table 2, we can see that the 4096×4096 resolution is too high and decreases average performance for the Dragon and Sponza models.

The Sponza scene renders slower than the Dragon scene, despite having less triangles. This is due to the fact that it consists of a large number of small triangle batches, while the Dragon model contains only two triangle batches. Performance of the Hairball scene is relatively low, mostly due to the fact that traced light rays pass through a large amount of triangles (see Fig. 5).

The worst case scenario for our approach is when all view samples fall into one shadow map texel and we create one linked list of all view samples. However, it is very unlikely to happen in a typical setting of video games or similar 3D applications. Employing a fitting scheme further decreases the probability.

5   CONCLUSION

We have presented a novel algorithm for rendering hard-edged alias-free shadows in real-time. Linked list shadow mapping is a view-sample mapping algorithm adapted to GPU-supported operations. It simulates irregular rasterization through OpenGL shaders. We have provided a brief analysis of the introduced algorithm, together with some performance improvements and results.

A lot of tests have to be done to compare this algorithm to existing alias-free hard shadow algorithms, such as shadow volumes or similar modifications to shadow mapping. The interesting factor will be performance.

At the moment, the performance of the algorithm is not viable for real-time display of shadows. A hybrid approach combining standard shadow mapping with linked list shadow mapping might inherit the alias-free guarantee while retaining plausible frame rates. A more thorough analysis of the irregular rasterization step is required. Alternative data structures instead of a regular grid (such as a kD-tree) might improve the speed of irregular rasterization, at the cost of a more complex spatial data structure creation.

ACKNOWLEDGEMENT

This work was funded by Comenius University, grant no. UK/199/2014. We also thank Frank Meinl for the Sponza model, the Stanford 3D Scanning Repository for the Dragon model and Samuli Laine and Tero Karras for the Hairball model.

REFERENCES

Aila, T., & Laine, S. (2004). Alias-free shadow maps. In *Proceedings of the Fifteenth Eurographics Conference on Rendering Techniques*, EGSR'04, (pp. 161–166). Aire-la-Ville, Switzerland, Switzerland: Eurographics Association.

Akenine-Möller, T., & Aila, T. (2005). Conservative and tiled rasterization using a modified triangle set-up. *Journal of Graphics, GPU, and Game Tools*, 10(3), 1–8.

Crow, F.C. (1977). Shadow algorithms for computer graphics. In *Proceedings of the 4th annual conference on Computer graphics and interactive techniques*, SIGGRAPH '77, (pp. 242–248). New York, NY, USA: ACM.

Deering, M., Winner, S., Schediwy, B., Duffy, C., & Hunt, N. (1988). The triangle processor and normal vector shader: a VLSI system for high performance graphics. *SIGGRAPH Comput. Graph.*, 22, 21–30.

Eisemann, E., Schwarz, M., Assarsson, U., & Wimmer, M. (2011). *Real-Time Shadows*. Natick, MA, USA: A.K. Peters, Ltd., 1st ed.

Ferko, M. (2013). Resolution estimation for shadow mapping. In *Theory and Practice of Computer Graphics*, (pp. 109–114). The Eurographics Association.

Fernando, R., Fernandez, S., Bala, K., & Greenberg, D.P. (2001). Adaptive shadow maps. In *Proceedings of the 28th annual conference on Computer graphics and interactive techniques*, SIGGRAPH '01, (pp. 387–390). New York, NY, USA: ACM.

Giegl, M., & Wimmer, M. (2007a). Fitted virtual shadow maps. In *Proceedings of Graphics Interface 2007*, (pp. 159–168). ACM.

Giegl, M., & Wimmer, M. (2007b). Queried virtual shadow maps. In *Proceedings of the 2007 symposium on Interactive 3D graphics and games*, (pp. 65–72). ACM.

Hasselgren, J., Akenine-Möller, T., & Ohlsson, L. (2005). Conservative rasterization. In *GPU Gems 2*, (pp. 677–690).

Johnson, G.S., Lee, J., Burns, C.A., & Mark, W.R. (2005). The irregular z-buffer: Hardware acceleration for irregular data structures. *ACM Trans. Graph.*, 24(4), 1462–1482.

Johnson, G.S., Mark, W.R., & Burns, C.A. (2004). The irregular z-buffer and its application to shadow mapping. Tech. rep., Department of Computer Sciences, The University of Texas at Austin, Austin, TX.

Lefohn, A.E., Sengupta, S., & Owens, J.D. (2007). Resolution-matched shadow maps. *ACM Trans. Graph.*, 26(4), 1–17.

Saito, T., & Takahashi, T. (1990). Comprehensible rendering of 3-D shapes. In *Proceedings of the 17th annual conference on Computer graphics and interactive techniques*, SIGGRAPH '90, (pp. 197–206). New York, NY, USA: ACM.

Thibieroz, N., & Gruen, H. (2010). OIT and indirect illumination using DX11 linked lists. In *GDC San Francisco 2010*.

Williams, L. (1978). Casting curved shadows on curved surfaces. *SIGGRAPH Comput. Graph.*, *12*, 270–274.

Current Issues of Science and Research in the Global World – Kunova & Dolinsky (Eds)
© 2015 Taylor & Francis Group, London, ISBN 978-1-138-02739-8

Applications of 360° object photography in augmented reality

J. Lacko
Faculty of Informatics, Pan-European University, Bratislava, Slovakia

ABSTRACT: In the paper we describe the possibilities of applications of 360° object photography in augmented reality. We focus on the object photography acquisition stage, where we use a real objects for digitalization into the form of structured photographic set. We describe the whole process of the digitalization. In the rendering stage we combine techniques of computer vision for localization of captured object (marker) and image based rendering methods for rendering of the digitalized object. We also focus on the different aspects and problems of rendering, especially on the lighting. By using this kind of the objects we can group them into scenes. In this stage of the process, we are limited by the camera positions in the process of sensing or digitalization. In the paper we describe also the scene formation from various objects. In the paper we will focus on the applications of augmented reality with object photography in the so called vireal museums (virtual objects in the real museums) and other application domains.

1 INTRODUCTION

There are many applications which are using techniques of augmented reality (Bimber & Raskar 2005) for showing 3D objects as the combination of real world and 3D models in virtual space. In most cases the objects are true 3D models (mesh or volumetric). In our work we will focus on using 360° object photographs (VR Photography 2014) (in our work we will use the name object panoramas) as the data for impression of 3D models. The object panoramas are photographed objects from different angles in the horizontal plane and also in the vertical angles (elevations). This is a cheap solution for making an impression of real 3D object in the image space. In the paper we will discuss about digitization process and possibilities of presentation such kind of data in different application areas.

2 MOTIVATION

In years 2007–2013 there was in Slovak republic massive digitization campaign as the part of the Operational Programme Informatisation of Society (OPIS 2014a) supported by European structural funds. Under the priority axis 2 Development repository institutions and renewal of their national infrastructure and measure 2.2 Digitisation of the content of repository institutions, archiving and provision of access to digital data (OPIS 2014b), there were digitized about 3 000 000 of various objects like paintings, museum artifacts, books, movies, castles and audio files. These data are stored and need to be present to professionals and public. From that amount, there are about 20 000 of different objects, digitized in the form of object panorama. Augmented reality is popular way how to present the data to public by using their information technology equipment like smartphones or tablets.

3 DIGITIZATION PROCESS

3.1 *Object panorama definition*

Object panorama is aligned set of object photographs with respect to the horizontal and vertical camera position. In horizontal way, there is usual to have about 10° camera angle step between two shots. For vertical camera position there are usual 4 camera positions (elevations)—5° or 0° from horizontal plane of the object, 30°, 60° and 80–90°, which depends on possibilities of the object attachment. As we can see on the Figure 1, there is a part of object panorama of František II. Rákoczi statue in one elevation with horizontal camera angle of 30°.

3.2 *Setup of digitization equipment*

On the Figure 2, we can see the setup for small 3D objects digitization into the form of object panorama for museums. There are 4 cameras, turning table, light sources and background, which color can be changed to be contrast enough with comparison to object.

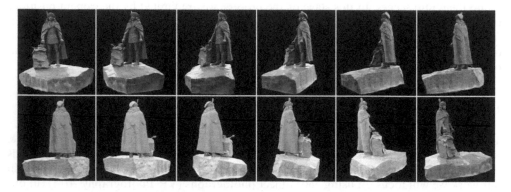

Figure 1.   Part of the object panorama of František II. Rákoczi statue.

Figure 2.   Digitization setup.

Figure 3.   Lighting artifacts due to object rotation.

The digitization process consists of these steps:

1. Align the object vertical axis into the middle of the turning table.
2. Change the camera positions into proper elevations.
3. Shot the image from 4 cameras synchronously and rotate the table in 10°
4. Repeat 3. until the object turn around the vertical axis.

The usual set of photographs consists of 144 photos, which we can reduce into 109 photos, in the case that fourth camera is placed directly above the object in the 90° position, because all 36 photographs from this camera are invariant to rotation. The set of the images can be used as secondary input dataset (Lacko 2010) for photogrammetry.

3.3 *Lighting setup*

When we want to capture the images in the correct lighting conditions, we need to use diffuse lighting and diffusers to reduce the environmental artifacts. The most problems are with the specular and transparent objects. In our setup, we are not able to turn the lights with the objects together, so there we can see some artifacts in the images. As we can see on the Figure 3, when we rotate the objects and positions of the lights are stable, there are highlights and shadows in other part of the object in each image. If we use traditional way of presentation without the real context of augmented reality, there the user can imagine that light is fixed, but when we will use the images in the augmented reality and we will turn around the object, these artifacts become more visible to the user. The solution of this problem could be the light rotation together with turning table while object is digitized.

4 PRESENTATION

For the presentation of the objects we use dataset consists of JPEG (JPEG Standards 2014) images in the server side of the application. The input data from digitization has 80MPix resolution which is done by Phase One cameras. For presentation purposes in the augmented reality, we don't need such high resolution, so we can optimize the input images by using level of detail techniques.

4.1 *Augmented reality algorithm*

The augmented reality algorithm, we use in our work consists of these steps:

1. Camera image evaluation (marker finding)
2. Computation of the marker position in the 3D space

3. Object panorama alignment with marker
4. Selection of most suitable image from object panorama
5. Rendering of frame.

We repeat the whole process until camera of equipment is sensing. As the marker we can use any image, but for the correct computation of the marker orientation in the 3D space we don't want images which are symmetric, because we need to easily determine the "north" in the image, which is the beginning position for very first image in the set.

4.2 *Masking*

JPEG image format doesn't support transparency. If we use the image from the object panorama as it is, we will see also the background, which was set in the digitization algorithm. For better results of human perception, we need to solve this problem. First possibility is to use another image format like PNG (Roelofs 2014). But this format is much bigger (memory) than JPEG. Benefit of this image format is that it has 8 bit alpha channel for semitransparent images. Another solution is to use a mask and combine original image with the mask. For this solution we need to have fulfilled some prerequisites. Firstly, the background of the image must be contrast enough to the object and border of the image can't consist of the same pixel colors as the background. This kind of condition, we can use when we use snake techniques of object separation. Secondly, if we want to use for masking more similar techniques like threshold, we don't want to have background color in the whole object part of the image. The pixels in the object border are little blurred for more realistic look.

For our purpose we use the second approach and on the Figure 4, there is a combination of original image and its mask shown.

4.3 *Scenes*

The object panoramas can be combined into the scenes. For each object panorama we can define its position in the 3D space, rotation and scale. By using these transformations we can

Figure 4. Photograph and mask.

Figure 5.   Scene consist of object panoramas.

make the scenes around the center point (0,0,0). When we render the scene, we rotate the camera around the center point and position of the object is defined relatively to the marker position. In general the center point of the scene is in the middle of the marker. We can combine together not only object panoramas, but also 3D models. In the rendering stage there is proper image from object panorama computed separately for each object. We can see the scene on the Figure 5.

4.4   *Zooming objects*

Scaling in the augmented reality as the one of the basic transformations in the 3D space, can be realised, similar as the other transformations, in two ways. First one is done, when the position of the marker is fixed and the camera is moving closer to the marker, then the object is zoomed. Another possibility is to zoom object by its scaling interactively when camera and marker position is fixed. We can scale the object by touch gestures.

5   APPLICATIONS

Our method can be applied for presentation of various kinds of object panoramas. The application areas are vireal museums (Lacko et al. 2011) (which are virtual objects in the real museums), for marketing/product presentations (combination with e-shops). Presentation of cultural heritage objects in museums is the way how to show to public various objects from different museums. We can make various kind of presentation scenarios, e.g. presentation by using tablets, or multitouch augmented reality system (Novotný et al. 2012). Example of using the object panoramas in real scenarios are on the Figures 6–7. Another possibility is to use not only static image, but the video. 360° video must be captured by multiple camera rig and must be synchronized. We need the mask for each frame of the video and for illusion of 3D object we don't need to generate 3D models of the captured objects.

Figure 6.   Object panorama in augmented reality.

Figure 7.   Detail of object panorama in augmented reality.

6 FUTURE WORK

In the future we plan to add precomputed or precaptured lighting scenarios based on actually sensed lighting conditions. If we have 3D models it is possible to compute the shading of the object directly per vertex. But in the image space we need to capture the object into different object panoramas with different lighting setup and in the rendering stage we need to choose proper object panorama.

To enhance the perception of the object and its position in the space, it is important to add the shadow. We can add shadow to masked object panorama which is precomputed e.g. from top view and placed under the panorama.

The cameras in the tablets or smartphones have the limited depth of field. We can compute it from camera parameters and also from captured images of real world. If our presented objects are bigger than computed depth of field, we can apply blur into the object parts which are very close to the camera or too far from the camera.

7 CONCLUSION

In our paper we propose the method of combination of augmented reality and object panoramas. We described the process of object panorama acquisition and its rendering and also the possibilities of their applications in the real world. As we can see there are various problems which can be easily solved if we will use instead of object panoramas 3D models, but the most of the data from massive digitization of cultural heritage objects are in the form of object panoramas.

REFERENCES

Bimber, O. & Raskar, R. 2005. Spatial augmented reality: merging real and virtual worlds. A.K. Peters.
JPEG Standards. 2014. *http://www.jpeg.org/*. Page accessed June 6 2014.
Lacko, J. 2010. Inverse problem solver. Dissertation thesis.
Lacko, J., Novotný, M. & Samuelčík, M. 2011. Koncept vireálneho múzea. *Informačno-komunikačné technológie – využitie v prezentačnej činnosti múzeí. Proc. Banská Štiavnica, 12–13 October 2011.* Slovenské technické múzeum.
Novotný, M., Lacko, J. & Samuelčík, M. 2012. MARS: Multi-touch augmented reality system and methods of interaction with it. *Information Technology Applications* (ISSN 1338–6468) 2/2012: 30–37.
OPIS. 2014a. http://www.opis.gov.sk/data/files/2440_3907.pdf. Page accessed June 8 2014.
OPIS. 2014b. http://www.opis.gov.sk/data/files/2446_4516.pdf. Page accessed June 8 2014.
Roelofs, G. 2014. Portable network graphics. *http://www.libpng.org/pub/png/*. Page accessed June 6 2014.
VR Photography. 2014. *http://en.wikipedia.org/wiki/VR_photography.* Page accessed June 8 2014.

Current Issues of Science and Research in the Global World – Kunova & Dolinsky (Eds)
© 2015 Taylor & Francis Group, London, ISBN 978-1-138-02739-8

The simulation of the scattering characteristics for cavities with complex shape

Igor Lvovich & Frank Schindler
Pan-European University, Bratislava, Slovakia

Andrey Preobrazhensky
Voronezh Institute of High Technologies, Voronezh, Russia

ABSTRACT: In the paper we consider the processes of scattering of electromagnetic waves on the cavities with complex shape. The particularities of the method of moments are given. The advantages of the method of integral equations are shown. In solving the integral equation by the method of moments as basis functions chosen us piecewise constant functions and test functions-δ-Dirac function. The universality of the considered algorithm is due to the fact that for the calculation of the scattering characteristics of perfectly conducting hollow structures of various forms necessary to change the outline of a structure when programming with regard to the necessary steps of the partition—the main steps of the algorithm are the same. Some results of calculations of some types of hollow structures are shown.

1 INTRODUCTION

One of the tasks of great importance in the development of objects of engineering, is the study of electromagnetic wave scattering on hollow structures of different sections and shapes in radar range of wavelengths in the resonance region, for example. In this case, the characteristic dimensions of apertures hollow structures are from one to ten wavelengths of the incident electromagnetic wave. When modeling of electromagnetic wave scattering on hollow structures with use of the method of integral equations are two possible approaches: the surge and diffraction.

2 THE METHODS OF CALCULATION THE SCATTERING CHARACTERISTICS OF THE OBJECTS

In many cases, hollow structures are not bodies of rotation and can be complex. In this case it is necessary to solve the integral equation with account of peculiarities of each part of the surface of patterns. Of course, will increase significantly time accounts, and volume of the required machine memory. But this approach has some advantages:

1. On the basis of the techniques discussed the opportunity to conduct simulation of electromagnetic wave scattering on hollow structures of various forms of cross-sections, with regard to the radio absorbing coatings on the surface of this structure.
2. Unlike existing waveguide techniques the opportunity to analyze the electromagnetic field reflected from the outer surface of the hollow structure can be used. Next, you will also have comparison of results of calculations on the basis of the techniques discussed with the results of calculations based on modal method.
3. The size of the structure may be that one element can suffice machine resources, however, can then be used approaches to assess the characteristics of several such items grouped.

The electric currents $J_z(r)$ on the circuit structure are obtained by solving the integral equation method of moments. During this process we have the discretazation of circuit structure.

It can be noted that the advantage of the method of moments is the possibility of calculations with definite precision. The integral equations are exactly in principle and the method of moments provides a direct numerical solution of these equations. All features of scattering of the waves, that is the appearance of surface waves, moving waves, diffraction at the edges, or edges, etc. recorded in the integral equations and "automatically" included in an explicit decision by the moment method. A second advantage of the method of moments is its practical application to geometrically complex scattering objects.

The factors to be considered when choosing a basis and test functions are diverse and complex. On the one hand it is desirable to choose basis functions which are similar in behavior to the real distribution of the currents on the other hand, it is desirable to have the functions which have easily calculated by the integral circuit. Typical examples of the choice of basis and test functions are given in.

In solving the integral equation by the method of moments as basis functions chosen us piecewise constant functions and test functions-δ-Dirac function.

The above procedure going to the linear algebraic integral equation method is also called piecewise constant basis with point wise stitching (Bogolyubov-Krylov method).

It is essential that initially the integral equation is not in any special form. And this is due to an extremely broad universality of the method that allows to solve the problem of diffraction for a wide variety of circuits hollow structures.

Of course, for each specific geometry of the hollow structure we can choose a more economical and efficient method for the numerical solution of the integral equation, which is usually offered in some journal articles in solving the particular diffraction problem (or class of problems). Everything here is determined by the best choice of basic functions, highlighting the features of the solution of the equation (for reducing the dimension of the linear algebraic equation). Therefore, as the lack of the Bogolyubov-Krylov it can be noted the large size of solvable systems of linear equations. However, for example, in it was shown that in diffraction problems with moderate accuracy requirements solutions simple piecewise constant approximation is most effective when compared with the more accurate linear five-point parabolic approximations or splines.

For fast convergence and high precision solutions magnitude plot sampling $\Delta \ell$ chosen from the condition:

$$\Delta \ell \leq \lambda / 5 \qquad (1)$$

The $\Delta \ell$ quantity may vary depending on the area of the surface sampling structure. If the shape of the outline structure is quite complicated, then to find the solution of the integral equation with the required precision is necessary in areas where the curvature of rapidly changing integration step chosen sufficiently small that might be overkill in other areas of integration. As a result of numerical simulation, it was found that in order to make the error in the calculation of the current in this example does not exceed 5%, the calculations necessary to choose the value of the sampling interval $\Delta \ell \leq \lambda / 10$.

In constructing the algorithm to compute $[J_q]$ Bogolyubov-Krylov method is necessary to ensure stability of the numerical solutions to as the number of basic functions (increasing the dimension of the matrix $[A_{pq}]$ and the column vector $[U_p]$) series, approximating $J(z)$, converges to the exact value the current distribution on the surface of the hollow structure. Practical criterion validity of the numerical solution of integral equation is a constant current $J(z)$ as the dimension $[J_m]$.

However, for a fixed dimension of system of algebraic equations' roots are usually determined, approximately due to inaccurate job $[A_{pm}]$ and $[U_p]$, and also due to rounding errors in the solution equation on a computer. Therefore, to obtain a stable solution of integral equation is necessary to ensure a well-conditioned matrix reversal generalized impedance.

Quantitative index of the matrix is the condition number

$$\nu(m) = \|[B_{pm}]\| \, \|[B_{pm}]^{-1}\|, \tag{2}$$

where $\|[B_{pq}]\|$—norm of the matrix $[B_{pq}]$; $[B_{pq}] = [A_{pq}] + [\Delta A_{pq}]$; $[\Delta A_{pq}]$—matrix of errors which arise due to inaccurate job $[A_{pq}]$; $[B_{pq}]^{-1}$—matrix reverse $[B_{pq}]$.

Quantitative measure of error $[\Delta J_m]$ calculations $[J_m]$, arising due to rounding errors when performing arithmetic operations on a computer is the number $\varepsilon(q)$, defined by the inequality:

$$\|[\Delta B_{pq}]\| \leq \varepsilon(q) \cdot \|[B_{pq}]\|, \tag{3}$$

where $[\Delta B_{pq}]$—matrix of the residuals, which characterizes the error solving the linear due to rounding of numbers in computer calculation.

From (2) and (3) it follows that in calculating $[J_m]$ must control the amount of $\varepsilon(q) \cdot \nu(q)$, to an increase in m, it remained relatively low. Smallness works $\varepsilon(q) \cdot \nu(q)$ ensures the sustainability of convergence of the numerical solutions. With increasing $\nu(q)$ is necessary to increase the accuracy of calculating the matrix elements $[A_{pq}]$ (decrease $\varepsilon(q)$). To avoid a sharp increase in the condition number, it is necessary, as the analysis, with piecewise constant approximation to implement a uniform sampling of the lens surface.

In solving the integral equation method of moments necessary to correctly describe the singularity of the kernel of integral equations. Thus, the coincidence of the coordinates of the observation point and integration, that is when $R \to 0$ two-dimensional Green's function.

Obtained on the basis of solving the linear components $[J_m]$ allow us to determine the scattered field.

1. The scattered electromagnetic field is determined on the technique based on the Kirchhoff integral.
2. Calculation of RCS by scattered electromagnetic field is carried out by the approach that can connect 3D and 2D solutions.

The methods of calculating the scattering characteristics of two-dimensional hollow structures can be used to evaluate the scattering characteristics of three-dimensional hollow structures of rectangular cross section, just as it is done in the framework of the combined boundary—integral modal method. In this algorithm for calculating EPR hollow structures of rectangular cross section consists of the following major steps:

1. Three-dimensional structure of a hollow rectangular cross-section of complex shape is associated with a two-dimensional structure;
2. For two-dimensional structure is written the integral equation;
3. Currents are determined by the two-dimensional surface structure;
4. Calculated two-dimensional RCS;
5. Using approximate formula is converted a two-dimensional to three-dimensional RCS

$$\sigma(3D) = (2b^2/\lambda)\sigma(2D)$$

where b—the size of the aperture of the hollow three-dimensional structure in the direction of the axis y.

So, the basic provisions of the above formulated the algorithm for calculating two-dimensional electromagnetic wave scattering characteristics of hollow structures containing radioabsorbing coatings. The main advantages of this algorithm are as follows:

1. Within the framework of the considered algorithm hollow structure is considered as a body of complex shape, so unlike waveguide approach (e.g., boundary—integral modal method) there is consideration of reflection from the outer region of the structure that allows us to analyse the scattered electromagnetic field across the sector angles of incidence of a plane electromagnetic wave. Above algorithm is better to use when calculating the

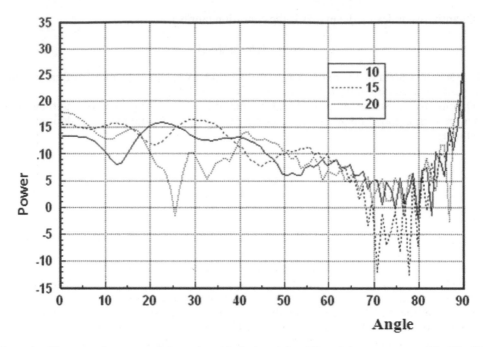

Figure 1. The scattering power of the cavity with the length L = 20λ and the aperture a = 10λ, 20λ, 30λ.

radioabsorbing materials hollow structures with small dimensions (a few wavelengths—the resonance region), in which poorly beam methods;

2. A method of integral equations previously used to calculate the characteristics of two-dimensional scattering bodies for various complex shapes are mainly convex bodies such as cylinders, wedges, etc. The analysis of the possibility of using an algorithm based on the method of integral equations for calculating the radar characteristics of hollow structures of complex shape containing radioabsorbing coatings on the inner walls.

3. The universality of the considered algorithm is due to the fact that for the calculation of the scattering characteristics of perfectly conducting hollow structures of various forms necessary to change the outline of a structure when programming with regard to the necessary steps of the partition—the main steps of the algorithm are the same.

3 THE RESULTS OF CALCULATION THE SCATTERING CHARACTERISTICS

Figure 1 shows the results of calculations of some types of hollow structures are given.

4 CONCLUSIONS

In the paper we consider the approach for calculation of the scattering characteristics for cavities with complex shape. Some results of calculation is given.

REFERENCES

Baranov A. 2012. *The problems of functioning of mesh networks*. Bulleting of the Voronezh Institute of high technologies. No 9. pp 49–50.

Bashkatov A. 2013. *The analysis of scattering characteristics of uwb-signals on the subject is complex*. Modeling, optimization and information technologies, No 3, p. 8.

Ling H. 1990. *RCS of waveguide cavities: a hybrid boundary-integral/modal approach.* IEEE Transactions on Antennas and Propagations 38(9). pp. 1413–1420.

Lvovich I., Preobrazhensky A., Yurov R., Choporov O. 2006. *Software complex for automated analysis of scattering characteristics of objects with application of mathematical models.* Management systems and information technologies. No 2 pp. 96–98.

Lvovich Ya., Lvovich I., Preobrazhensky A. 2010. *The problems of evaluating the characteristics of scattering of electromagnetic waves by a diffraction structures in their design.* Bulleting of the Voronezh Institute of high technologies. No 6. pp. 255–256.

Lvovich I., Preobrazhensky A. *The calculation of the characteristics of metal-dielectric antenna.* Vestnik VSTU. 1(11). pp. 26–29.

Miloshenko O. 2012. *The methods of characteristic estimation of radio wave propagation in the systems of the mobile radio communication.* Bulleting of the Voronezh Institute of high technologies. No 9. pp. 60–62.

Mitra, P. (ed.), 1977. *Computational methods in electrodynamics,* Moscow, Mir, 485 p.

Mitra, P. (ed.), 1977. *Computational methods in electrodynamics,* Moscow, Mir, 485 p.

Motin D. 2013. *On simulation of coverage area of service in a wireless communication system.* Modeling, optimization and information technologies, No 1, p.13.

Preobrazhensky A.P. 2007. *Modeling and algorithmic analysis of diffraction structures in cad radar antennas.* Voronezh, Nauchnaya kniga, 248 p.

Preobrazhensky A.P., Choporov O.N. 2004. *Forecasting algorithms of radar characteristics of objects when restoring radar images.* Management systems and information technologies. No 5 pp. 85–87.

Preobrazhensky A.P., Choporov O.N. 2004. *The method of prediction of radar characteristics of objects in the wavelength range using the measurement results of scattering characteristics of the discrete frequencies.* Management systems and information technologies. No 2. pp. 98–101.

Preobrazhensky, A. 2007. *Simulation and algorithmic analysis of diffraction structures in CAD radar antenna.* Voronezh. Scientific book. 248 p.

Rozzi T. 1990. *Equivalent network of transverse dipoles on inset dielectric guide: application to linear arrays.* IEEE Transactions on Antennas and Propagations 38(3). pp. 380–383.

Ufimtsev, P. 1962. *Method of edge waves in physical theory of diffraction.* Moscow, Soviet radio, 243 p.

Vasiliev, E. 1987. *Excitation of rotating bodies* Moscow. Radio and communication. 270 p.

Wasiljev E., Makkaveev V., Gorelikov A. 1982. *On the application of some quadrature formulas for the solution of integral equations of the second kind.* Machine design of devices and systems of the microwave.—Vol. 6.—p. 68–84.

Yackevich V., Karshakevich S. 1981. *The sustainability of the process of convergence of the numerical solution in electrodynamics.* Izvestiya Vuzov Seriya Electronics. – 24(2)—pp. 66–72.

Zhulyabin D. 2014. *Models of channels for wireless communication systems.* Modeling, optimization and information technologies, No 1, p.1.

Current Issues of Science and Research in the Global World – Kunova & Dolinsky (Eds)
© 2015 Taylor & Francis Group, London, ISBN 978-1-138-02739-8

The possibilities of calculation the scattering characteristics on parallel approach

Igor Lvovich & Eugen Ružický
Pan-European University, Bratislava, Slovak Republic

Oleg Choporov
Voronezh Institute of High Technologies, Voronezh, Russia

ABSTRACT: In the paper we consider the processes of scattering of electromagnetic waves on the objects with complex shape. Different types of methods, depending from the wavelength are analyzed. The main steps of the solution of the problem of scattering of electromagnetic waves on the object with complex shape are discussed. The possibilities of use the parallel approach for calculation is discussed. Based on integral equations method developed an algorithm for the numerical solution of the problem, with the help of parallel algorithm of the method the Gauss described the process of parallelization. Some results of calculations are given.

1 INTRODUCTION

The scatterers of radar signals (radar targets of space, air, land and water based) and converters of energy of Electromagnetic Waves (EMW) (banners, the means of reducing the visibility of radio waves, antenna device SHF and EHF ranges of waves), as a rule, are characterized by large electric size, complex geometry, the presence of absorbing and non-linear elements.

Analysis and synthesis of the above electrodynamics objects based on a rough idea about running in them physical processes carry the risk of significant and difficult-to-control of estimation errors of their main characteristics, which, as a rule, very quickly change when you change the frequency, type of polarization and angle of incidence of the electromagnetic wave.

Measurement of the main characteristics of the radar targets (polarization matrix, effective surface scattering in monostatic and bistatic modes scattering) in a wide frequency band and extensive angular sector requires either certified specially equipped antenna polygon or certified anechoic chamber (the cost of which may reach several million dollars), and also the big expenses of time and means. To achieve this goal it is necessary to solve a number of tasks:

1. To make the model of scattering of electromagnetic wave on a 2D perfectly conducting the objects of arbitrary shape;
2. To make the algorithm of numerical solution of the problem of scattering of electromagnetic waves on the basis of the method of integral equations;
3. To develop the software tool on the basis of developed algorithm allowing to calculate the characteristics of scattering of electromagnetic waves (effective area scattering) for E-polarized incident electromagnetic wave, with application of parallel computing.

2 THE METHODS OF CALCULATION THE SCATTERING CHARACTERISTICS OF THE OBJECTS

The methods can be divided into three classes. The first is asymptotic methods. In turn, they can be divided into two groups: asymptotic and heuristic. The difference here lies in the extent of their mathematical validity. The first is more justified—can be attributed geometrical optics, geometrical theory of diffraction, the second physical optics (so-called Kirchhoff approximation) and the boundary wave method of P. Ya. Ufimtsev.

The second class is rigorous methods. Those are the methods through which you can obtain the solution arbitrarily close to accurate.

These measures include a method of separation of variables, the method of integral transformations, and the method of integral equations.

Finally, to the third class are a hybrid methods. As a rule, in the framework of this approach, using one or another of the approximate method are the points (or fields) that are then substituted into various electrodynamics operators (or specified using the latter). An example of such methods is the method of stationary functional J. Schwinger, various kinds of combined methods, combining a variety of methods and others.

There are three characteristics of the area, which can be the size of the lens L: quasi-static (Rayleigh region), when $L/\lambda \ll 1$; resonance region, when $L/\lambda \sim 1$; quasi-optical region, when $L/\lambda \gg 1$ (the length of the electromagnetic wave).

In the quasi-static region the solution of the problem is obtained from the solution of the wave equation (Helmholtz equation), but the solution in analytical form does not always have to resort to numerical solution.

In the resonance region (as the most difficult for research) is very often used method of separation of variables or method of integral equations. In the quasi-optical field is applicable two types of methods: radiation and waveguide. Radiation methods, above all, is the geometric optics and refinement: geometric theory of diffraction, lets spread the geometrical methods in diffraction problem; complex geometrical optics, allowing to calculate fields in the field of refractive shadows, the parabolic equation method, also widen the scope of radiological methods.

Wave methods include the method of physical optics (Kirchhoff approximation) with amendments, of which we mention the boundary wave method is used to locate the amendments to the field of radiation associated with construction of the approximate (quasi-optical) of own functions.

The wording in the form of the integral equation or system of integral equations, usually reduces the dimensionality of the problem and, secondly, reduces the initial boundary problem in unbounded region to the task in a limited area (surface and in the volume of the lens).

Under this method, points, current on the surface of the body, is determined on the basis of the solution of the integral equation. Calculation of scattered electromagnetic field is carried out on the basis of found currents. This method is effectively used for calculation of scattering characteristics of bodies, which sizes lay in the resonance region (the two-dimensional problem), and bodies whose sizes are several wavelengths (three-dimensional problem). With increasing body size increases sharply necessary for calculations of machine time, the amount of RAM. In the framework of the method, it is possible calculation of scattering characteristics of a perfectly conducting bodies with radio absorbing coatings.

3 THE USE METHOD OF MOMENTS IN PARALLEL CALCULATION OF SCATTERING CHARACTERISTICS

Very often when a numerical study of the problem of scattering of electromagnetic waves by the method of moments is used. This General approach to the tasks of the radiation consists essentially in the reduction of the investigated integral equation to a system of linear algebraic equations with N unknowns, which usually represent the coefficients of some expansion for current.

The problem solution can be obtained in four steps:

1. The vector of current J is decomposed into a number of basic functions J_n in the field of definition of the operator L_{on}.
2. Determined appropriate internal work and establishes a system of weighting functions.
3. Calculated inner product and thereby equations are the matrix mind.
4. The solution of the matrix equation is found. After determining the unknowns (i.e. current) it is easy to determine such characteristics as the chart radiation in the far-field zone and the complex resistance (impedance) antenna.

During the calculation of the currents on the surface of the object we can use the parallel approach. Introduced to the original values and is a matrix, the matrix is filled with elements, the number of elements of the matrix depends on the size of the matrix. The size of the matrix depends on the number of sampling, as the number of spatial discretization is great, then the matrix size should be large enough. It follows that the computation time of this matrix will grow depending on the number of matrix elements. In order to decrease the time to perform the calculations, we apply the method of parallel computing by creating threads. At the stage of formation of the matrix, to create new threads to divide the matrix in two parts (for 2 core processor) and the process of forming a matrix wakes take place in two independent threads that will allow more efficient use of computing power, computers, and get gain time, because the calculation process will occur on two cores. We use Gauss Method with application of technology of parallelization during the calculations.

In computing practice, for example, when solving boundary value problems for partial differential equations by the method of finite differences and finite elements, the systems of algebraic equations can be obtained with sparse (tape) matrices A.. The solution of such problems can be obtained by using the method the Gauss and above other direct methods. Developed many proprietary direct-based methods tape systems, for example, block methods, the method of cyclic reduction, etc. Note that when solving tape the systems of algebraic equations as serial and parallel computing systems for storage matrix A is advisable not to use a two-dimensional (n*n) array, and the number of one-dimensional arrays for storing nonzero diagonals of the matrix A, or single-dimensional array, in which the diagonal of the matrix A are stored sequentially.

Based on integral equations method developed an algorithm for the numerical solution of the problem, with the help of parallel algorithm of the method the Gauss described the process of parallelization. Using the considered algorithm we developed the program. The calculation is performed for resonance region, i.e. the size of the body is about several wavelengths. In the calculations of scattering cross-section is normalized by the square of the wavelength.

On the Table 1 we can see the time calculations in direct and parallel modes. After completion of calculations is displayed on screen graph of effective work. Blue denotes work in simple mode, red in parallel, seen a noticeable difference and we can conclude that at increasing the dimension of the matrix is used in parallel mode requires less time for dawns than usual.

Table 1. Time of calculation.

Dimension of matrix	500	5000	6000	7000	8000
Time of usual calculation, sec	5	12	19	25	35
Time of parallel calculation, sec	5	27	40	57	73

4 CONCLUSIONS

In the paper we consider the possibility of calculation of scattering characteristics of the objects with complex shape. It was shown, that the time, necessary for calculation can be decreased in several times.

REFERENCES

Baranov A. 2012. *The problems of functioning of mesh networks.* Bulleting of the Voronezh Institute of high technologies. No 9. pp. 49–50.

Bashkatov A. 2013. *The analysis of scattering characteristics of uwb-signals on the subject is complex.* Modeling, optimization and information technologies, No 3, p.8.

Golovinov S., Preobrazhensky A., Lvovich I. 2010. *Modeling of the propagation of millimeter waves in urban areas on the basis of the combined algorithm.* Telecommunications. No 7. pp. 20–23.

Hujanen A., Holmberg J., Sten J. 2005. *Bandwidth limitation of impedance matched ideal dipoles.* IEEE Transactions on Antennas and Propagations, 53(10), pp. 3236–3239.

Lvovich Ya., Lvovich I., Preobrazhensky A. 2010. *The problems of evaluating the characteristics of scattering of electromagnetic waves by a diffraction structures in their design.* Bulleting of the Voronezh Institute of high technologies. No 6. pp. 255–256.

Lvovich Ya., Lvovich I., Preobrazhensky A., Golovinov S. 2012. *Study of the method of ray tracing for the design of wireless communication systems.* Electromagnetic waves and electronic systems. No 1. pp. 32–35.

Miloshenko O. 2012. *The methods of characteristic estimation of radio wave propagation in the systems of the mobile radio communication.* Bulleting of the Voronezh Institute of high technologies. No 9. pp. 60–62.

Mitra, P. (ed.), 1977. *Computational methods in electrodynamics*, Moscow, Mir, p. 485.

Motin D. 2013. *On simulation of coverage area of service in a wireless communication system.* Modeling, optimization and information technologies, No 1, p. 13.

Preobrazhensky A.P. 2007. *Modeling and algorithmic analysis of diffraction structures in cad radar antennas.* Voronezh, Nauchnaya kniga, p. 248.

Ross, D., Volakis, J., Anastassiu, H. 1995. *Hybrid finite element-modal analysis of jet engine inlet scattering.* IEEE Transactions on Antennas and Propagations, 43(3), pp. 277–285.

Shirman, Ya., Losev, Yu., Minervin, N., Moskvitin, S., Gorshkov, S. and other. 1998. *Radioelectronic systems: principles of construction and theory.* Moscow, CJSC «Maquis», p. 832.

Ufimtsev, P. 1962. Method of edge waves in physical theory of diffraction. Moscow, Soviet radio, p. 243.

Zhulyabin D. 2014. *Models of channels for wireless communication systems.* Modeling, optimization and information technologies, No 1, p. 1.

The problems of estimation of characteristics in Wi-Fi communications

Yakov Lvovich, Andrey Preobrazhensky & Alyona Kurotova
Voronezh Institute of High Technologies, Voronezh, Russia

ABSTRACT: This paper describes the methods for estimation the propagation character-istics of electromagnetic waves in Wi-Fi communications. Using mathematical relationships below, an algorithm was built for estimating the degree of attenuation of radio waves in the urban environment. Considered algorithm ignores natural interference caused by the propa-gation of electromagnetic waves (including broadband). The program was created for solving the problem of wave propagation. The testing software were hold. The system requirements for the operation of this software were present.

1 INTRODUCTION

The questions of modeling and design of wireless networks are of great and increasing interest.

In the era of passing information, this phenomenon comes in even the most remote cor-ners of our planet. We often hear in the media on new projects on information society. For example, you can hear from our politicians, what happens supply computer equipment and Internet services of rural schools and schools of small towns. But this process is very slow. Much faster the development of computer technology.

The aim is to build a software tool to simulate the propagation of waves in the network Wi-Fi.

To achieve this goal following problems were solved:

– As a result of analysis of theoretical sources identified the theoretical basis of using a wire-less network Wi-Fi.
– The possibilities and characteristics of the main ways of modeling network Wi-Fi, consid-ered their classification.
– A review of existing methods for optimizing the network Wi-Fi.
– The model of network Wi-Fi, and held its optimization for future use.
– Implemented processing, systematization and generalization of the study results.

2 THE MODELS FOR ESTIMATION OF FADING IN WIRELESS NETWORKS

When the wave propagates from transmitter to receiver her way very diverse: from their line of sight to strongly closed obstacles, buildings, trees, and path. In addition, unlike wire com-munications, where considered constant parameters in the wireless radio channels have a substantially random parameters that are difficult to analyze. Simulation of radio—the most difficult task of designing radio. Analysis mainly performed statistically using the experimen-tal data, it is sometimes performed for the same or similar system.

Typically, the radio propagation modeling based on prediction average received signal level at a given distance from the emitter as well as determining its dispersion values depending on the situation on the road. The calculation to determine the radio coverage area of the transmitter. In this paper we consider the construction of subsystem evaluation network coverage Wi-Fi.

Using mathematical relationships below, an algorithm was built for estimating the degree of attenuation of radio waves in the urban environment. Transmission loss inside the buildings (including the walls and floor decks) at 2400 MHz are estimated as follows:

$$L[\partial B] = 40 + 35 \lg(R[\kappa M]).$$

Transmission loss at the base—line caller determined on the basis of the following relation

$$L[\partial B] = 65 + 40 \lg(R[\kappa M]) + 40 \lg(f[M\Gamma s]) - 4G[dBi]$$

where R—the distance from the base to the subscriber, G—gain antenna, f—frequency.

At the output of the subsystem is determined by the required access points.

Within this algorithm is based on several possible base stations. In this case, the interference of radio waves will form a complex picture of the power distribution propagating electromagnetic waves. The algorithm is intended to account for the phenomena of electromagnetic wave propagation in urban areas is because there is a large number of obstacles, causing attenuation of the desired signal. With increasing frequency, the transmission signal attenuation of the propagating electromagnetic wave is increased.

The propagation of radio waves of the energy will be spent in reflection. Most of the rays while going past the party, as it is in a restricted area.

Positioning accuracy is connected in order to maximize the power delivered from the base station is determined by the sampling pitch angle.

Considered algorithm ignores natural interference caused by the propagation of electromagnetic waves (including broadband).

Above algorithm is universal, it is automatically considered a wide range of possible input parameters.

The advantage of this approach is that it allows you to evaluate ("computer experiment") performance in wireless communication systems without a real "full-scale" experiment.

Generally, the algorithm can be refined models involving propagation of electromagnetic waves—diffraction edge waves, creeping waves, diffuse reflection, the reflection from the earth's surface. In addition, can be refined multiple reflections of waves inside the building, but, apparently, their contribution to extending the main field at least 10 dB less than the waves having a single reflection. All these additional refinements in the future may be presented as optional modules with the corresponding attenuation characteristics of electromagnetic waves.

The result is a software product that allows a visual assessment of propagation. In developing the program, and in particular the interface to consider many requirements such as simplicity and convenience. It was added specifically to people who have never worked with such systems could easily deal with the product.

In this paper, the software subsystem implemented optimize the location of access points, Wi-Fi on the specified criteria is developed. Based on the results of computer simulation were identified network coverage Wi-Fi. At the output of the subsystem is determined by the required access points.

3 CONCLUSIONS

When the wave propagates from transmitter to receiver her way very diverse: from their line of sight to strongly closed obstacles, buildings, trees, and path. In addition, unlike wire communications, where considered constant parameters in the wireless radio channels have a substantially random parameters that are difficult to analyze. Simulation of radio—the most difficult task of designing radio. Analysis mainly performed statistically using the experimental data, it is sometimes performed for the same or similar system.

Typically, the radio propagation modeling based on prediction average received signal level at a given distance from the emitter as well as determining its dispersion values depending on the situation on the road. The calculation to determine the radio coverage area of the transmitter. In this paper we consider the construction of subsystem evaluation network coverage Wi-Fi.

Using mathematical relationships below, an algorithm was built for estimating the degree of attenuation of radio waves in the urban environment. Transmission loss inside the buildings (including the walls and floor decks) at 2400 MHz are estimated as follows:

$$L[dB] = 40 + 35 \lg(R[km]).$$

Transmission loss at the base-line caller determined on the basis of the following relation

$$L[dB] = 65 + 40 \lg(R[km]) + 40 \lg(f[MHz]) - 4G[dB],$$

where R—the distance from the base to the subscriber, G—gain antenna, f—frequency.

At the output of the subsystem is determined by the required access points.

Within this algorithm is based on several possible base stations. In this case, the interference of radio waves will form a complex picture of the power distribution propagating electromagnetic waves. The algorithm is intended to account for the phenomena of electromagnetic wave propagation in urban areas is because there is a large number of obstacles, causing attenuation of the desired signal. With increasing frequency, the transmission signal attenuation of the propagating electromagnetic wave is increased.

The propagation of radio waves of the energy will be spent in reflection. Most of the rays while going past the party, as it is in a restricted area.

Positioning accuracy is connected in order to maximize the power delivered from the base station is determined by the sampling pitch angle.

Considered algorithm ignores natural interference caused by the propagation of electromagnetic waves (including broadband).

Above algorithm is universal, it is automatically considered a wide range of possible input parameters.

The advantage of this approach is that it allows you to evaluate ("computer experiment") performance in wireless communication systems without a real "full-scale" experiment.

Generally, the algorithm can be refined models involving propagation of electromagnetic waves—diffraction edge waves, creeping waves, diffuse reflection, the reflection from the earth's surface. In addition, can be refined multiple reflections of waves inside the building, but, apparently, their contribution to extending the main field at least 10 dB less than the waves having a single reflection. All these additional refinements in the future may be presented as optional modules with the corresponding attenuation characteristics of electromagnetic waves.

The result is a software product that allows a visual assessment of propagation. In developing the program, and in particular the interface to consider many requirements such as simplicity and convenience. It was added specifically to people who have never worked with such systems could easily deal with the product.

In this paper, the software subsystem implemented optimize the location of access points, Wi-Fi on the specified criteria is developed. Based on the results of computer simulation were identified network coverage Wi-Fi. At the output of the subsystem is determined by the required access points.

REFERENCES

Baranov A. 2012. *The problems of functioning of mesh networks*. Bulleting of the Voronezh Institute of high technologies. No 9. pp 49–50.
Bashkatov A. 2013. *The analysis of scattering characteristics of uwb-signals on the subject is complex*. Modeling, optimization and information technologies, No 3, p. 8.

Dmitriev, A., Efremova, E., Kletsov, A., Kuzmin, L., Laktyushkin, A., Yurkin, V. 2008. *Wireless Ultrawideband Communications and Sensor Networks*. Journal of communications technology and electronics, 53(10): 1278–1289.

Kochin B., Vorotnitsky Yu., Strigalev D. 2012. *Methodology for quick evaluation of the power of WI-FI signal when passing obstacles within the building* (http://elib.bsu.by/bitstream/123456789/22199/1/%D 0%9 A%D0%BE%D1%87%D0%B8%D0%BD%20%D0%92_%D0%9F.pdf).

Lvovich I., Preobrazhensky A., Yurov R., Choporov O. 2006. *Software complex for automated analysis of scattering characteristics of objects with application of mathematical models*. Management systems and information technologies. No 2 pp. 96–98.

Lvovich Ya., Lvovich I., Preobrazhensky A. 2010. *The problems of evaluating the characteristics of scattering of electromagnetic waves by a diffraction structures in their design*. Bulleting of the Voronezh Institute of high technologies. No 6. pp. 255–256.

Miloshenko O. 2012. *The methods of characteristic estimation of radio wave propagation in the systems of the mobile radio communication*. Bulleting of the Voronezh Institute of high technologies. No 9. pp. 60–62.

Motin D. 2013. *On simulation of coverage area of service in a wireless communication system*. Modeling, optimization and information technologies, No 1, p. 13.

Preobrazhensky A.P., Choporov O.N. 2004. *The method of prediction of radar characteristics of objects in the wavelength range using the measurement results of scattering characteristics of the discrete frequencies*. Management systems and information technologies. No 2. pp. 98–101.

Preobrazhensky A.P., Choporov O.N. 2004. *Forecasting algorithms of radar characteristics of objects when restoring radar images*. Management systems and information technologies. No 5 pp. 85–87.

Preobrazhensky A.P. 2007. *Modeling and algorithmic analysis of diffraction structures in cad radar antennas*. Voronezh, Nauchnaya kniga, 248 p.

Ryzhov, A., Lazarev, V., Mohseny, T., Nikerov, D., Andreev, Yu., Dmitriev, A., Chubimsky, N. 2012. *Weakening of UWB chaotic signals range of 3–5 GHz when passing through walls*. Journal of Radio-electronics. No 5. (http://jre.cplire.ru/jre/may12/1/text.html).

Zhulyabin D. 2014. *Models of channels for wireless communication systems*. Modeling, optimization and information technologies, No 1, p. 1.

Intelligent interferer in LTE and LTE-A

T. Páleník, A. Ralbovský & M. Rakús
*Faculty of Electrical Engineering and Information Technology, Institute of Telecommunications,
Slovak University of Technology, Bratislava, Slovakia*

J. Doboš
Faculty of Informatics, Pan-European University, Bratislava, Slovakia

A. Silva
Department of Electronics and Telecommunications, University of Aveiro, Portugal

M. Rupp
Institute of Telecommunications, Technische Universitaet Wien, Vienna, Austria

ABSTRACT: LTE (Long Term Evolution) is a 3GPP standard for high-speed low-latency mobile communication based on OFDM technology. LTE-A (LTE Advanced) is an enhanced version of the LTE. Although security is an integral part of both standards, it is mainly focused on preventing eavesdropping. Several vulnerabilities have been found in LTE standard, that allow performing DoS attacks to make regular communication impossible. In this paper, we analyze weaknesses of LTE as well as LTE-A and perform computer simulations to quantify the influence of an intelligent interferer on the quality of communication.

1 INTRODUCTION

Wireless communications systems are constantly evolving and new standards providing better parameters are being introduced every few months. A major step in mobile technologies was introduction of Long Term Evolution (LTE, also referred to as 4G), which was intended as a successor of the UMTS (3G). Primary motivation for developing LTE was to enable network operators to increase their network capacity and provide the end users with higher data rates as well as lower delays than were possible with UMTS. However, large portions of radio spectrum are not today available (or very high costs are associated with it), therefore achieving high spectral efficiency was a key issue in order to fulfil those requirements. LTE-A is an enhanced version of LTE providing increased peak data rate, higher spectral efficiency and increased number of simultaneously active subscribers. The paper is organized in the following way: section II provides a brief simplified overview of the LTE physical layer with focus on the identified weaknesses regarding interference. Section III describes an intelligent interferer strategy and provides a quantitative evaluation of the effect of such interference on system performance. Section IV discusses the possibilities of acceleration of such simulation by means of massively parallel processing utilizing the GPU. Section V then concludes the paper.

2 OVERVIEW OF LTE PHYSICAL LAYER

Physical layer of LTE is completely different from that of UMTS. Instead of CDMA used in UMTS, OFDMA and SC-FDMA techniques are used in LTE for downlink and uplink, respectively.

2.1 *LTE frame structure*

An LTE frame is a basic data structure used in LTE, which consists of several smaller units and blocks. Two types of frame defined in LTE specification (ETSI TSGR 2014), depending on duplex type. The specification defines both Frequency Division Duplex (frame structure of Type 1) as well as Time Division Duplex (frame structure of Type 2), however vast majority of operators prefer FDD as it fits well into existing and perspective spectrum assignments (TSI 2010) Therefore throughout this paper, we deal only with FDD and unless specified, by the LTE frame we mean the frame of Type 1.

Structure of frame for FDD consists of 10 equally long subframes or 20 slots, respectively (two consecutive slots form one subframe). It is graphically illustrated in Figure 1. According to specification (ETSI TSGR 2014): "For FDD, 10 subframes are available for downlink transmission and 10 subframes are available for uplink transmissions in each 10 ms interval. Uplink and downlink transmissions are separated in the frequency domain. In half-duplex FDD operation, the UE cannot transmit and receive at the same time while there are no such restrictions in full-duplex FDD".

2.2 *Resource Block*

Resource Block is a smallest unit of radio spectrum that can be allocated to an individual. It consists of a mesh of 12 carrier frequencies and 6 or 7 time symbols depending on length of the Cyclic Prefix. Individual fields defined by an exact frequency and time position are called Resource Elements. One symbol is carried by one Resource Element. Except the user data itself, also physical signals not carrying any useful data, called Downlink Demodulation Reference Signals (pilot symbols) are transmitted on defined positions in the Resource Block in order to allow coherent demodulation. They are described in detail later in this paper.

2.3 *Physical channels in LTE*

LTE standard operates with a concept of physical channels. They are separated and clearly distinguishable data streams (usually by frequency or position in LTE frame) used for transmission of different types of information. Complete list physical channels as well as their detailed description can be found in (ETSI TSGR 2014). For the purposes of this paper, only Physical Downlink Shared Channel (PDSCH) is interesting. It is used to carry user data.

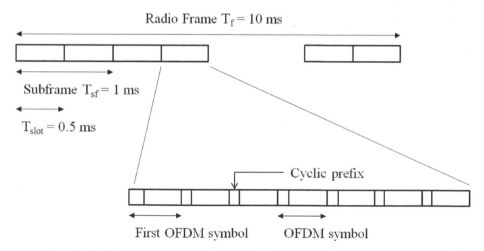

Figure 1.   LTE frame structure type 1 (FDD) with Cyclic Prefix (TSI 2010).

2.4 *Physical signals*

In addition to signals from higher layers carrying useful data, some supplementary signals are added in order to enable communication. The most important ones are synchronization signals are reference signals. The purpose of reference signals is to facilitate the receiver to estimate channel state. In this paper we focused on Demodulation-Reference signal, also called UE-specific Reference Signal, which is intended to assist the UE at estimating channel input response in order to correctly perform equalization. It is contained in four Resource Elements in a Resource Block. The positions of the Resource Elements containing symbols of this reference signal are selected in such a manner, that one RE lies on every third frequency. This enables the receiver to perform relatively precise channel estimation by interpolating these values and equalization in frequency domain. The reference signal consists of known symbols transmitted at a well-defined OFDM symbol positions in the slot. There is one reference signal transmitted per downlink antenna port and an exclusive symbol position is assigned for a reference signal.

Our assumption is that by jamming reference signals, it would be possible to increase the overall error rate and therefore make the communication impossible while the overall energy of the jamming waveform would remain the same.

2.5 *Quantitative verification of interference effects*

To verify the impact of intelligent interferer on LTE system by quantitative means, we created a set of MATLAB scripts operating as an LTE simulator.

All simulations were performed by Monte Carlo method (Jeruchim 2007). We simulated transmission of random binary data, which were first modulated using Quadrature Amplitude Modulation (QAM) of appropriate order, mostly 4-QAM. After transmission through the selected channel model and performing other operations depending on the chosen scenario, received data were bit-after-bit compared with the original data and number of errors was computed. Then the resulting BERs (Bit Error Rates) were.

In simulations discussed below we consider scenario of one interferer located in the vicinity of the UE as well as base station and capable of sending any signal in the frequency range used by the eNodeB and UE. In the real scenario, the interferer would also have to be able to intercept actual signals transmitted by eNodeB for the UE in order to synchronize to the network and figure out correct frequencies.

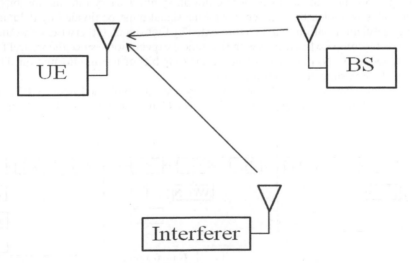

Figure 2.   Principle of jamming in simulation scenarios.

AWGN is generated according to equation (1), where T_{sym} is the signal's symbol period and T_{samp} is the signal's sampling period:

$$\frac{E_s}{N_0}[dB] = 10\log\left(\frac{T_{sym}}{T_{samp}}\right) + SNR[dB] \qquad (1)$$

The relationship between E_s/N_0 and E_b/N_0 is determined by formula (2), where k is the number of information bits per symbol.

$$\frac{E_s}{N_0}[dB] = \frac{E_B}{N_0}[dB] + 10\log_{10}(k) \qquad (2)$$

Since all simulations operate in discrete-time (every symbol is represented by one complex sample from the I-Q plane), time duration of one symbol is equal to time duration of noise sample. Therefore according to (1) this ratio is equal to SNR. All data are transmitted in baseband.

2.6 LTE simulator

In order to make implementation simpler and keep the computational demands on reasonable level, without a substantial loss of usability, our LTE simulator includes some simplifications:

1. Transmission only for LTE downlink data shared channel is emulated.
2. Only single-antenna scenarios are considered.
3. Only full Frequency Division Duplex is emulated, therefore only frame structure of Type 1 is used.
4. Custom interpolation method is applied on demodulation reference signals, described in detail later in this chapter.
5. Demodulation reference signals are of constant value.
6. Perfect synchronization is assumed and all symbols except those reserved for demodulation reference signals are used to transmit data.
7. No scrambling and error coding is used and only error rate of lowest layer is evaluated.

Despite these simplifications, however, the results produced by the simulator should be close enough to reality. In scenarios, where not all symbols carry information bits (Cyclic Prefix and pilot symbols), the total energy of the signal must be divided by the number of bits of the useful information that are represented by that signal. Otherwise, we would not be able to correctly compare the results with theoretical expectations, since theoretical formulas are constructed for E_b/N_0 parameters considering only bits of useful information. The block diagram of the link simulator is shown in Figure 3.

The acronyms should be well known, but they are also explained here: ARQ stands for the Automatic Repeat and reQuest, ECC for Error Control Coding, CM for Constellation

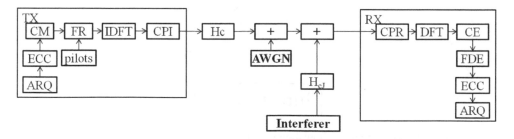

Figure 3. Block diagram of full LTE simulator with generic interferer.

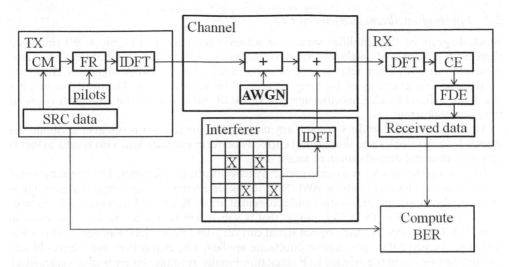

Figure 4. Diagram of simplified simulation of intelligent interferer in LTE system.

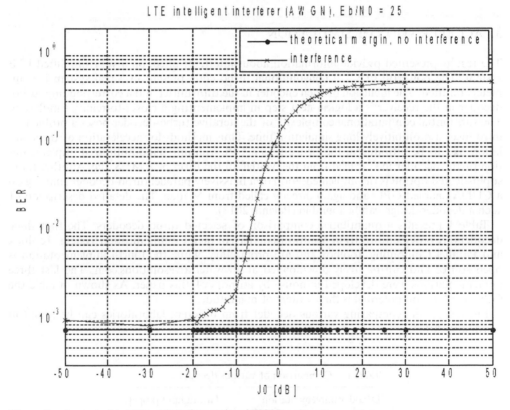

Figure 5. Impact of intelligent interferer using AWGN.

Mapping—the modulation in digital domain, FR for Framing, (I) DFT for (Inverse) Discrete Fourier Transform, CPI/R for Cyclic Prefix Insertion/Removal, CE for Channel Estimation, FDE for Frequency Domain Equalization, The channel model is assuming a Rayleigh-fading multipath channel described by channel matrix Hc and (Hcj respectively) with AWGN noise.

Block diagram of the simplified simulation scenario is depicted in Figure 4. We simulated therein an influence of intelligent interferer jamming only selected symbols in LTE frame structure—demodulation reference signals, also called pilots. It can be highly effective even though the overall energy of jamming signal remains relatively low. The positions of pilot signal are defined by LTE specification (ETSI TSGR 2014) and do not change throughout the communication.

Although the attacker does not jam any useful data, the receiver is not able to obtain the channel characteristics and thus cannot correctly perform equalization. This results in incorrect detection and demodulation of useful data.

To perform the attack, we assume perfect synchronization of jammer. The jamming signal in frequency domain consists of AWGN in Resource Elements where demodulation reference signal symbols are transmitted and zeroes in all other Resource Elements. Such resource blocks are grouped to form LTE Frame that is subsequently transformed to time domain using IFFT and added to the original signal carrying data in our LTE simulator. This addition is performed before any channel effects are applied. The transmitter and receiver blocks remain the same as in the original LTE simulator. Finally, resulting bit error rate is calculated. Similarly to the previous scenarios, effects of intelligent interferer are investigated at E_b/N_0 equal to 25dB, what is value, at which BER of 10^{-3} is reached.

3 ACCELERATION OF SIMULATION

The results presented provide information about interferer effect on a very simplified LTE system. The powerful ECC—a convolutional turbo code present in a real system is completely omitted. The inclusion of such encoder in the transmitter and a corresponding turbo-decoder in the receiver is a logical next step in implementing a closer-to-reality simulation. The substantial computational complexity of the iterative turbo-decoder poses a problem: it introduces a prohibitively long simulation time. Thus methods for acceleration of the simulation had to be introduced: the first one is a CPU-running hand-optimized implementation of the decoder in C, called from the Matlab in the form of a MEX file. Another more sophisticated option is the implementation of a massively parallel turbo-decoder running on a CUDA-enabled GPU and again directly called from Matlab. The detailed description of such a decoder design can be found in (Palenik 2013).

Table 1 presents a quantitative comparison of selected turbo decoders. The first three decoders are provided directly by Mathworks as part of their Communication Systems Toolbox and Parallel Computing Toolbox and are proprietary, while the fourth implementation is open source and can be freely downloaded at www.iterativesolutions.com. The last three implementations were developed in-house by authors of this paper. As shown in table the GPU acceleration potential is three orders of magnitude.

Further results comparing various decoder from different laboratories can be found in (Cavallaro 2011).

Table 1. Comparison of various decoders.

Decoder implementation	Throughput [kbps]
Mathworks CPU	225
Mathworks GPU	107
Mathworks GPUopt	339
IS CML	196
Matlab CPU native	1.17
Matlab CPU MEX	167
Matlab GPU exp	1424

4 CONCLUSION

In this paper we analyzed impact of specific kind of interference in LTE system. By the means of simulation we found, that LTE is susceptible to DoS attacks on physical layer. Such an attack can be performed by focusing jamming energy on demodulation reference signals, and causes severe disruption of communication. The effect of the interferer was quantitatively evaluated using simplified simulations, which show that even a very small ESD of the interfering signal results in a sharp increase in communicating system's error ratio, effectively increasing the number of errors that occur during the transmission hundred-fold. On the other hand, an accelerated massively-parallel system block necessary to implement a more realistic ECC-enabled complex simulation was also introduced.

ACKNOWLEDGEMENT

This work was supported by, Slovak Research and Development Agency under contracts SK-AT-0020-12 and SK-PT-0014-12, by Scientific Grant Agency of Ministry of Education of Slovak Republic and Slovak Academy of Sciences under contract VEGA 1/0518/13, by EU RTD Framework Programme under ICT COST Action IC 1104 and by Visegrad Fund and National Scientific Council of Taiwan under IVF–NSC, Taiwan Joint Research Projects Program application no. 21280013 "The Smoke in the Chimney—An Intelligent Sensor—based TeleCare Solution for Homes".

REFERENCES

Cavallaro, J., Wu, M., Sun, Y., Wang, G., 2011. Implementation of a High Throughput 3GPP Turbo Decoder on GPU. *Journal of Signal Processing Systems (JSPS)*, 2011. New York: Springer Publishing.

ETSI TSGR. 2014. *3GPP: Evolved Universal Terrestrial Radio Access (E-UTRA); Physical channels and modulation (3GPP TS 36.211 version 11.5.0 Release 11)*. Sophia Antipolis Cedex: European Telecommunications Standards Institute.

Jeruchim, M.C., Balaban, P., Shanmugan, K.S. 2007. *Simulation of Communication Systems: Modeling, Methodology and Techniques*. NewYork: Kluwer Academic.

Palenik, T., Farkas, P. 2013. *Utilizing massive parallelism in decoding of modern error-correcting codes for accelerating communication systems simulations,* Kosice: FEEI TU.

TSI. 2010. *LTE in a Nutshell—Physical layer*, Toronto: Telesystem Innovations Inc.

The smoke in the chimney—international cooperation at Pan-European University

J. Doboš & E. Ružický
Faculty of Informatics, Pan-European University, Bratislava, Slovakia

T. Páleník & P. Farkaš
Slovak University of Technology, Bratislava, Slovakia

Y.C. Tseng & Y. Ren
National Chiao Tung University, Hsinchu, Taiwan

A. Vidács & L. Vajda
Budapest University of Technology and Economics, Budapest, Hungary

M. Šimek
Brno University of Technology, Brno, Czech Republic

ABSTRACT: This paper describes our project: "The Smoke in the Chimney—An Intelligent Sensor-based TeleCare Solution for Homes" (TeleCalmPlus), that is currently being implemented in the Faculty of Informatics in cooperation with partner universities from Taiwan, Hungary and Czech Republic. It is one of two projects which were selected for funding by International Visegrad Fund and National Scientific Council of Taiwan from all submitted proposals reacting on call in year 2012. The basic idea is to design a remote monitoring system for assisting the care of senior citizens utilizing cutting edge sensor networks technologies at their home, a cloud infrastructure for intense data-processing tasks, and a simple app for Android smartphones. This paper was compiled from the detailed project description (Vidacs 2013). Please refer to this document for more information.

1 INTRODUCTION

While in recent years much has been done in Europe and all over the world to raise awareness of the challenges that are caused by the ageing population and the sharply increasing number of elderly people. The evolution of information technology and telecommunications, especially the spread of embedded systems and data communications networks opens new possibilities how to deal with this situation. One of the promising applications, targeted in this project, is assistive application, among other systems monitoring and helping the elderly, the handicapped and those who need rehabilitation or depend on nursing care. Besides their social importance, assistive applications have great economic significance as well: they create new market opportunities for the economy.

The TeleCalmPlus represents a system based on the analogy of "smoke in the chimney". In the past family members and relatives were living near each other, and were able to observe the health state of each other on a daily base. The easiest way to do this was to look early in the morning on the chimney of the relatives or family members, checking for smoke. If there was smoke in the chimney, than everything was fine (they started their day in a normal behavior). If no smoke was observed than something was fishy, and attention was needed. Nowadays families live far away from each other, not being able to check the smoke in the chimney anymore. Our goal is to allow people to do something similarly simple with the help of informatics.

The system consists of simple devices placed in the users' home, a server for data storage/evaluation and an Android application for the relatives. The information from the user devices are stored on the server and the data collected can be viewed remotely. Based on the activities and physiological information gathered from the devices, the TeleCalmPlus system gives an overall-evaluation of the users' health/daily living condition. The users' condition represented by state (green, yellow, red) can be viewed remotely on the Android based devices with internet connection. The user can define whom to give permission to view his/her everyday activities and physiological condition. The main advantage of using this system is, that it will signalize continuously if everything is alright. Meaning people don't need to wonder, why no alarm is coming for a longer time.

2 OUR PARTNERS

The consortium consists of four research organizations from three Visegrad Four countries and one Taiwanese partner. It is a truly beneficial to have a joint research project between these partners because they are considered to be experts in different fields, namely technological, dissemination, system programming, telecommunication, software engineering know-how etc.

Budapest University of Technology and Economics (BME) (http://telecalmplus.tmit.bme.hu/) is the largest school of technology in Hungary, with eight faculties, over 3000 employees and 20 000 students. Its biggest faculty, the Faculty of Electrical Engineering and Informatics (https://www.vik.bme.hu) has ten departments and currently twelve so called knowledge centers. The latter cooperate closely with the departments and some of the faculties. The Department of Telecommunications and Media Informatics (BME-TMIT) and the Healthcare Technologies Knowledge Centre (BME-EMT) will represent BME in the planned project.

Brno University of Technology (BUT) (http://wislab.cz/about) is the second largest and the second oldest technical university in the Czech Republic. The Faculty of Electrical Engineering and Communication is the third largest faculty of BUT. The Faculty presently consists of 12 departments with about 190 teachers, 360 PhD students, and 4000 students in basic and advanced programs. The funding of research work comes from Czech and international research projects, and grants provided by the Czech Science Foundation, Ministry of Education and especially by the Ministry of Industry and Commerce. Particular attention is currently paid to projects for the 7-th Framework Programme and Operational Programs under EU Structural Funds. The WISLAB laboratory was formed under the BUT in order to group the researchers focusing on the recent research and development in the field of the wireless ad-hoc and sensor systems. The laboratory staff regularly publishes results in the prestigious international research journals such as Wireless Personal Communication, Telecommunication Systems, etc.

National Chiao Tung University (NCTU) is a public university located in Hsinchu, Taiwan (Republic of China). NCTU is Taiwan's oldest university, with eight colleges, over 500 academic staffs and 13,000 students. NCTU is one of the most prestigious universities in Taiwan and the world. According to Academic Ranking of World Universities, NCTU ranks 47-th in Computer Science in 2012. The Department of Computer Science (http://www.ccs.nctu.edu.tw/en/department/index.php) will represent NCTU in the planned project.

3 TELECALMPLUS SYSTEM OVERVIEW

The main system components proposed:

1. Intelligent home environment:
   This component is responsible for the collection of the measurement results from the various data sources. The hardware devices are installed in the home environment and lightweight software has to run on the personal computer of the relative.

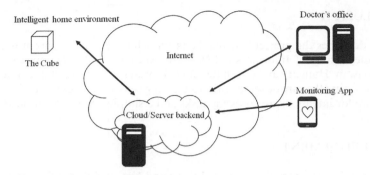

Figure 1.   TeleCalmPlus system overview.

2. Server backend:
   Measurement results are periodically uploaded to a central database where simple predictions are made in order to determine the health status of the relatives. The measurement and prediction results are made available for query through web services.
3. Smartphones and tablets (observer):
   These devices are used by the relatives of the elderly who want to get the prediction result. In its simplest form, the result is presented on the main screen of the device as an icon with three possible colors: (1) green—everything is all right, (2) yellow—the relative should pay attention as something is not in conformance with the normal behavior, and (3) red—the relative should check on the monitored individual.

3.1   Technical specification of the components

3.1.1   Intelligent home environment
* The component consists of the sensor network, the blood pressure meter and the Skype monitoring module at the current state of development. The sensor network module with the various device managers was developed under the TeleCalm project for which BME EMT was a project partner.
* The manager programs for the sensor network are written in C++ (with Boost and Qt). This subsystem can be installed onto the *nix system. The network consists of Netvox devices (motion, switch, etc. sensors) connected with ZigBee.

3.1.2   Server backend
* The measurement results are periodically uploaded to a central server through HTTP connection by GET queries.
* The server runs a MySQL database to store the results.
* The prediction is made by a standalone Java SE application which connects to the MySQL database through JDBC.
* Web services are provided to query the prediction results and the necessary artifacts are created by JAX-RS.

3.1.3   Smartphones and tablets
* The observer component is written as an Android application and the required minimum SDK version is 8 (Android 2.2). It is implemented to support multiple screen sizes and resolutions in order to utilize the bigger screens of the tablet devices too.
* The application is an Android Widget and uses the Alarm Service for the periodic query of the prediction results. The most up-to-date measurement results are stored on the device and can be viewed offline too.
* The measurement results are sent in JSON format and the Google Gson library is used for the serialization and de-serialization.

4 CONCLUSION

This paper describes the project The Smoke in the Chimney that is currently being implemented in the Faculty of Informatics of Paneuropean University in cooperation with partners from Taiwan, Hungary and Czech Republic. It presents a design of a remote monitoring system for assisting the care of senior citizens utilizing sensor networks at their home, a cloud infrastructure for intense data-processing tasks and a simple app for Android smartphones.

ACKNOWLEDGEMENT

This work is mainly supported by Visegrad Fund and National Scientific Council of Taiwan under IVF–NSC, Taiwan Joint Research Projects Program application no. 21280013 "The Smoke in the Chimney—An Intelligent Sensor—based TeleCare Solution for Homes and also partially by Slovak Research and Development Agency under contracts SK-AT-0020-12 and SK-PT-0014-12, by Scientific Grant Agency of Ministry of Education of Slovak Republic and Slovak Academy of Sciences under contract VEGA 1/0518/13 and by EU RTD Framework Programme under ICT COST Action IC 1104.

REFERENCE

Vidacs, A. et al., 2013. Detailed Project Description, The Smoke in the Chimney—An Intelligent Sensor-based TeleCare Solution for Homes (TeleCalmPlus), Budapest: BME.

Role of the process-driven management in informatics and vice versa

Václav Řepa
University of Economics, Prague, Czech Republic

Jiří Voříšek
Pan-European University, Bratislava, Slovakia

ABSTRACT: This paper is about the relationship between two highly actual and important phenomena: business process management and information system development. We shortly explain the essence of so-called process-driven management and show that it leads to the unity of infrastructures of the organization via information technology. This fatal relationship is then discussed with use of the information system maturity model according to Richard Nolan and some important general conclusions are made.

1 INTRODUCTION

During almost 20 years of existence of this approach, thinking in terms of business processes became a regular part of organization management practice. Nevertheless, the Business Process Reengineering and Process-driven Management means much more than it is regarded in ordinary managerial praxis. First of all it is a real paradigmatic change in the theory of management. The essential factor working as a trigger as well as an enabler of this managerial style is information technology.

The paper draws attention to the close relationship between the process-driven management and information system and tries to show how these two phenomena are converging together into the single organizational-technological complex called process-driven organization.

2 PROCESS-DRIVEN MANAGEMENT

The first complete explanation of the idea of process management as a style of managing an organization has been published in [Hammer, Champy 1993]. The authors excellently explain the historical roots, as well as the necessity, of focusing on business processes in the management of the organization. The major reason for the process-orientation in management is the vital need for the dynamics in the organization's behavior. It has to be able to reflect all substantial changes in the technology as well as in the market as soon as possible. The only way to link the behavior of the organization to the changes in the market and technology possibilities is to manage the organization as a set of processes principally focused on customer needs. As customer needs, as well as requirements driven with the technology possibilities, are constantly changing the processes in the organization should change as well. That means that any process in the organization should be linked to the customer needs as directly as possible. Thus, the general classification of processes in the organization distinguishes mainly between:

- *Key processes*, i.e. those processes in the organization which are linked directly to the customer, covering the whole business cycle from expression of the customer need to its satisfaction with the product/service.
- *Supporting processes*, which are linked to the customer indirectly—by means of key processes which they are supporting with particular products/services.

While the term "key process" typically covers the whole business cycle with the customer—it is focused on the particular business case; the supporting process is typically specialized just to the particular service/product, which is more universal—usable in a number of business cases. This approach allows the organization to focus on the customers and their needs (by means of the key processes), and to use all the traditional advantages of the specialization of activities (by means of the supporting processes) at the same time. Key processes play the crucial role—by means of these processes the whole system of mutually interconnected processes is tied together with the customers' needs. Supporting processes are organized around the key ones, so that the internal behavior, specialization, and even the effectiveness of the organizations' activities are subordinated to customers and their needs.

2.1 *Nature of organizational behavior and its infrastructures*

Figure 1 illustrates basic factors of the organizational behavior and their mutual relationships.

The essence of any business is hidden in the primary function of the organization which has to be always oriented on the *goals outside of the organization*, in its environment. In other words, *essential meaning* of an organization is always given by its *value in the system* which it is a part of. In the market environment the system is represented by organization's customers. Therefore we speak about "*customer orientation*" although these principles are universal and valid for any organization in any system including even the public sector. The primary function is achieved by the behavior of the organization represented by its business processes. Thus business processes represent the nature of the organizational behavior. Any other aspect of the organization then represents some form of the support—some infrastructure. Figure 1 also shows basic kinds of infrastructures: information system and organizational system. While there has never been a doubt about the fact that information system is an infrastructure, in case of the organizational system such view is completely new, untraditional. Organization of the process-driven organization has not the form of predefined hierarchical structure but rather the set of roles associated to processes which define their mutual relationships which thus can be different in different time, under different circumstances.

The view of the behavior of a process-managed organization is quite different from the traditional one. Mainly, the key processes represent an unusual view of communication and collaboration within the organization. In traditionally managed organizations the organization structure reflects just the specialization of work; it is static, and hierarchical. The concept of key processes brings the necessary dynamics to the system—key processes often change according to the customer needs, while supporting ones are relatively stable (the nature of the

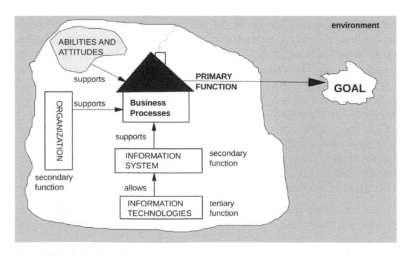

Figure 1.   Basic factors of the organizational behavior (source: author).

work is relatively independent of the customers' needs). At the same time, the key processes represent the most specific part of the organizations' behavior, while the supporting ones are more general and standard. Thus, the supporting processes are the best candidates for possible outsourcing while the key ones should be regarded, rather, as an essence of the market value of the organization. So, we have a system of different processes with different natures. To ensure the necessary communication among them, we need to have an interface which can enable overcoming of these differences. Similar problem can be seen also with the interface among the system of business processes and its supporting infrastructures: organizational structure, and information system. In both cases we have to harmonize different systems working differently, with different goals, and under different circumstances. The organization needs to be *flexible and ever prepared for a change*. Such a flexibility is not possible without the help of information technology via sophisticated information system.

3 INFORMATION SYSTEM AS A TOOL FOR ORGANIZATIONAL DEVELOPMENT

Figure 2 shows the model of the evolution of the organizational maturity through its information system. It is based on the work of Richard Nolan [Nolan 1974] who firstly expressed the evolution of the information system as a consequence of the growth of the organizational system via different levels of its maturity. This work later served as a base idea for other "maturity models" like CMM [Carnegie Mellon University Software Engineering Institute 1995] for instance. Nolan discovered the general dependency between the ability to use some kind of technology and the "maturity" of the organization as a whole which includes knowledge and attitudes of its people together with their experience with the previous—lower type of technology (information system). In fact, Nolan's concept of maturity model is a creative application of the famous idea of Abraham Maslow [Maslow 1954] which defines the hierarchy of "human needs" where fulfilling the needs on a lower level is a prerequisite for the needs of upper level; a general precondition for the transition to the upper level of the information system is always the perception of the necessity to solve the fatal problems connected

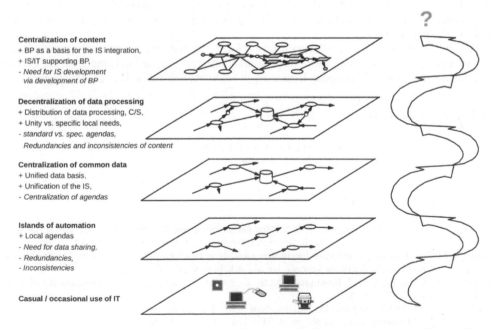

Figure 2. Evolutionary model of the organizational maturity (source: author according to [Nolan 1974]).

with the current level. The model shows how the organization grows from occasional use of information technology through islands of automation to the need for centralization of data (database system) and consequent need for their partial centralization which meets the idea of client-server technology.

Looking at the model at a glance one can find that the evolution process oscillates between two general aspects: centralization (integration) and decentralization (disintegration) in terms of the evolution spiral where the previously surpassed aspects will come back in the future in different, evolutionary higher, forms. According to this principle it can be supposed that the problems with the use of the client-server technology (representing some kind of decentralization) bring the need for some kind of centralization. As the figure shows the reason for such a need is the problem of necessary redundancies and consequent inconsistencies of the content of the organizational behavior following from this technology. The general solution of this problem is the integration (centralization) of the content via the *common definition of business processes* which exactly meets the idea of process-driven organization as well as of connected technology—*workflow management systems*. It can be also supposed that the future development will undoubtedly bring some problems with this way of centralization of the content and consequent need for some kind of decentralization in order to allow needed different interpretations of the process contents. Current trends to the *service-oriented approach* [OASIS 2006], [Cardoso, Voigt, Winkler 2009] to the infrastructural aspects of the organization can be possibly regarded as a symptom of this need.

The model shows that business processes represent the important tool for the integration of the information system. At the same time it also shows how the information technology serves as a tool for development of the organization itself via its information system.

4 CONCLUSIONS

The two actual phenomena discussed in this article: business process management and information system development are closely tied together. Moreover, they can be regarded as two opposite sides of the same coin. Information system based on the workflow technology is a fatal condition as well as a main development tool of the process-driven organization. On the other hand, the development of the information system in a process-driven organization has to be based on the permanent analysis and design of its business processes. One cannot exist without the other one.

The above discussed facts are the main reason for the high attention put on the topic of business process modeling and management in the study program Applied informatics at the Faculty of Informatics of the PanEuropean University in both, bachelor as well as master, degrees.

ACKNOWLEDGEMENT

This paper has been supported by the Faculty of Informatics and Statistics of the University of Economics, Prague in the project No. 400040.

REFERENCES

Cardoso, J., Voigt, K., Winkler. M. 2009. Service Engineering for the Internet of Services. Enterprise Information Systems, Lecture Notes in Business Information Processing (LNBIP), Vol. 19, pp. 15–27.

Carnegie Mellon University Software Engineering Institute 1995. Capability Maturity Model: Guidelines for Improving the Software Process, 1st ed., Addison-Wesley Professional.

Hammer, M., Champy, J. 1993. Re-engineering the Corporation: A Manifesto for Business Revolution. Harper Business, New York.

Maslow, A. 1954. Motivation and Personality, New York, Harper & Row.

Nolan R. 1974. Managing the Four Stages of EDP Growth, Harvard Business Review.

OASIS 2006. Reference Model for Service Oriented Architecture 1.0, OASIS Standard, 12 October 2006 (on: http://docs.oasis-open.org/soa-rm/v1.0/).

Statistical testing of random sequences

F. Schindler
Faculty of Informatics, Institute of Applied Informatics, Pan-European University, Bratislava, Slovakia

T. Szabó
Faculty of Central European Studies, Constantine the Philosopher University, Nitra, Slovakia

J. Bujda
Faculty of Electrical Engineering and Informatics, Slovak Technical University, Bratislava, Slovakia

ABSTRACT: In this paper we deal with random phenomena in informatics. We designed and tested some generators based on Linear Feedback Register (LFSR) to generate random sequences. To test the generated sequences we use the tests of FIPS 140-2, using Diehard battery of tests and basic statistical tests of the NIST test suite. Our paper is concluded by processing and evaluating the results.

1 INTRODUCTION

In computer science random numbers play a very important role. Let's recall the most important areas: their applications in cryptography, at the random choice for the computer games, in computer simulation, in methods of Monte Carlo, etc. From all this we may conclude that it is a quite broad domain of their usability. In this paper we will deal with random numbers from the point of view of cryptography. To mention a couple of examples of their usage in cryptographic algorithms we state here:

- use of identifiers in authentication steps for key distribution,
- key generation for symmetric ciphers,
- key generation for the RSA algorithm in cryptography with public keys,
- digital signature generation based on DSA algorithm (Levický 2010).

2 RANDOM NUMBERS GENERATORS

Random number generators that we deal with in this paper are created by means of an algorithm. It means they are deterministic random numbers generators. Resulting sequences coming out of them are periodic, e.g. generated numbers are not completely random, but only pseudorandom numbers. Though they look like sequences of random numbers.

In cryptography there are a couple of definitions for the concept of pseudorandom sequence generator or pseudorandom bit generator.

From the sequence of n bits every number is made. Everybody approaches this problem from a different point of view. Nevertheless all these definitions are equivalent. Our definition suggests the essence of them is as follows:

Definition 1 A pseudorandom bit generator is a deterministic algorithm, which takes in a true random bit sequence of the length k, and outputs a binary sequence of the length lk, which appears to be random. The generator input is called to be an initial seed, and its output is referred to as a pseudorandom bit sequence (Menezes, A. et al. 1996).

Freely said the pseudorandom generator is an efficient code segment or algorithm which outstretches a short random sequence into a long pseudorandom sequence (Goldreich 2000). This sentence includes three basic properties of pseudorandom generators:

- Efficiency: Each such generator must be efficient. This attribute of efficient computation means that the generator has to be implemented as deterministic algorithm executed in polynomial time.
- Stretching: This generator must outstretch its input sequence into a longer output sequence. To be more specific, from the input n bits it generates the corresponding output of (n) bits, where $(n) > n$.
- Pseudorandomness: The generator output must appear as random for each active observer, e.g. for any random initial seed each active procedure is to fail to recognize the generator output from real random sequence of the same length.

Very important generator characteristic in cryptography is its extemporaneousness. For the case of pseudorandom generator it implies that if we do not know initial seed it is impossible to foretell the next generated output sequence bit in spite of knowing all previously generated output bits. Exactly that the following definition is claiming:

Definition 2   A pseudorandom bit generator passes the next-bit test, if there is no polynomial time algorithm, which based on knowledge of first n output bits of generated sequence would know to foretell the (n+1)-th output bit with considerably higher probability than ½ (Menezes, A. et al. 1996).

According to this definition we can claim, that if the pseudorandom generator of binary sequences passes the next-bit test, then we call such a generator to be a cryptographically secure generator of pseudorandom binary sequences.

2.1   Cryptographic aspects of binary sequences

Pseudorandom binary sequences used for stream ciphers are very different from other sequences used for these purposes. This is due to the fact that sequences used for one stream cipher require other properties as those typically used for other stream ciphers. Therefore cryptographic sequences will differ from one another in some aspects. Though, in other aspects they will be the same. Such exigent sequence attributes are linear complexity, spherical complexity and differential sequence properties. In this paper we won't talk in more detail about individual attributes and related definitions and theorems. Further we will define a linear feedback register that we use in our design of generators.

2.2   Linear Feedback Register (LFR)

A great advantage of this register is its easy hardware realization, large period, and its easy analysis. Based on aforementioned properties it is often used in various generators.

Definition 3   A Linear Feedback Shift Register (LFSR) of length L consists of L stages (or delay elements) numbered 0, 1, ..., $L - 1$, each capable of storing one bit and having one input and one output; and a clock which controls the movement of data. During each unit of time the following operations are performed:

1. the content of stage 0 is output and forms part of the output sequence;
2. the content of stage i is moved to stage $i - 1$ for each i, $1 \le i \le L - 1$; and
3. the new content of stage $L - 1$ is the feedback bit sj which is determined by the following recursion: $s_j = (c_1 s_{j-1} + c_2 s_{j-2} + ... + c_L s_{j-L})$ mod 2, for $j \ge L$ (Menezes et al. 1996).

- LFR on Figure 1 is denoted by $<L, C(D)>$, where $C(D) = 1 + c_1 D + c_2 D^2 + ... + c_L D^L \in Z_2[D]$ is its characteristic polynomial.
- Every LFSR, whose level $C(D)$ is L is called nonsingular, provided $c_L = 1$.

Each linear feedback shift register must be initialized by values 0 or 1 for each its element. Then we say that $[s_{L-1}, ..., s_1, s_0]$ is initialization state (seed) of given LFR.

302

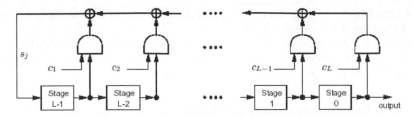

Figure 1.   A Linear Feedback Shift Register (LFSR) of length L. (Menezes et al. 1996).

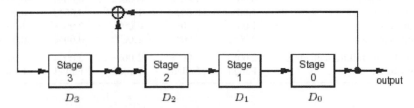

Figure 2.   Scheme LFSR $<4, 1 + D + D^4>$—example. (Menezes et al. 1996).

On Figure 2 LFSR is described as $<4, 1 + D + D^4>$. In such a way graphically depicted LFSR is more transparent than the general chart illustrated in Figure 1.

If the initial setting is the null vector the corresponding output sequence is infinite sequence of nulls. From nonzero initial register setting we get a sequence with the length not more than $2L-1$, where the equality holds if the generating polynomial is in Z_2 primitive. Therefore we will consider only this case here. The sequence length follows from the fact that during one period in the register there are changed off all possible states whose number is 2^L provided we do not count the null case.

Definition 4   If $C(D) \in Z_2[D]$ is a primitive polynomial of degree L, then $<L, C(D)>$ is called a maximum-length LFSR. The output of a maximum-length LFSR with non-zero initial state is called an *m*-sequence.

Theorem   (statistical properties of *m*-sequences) Let s be an *m*-sequence that is generated by a maximum-length LFSR of length L.

– Let k be an integer, $1 \le k \le L$, and let s be any subsequence of s of length $2^L + k-2$. Then each non-zero sequence of length k appears exactly 2^L-k times as a subsequence of s. Furthermore, the zero sequence of length k appears exactly 2^L-k-1 times as a subsequence of s. In other words, the distribution of patterns having fixed length of at most L is almost uniform.
– s satisfies Golomb's randomness postulates. That is, every *m*-sequence is also a *pn*-sequence. (Menezes et al. 1996).

3   DESIGN AND TESTING OF GENERATORS

The results and methods we are using in this section are described in Bujda (2009). We have designed five different pseudorandom generators by means of which we tested generated sequences according to statistical tests contained in FIPS 140-2, NIST STS a DIEHARD. During the generator design we used a combination of Linear Feedback Three Registers (LFSR) along with characteristic polynomials: $p_1 = x^{97} + x^6 + 1; p_2 = x^{50} + x^4 + x^3 + x^2 + 1; p_3 = x^{49} + x^9 + 1$.

Generators have been made by following ways (the corresponding seed has been chosen by the random function rand(), shift goes from the right to the left):

– G1—contains one LFSR, whose characteristic polynomial is p_1,
– G2—involves one LFSR with characteristic polynomial p_2,

- G3—whose base is the combination of generators created with polynomials p_1 and p_2 as above by means of bit operator **XOR**,
- G4—applies the combination of polynomials p_1, p_2 a p_3 via bit operator **XOR**,
- G5—employs the combination of polynomials p_1, p_2 and p_3 along with the majority function: $f(p_1, p_2, p_3) = (p_1 \otimes p_2) \oplus (p_1 \otimes p_3) \oplus (p_2 \otimes p_3)$.

Table 1. Test evaluation according to FIPS 140-2.

	G1	G2	G3	G4	G5
Monobit test	0,00%	1,00%	0,00%	0,00%	0,00%
Poker test	1,00%	0,00%	1,00%	0,00%	0,00%
Runs test	1,00%	5,00%	0,00%	2,00%	1,00%
Long runs test	0,00%	3,00%	1,00%	0,00%	0,00%
Number of failing sequences	2,00%	7,00%	2,00%	2,00%	1,00%
Conclusion	Failed	Failed	Failed	Failed	Failed

Table 2. Test Interpretation according to NIST STS.

	G1	G2	G3	G4	G5
Frequency (monobit) test	7,00%	91,00%	5,00%	1,00%	1,00%
Frequency test within a block	1,00%	90,00%	1,00%	1,00%	0,00%
	1,00%	92,00%	1,00%	1,00%	1,00%
	1,00%	81,00%	1,00%	0,00%	0,00%
	2,00%	83,00%	3,00%	1,00%	2,00%
	2,00%	94,00%	2,00%	1,00%	0,00%
	2,00%	97,00%	2,00%	1,00%	1,00%
	2,00%	88,00%	2,00%	2,00%	0,00%
	2,00%	92,00%	1,00%	3,00%	2,00%
	2,00%	92,00%	1,00%	1,00%	0,00%
Runs test	6,00%	92,00%	4,00%	1,00%	2,00%
Test for longest run of one in block	5,00%	91,00%	3,00%	2,00%	1,00%
	56,00%	91,00%	41,00%	2,00%	1,00%
	2,00%	91,00%	2,00%	3,00%	2,00%
Binary matrix rank test	2,00%	100,00%	1,00%	2,00%	2,00%
Discrete Fourier transform (spectral) test	3,00%	86,00%	4,00%	1,00%	2,00%
Non-overlapping template matching test	3,00%	91,00%	2,00%	1,00%	0,00%
Overlapping template matching test	3,00%	91,00%	4,00%	1,00%	1,00%
Maurer's univ. Stat.test	2,00%	91,00%	1,00%	0,00%	0,00%
Lempel-Ziv compression test	1,00%	80,00%	1,00%	0,00%	0,00%
	1,00%	81,00%	0,00%	0,00%	0,00%
	1,00%	82,00%	0,00%	0,00%	0,00%
	3,00%	97,00%	2,00%	0,00%	0,00%
	4,00%	91,00%	3,00%	0,00%	3,00%
Linear complexity test	100,00%	100,00%	67,00%	11,00%	10,00%
	100,00%	100,00%	69,00%	12,00%	12,00%
	100,00%	100,00%	72,00%	12,00%	10,00%
Serial test	4,00%	98,00%	2,00%	1,00%	2,00%
	4,00%	100,00%	1,00%	1,00%	2,00%
Approximate entropy test	3,00%	98,00%	1,00%	1,00%	0,00%
	4,00%	97,00%	3,00%	0,00%	0,00%
	4,00%	99,00%	3,00%	1,00%	2,00%
Cumulative sums test	7,00%	95,00%	4,00%	0,00%	1,00%
Random excursions test	25,00%	87,00%	17,00%	1,00%	1,00%
Random excursions variant test	34,00%	87,00%	23,00%	1,00%	1,00%
Number of failing sequences	100,00%	100,00%	73,00%	12,00%	12,00%
Conclusion	Failed	Failed	Failed	Failed	Failed

Table 3.   Test interpretation according to DIEHARD.

	G1	G2	G3	G4	G5
Birthday spacings test	12,00%	100,00%	10,00%	4,00%	1,00%
Binary matrix rank test	12,00%	100,00%	8,00%	2,00%	3,00%
(for 6 × 8, 31 × 31 and 32 × 32)	12,00%	100,00%	9,00%	4,00%	3,00%
	10,00%	100,00%	11,00%	7,00%	2,00%
Monkey tests on 20-bit words	12,00%	93,00%	11,00%	1,00%	6,00%
Monkey tests (OPSO, OQSO, DNA)	10,00%	100,00%	11,00%	6,00%	2,00%
Count the 1's test in stream of bytes	13,00%	90,00%	7,00%	3,00%	0,00%
Count the 1's test in specific bytes	100,00%	100,00%	32,00%	9,00%	9,00%
Parking lot test	9,00%	92,00%	7,00%	0,00%	0,00%
Minimum distance test	3,00%	93,00%	2,00%	6,00%	5,00%
Random 3D spheres test	7,00%	91,00%	5,00%	1,00%	2,00%
Squeeze test	13,00%	92,00%	12,00%	0,00%	2,00%
Overlapping sums test	15,00%	92,00%	12,00%	1,00%	1,00%
Runs test	17,00%	92,00%	9,00%	0,00%	0,00%
Craps test	24,00%	92,00%	15,00%	1,00%	2,00%
Number of failing sequences	100,00%	100,00%	43,00%	12,00%	11,00%
Conclusion	Failed	Failed	Failed	Failed	Failed

3.1   Generator testing

There is no reason to assume that some of the submitted generators would meet all testing packages, since our generators are rather simple. Our goal was to mutually compare the tests in order to study the change of their statistical properties. We decided to implement tests based on following packages: FIPS 140-2, NIST STS a DIEHARD (see FIPS PUB 140-2 and Marsaglia). Tests were made on testing sets containing 100 samples of randomly generated sequences. Some statistical tests from the packages NIST and DIEHARD require for adequate results at least random samples of the magnitude 10 MB (approximately 8.10^7 bits).

3.2   Testing results

Tests were successfully performed for all designed generators. Their results were processed in corresponding tables. They contain a percentage value expressing how many sequences did not pass the given test.

Tests coming from the package FIPS 140-2 were successfully mastered by all our generators. Their results are contained in Table 1. Only the generator G2 achieved during these tests limiting values, all others passed it with relative ease.

Test results from NIST STS are included in Table 2. Due to their severity none of the designed generators met their requirements. The worst results are in the case of the generator G2, which barely passed less severe test from the package FIPS 140-2. Generators G1 and G2 are made by means of only one linear feedback register. Due to this they did not pass the Linear Complexity test. The best results have been reached by generators G4 and G5, in which case three linear feedback registers we interconnected together. Generator G3 with two connected LFRs did not achieve by far such results as G4 and G5 did.

None of the generators met all the tests from DIEHARD, what we have also assumed because they are quite demanding. Even in this case generators G4 and G5 achieved better results.

4   CONCLUSIONS

In this paper we dealt with randomness from the point of view cryptography. It provides the reader a review about actual norms for random sequences as well as ways of their generation and testing.

REFERENCES

Bujda, J. 2009. Posudzovanie a použitie náhodnosti v informatike, Master's Thesis, Dept. of Applied Informatics, Bratislava: FEI STU.

FIPS PUB 140-2, Federal Information Processing Standards Publication, Security Requirements for Cryp-tographic Modules, U.S. Dept. of Commerce/N.I.S.T., 2001.

Goldreich, O. 2000. Pseudorandomness. In U. Montanari et al. (eds.): Automata, Lagnuages and Programming, ICALP 2000. Berlin: Springer.

Levický, D. 2010. Kryptografia v informačnej a sieťovej bezpečnosti. Košice: ELFA.

Marsaglia, G. DIEHARD Statistical Tests, http://stat.fsu.edu/geo/diehard.html.

Menezes, A. & van Oorschot, P. & Vanstone, S. 1996. Handbook of Applied Cryptography, Boca Raton: CRC Press Inc.

Database programming with MS SQL under .NET Framework

F. Schindler
Faculty of Informatics, Pan-European University, Bratislava, Slovakia

M. Šutka
Zymestic Solutions Inc., Bratislava, Slovakia

ABSTRACT: Programming databases is considered to be in today's world one of the most important topics. Among the prominent database systems it also belongs MS SQL, which thanks to its robustness and modularity is appropriate not only for large business systems, but also for simple database applications. Emphasis in these programs is paid to the choice of proper programming language and corresponding environment. One of the best platforms for doing this is .NET Framework containing the package of complex tools designed for work with MS SQL databases.

1 INTRODUCTION

In our modern times when information rolls at us from all sides at the light speed it is very important to store this data somewhere. Such a storage place or data warehouse is often called to be a database. When dealing with databases it is necessary to distinguish between their design and realization. Both of these concepts are mutually interconnected. MS SQL offers us not only the high security measure, but also high reliability and a quality realization of corresponding database design. An important role in all this is by itself the realization and the choice of proper environment including programming language. It is not an accident that in the last couple of years many different companies rework their database applications into .NET environment employing the programming language C#. From the family of programming languages C# is the youngest. Thanks to that it draws from rich experiences of similar languages. It is a product of the Microsoft. It is good to know that its compiler is freely accessible. There are also a couple of C# development environments. The most suitable is Visual Studio coming from Microsoft. Its part includes WebDeveloper for web applications generation. Visual Studio when compared with other competing development environments considerably simplifies the work with databases. It conveys database design visualization and interconnection of database tables by means ServerExplorer, which allows debugging stored procedures directly during the program run and etc. Data processing and interaction by itself provides .NET layer ADO.NET consisting of a few classes and name spaces. Those allow us to work with the database as a whole.

1.1 *ADO.NET*

The layer ADO.NET we may perceive as a complex name for all tools in .NET Framework that could be used during the data processing not only with MS SQL database, but also with other related databases. To realize the interconnection to the database one can exploit various providers. Depending on which database we want to work with we have to choose from one of them. Hence we choose one of the name spaces with which the execution of SQL commands is linked. When we make one project with MS SQL database and the other with Oracle we are often use two different name spaces that contain various and independent

Table 1. Name spaces in ADO.NET.

System.Data	Name space containing all basic classes
System.Data.OleDB	Classes for the provider OLE DB
System.Data.SqlClient	Classes for interconnection to MS SQL
System.Data.OracleClient	Classes for interconnection to Oracle database
System.Data.Odbc	Classes for connection to drivers ODBC
System.Data.Common	Classes shared among individual connection providers
System.Data.SqlTypes	Classes and structures for MS SQL server

classes. Identifiers of all these classes are easy to remember and therefore there is no problem to work with them. To see that we attach Table 1.

Classes contained in all these name spaces are mutually shared and depend on the used database. Such classes have usually alike names, but inside they function on the same principle (McClure et al. 2005).

2 INTERCONNECTION WITH DATABASE

We may directly access to a database which is placed on the server or to a locally accessible file (*.mdf). To the database source we are able to interconnect from VisualStudio via ServerExplorer without necessity to install further software. If we want to work with the file we should place it into the directory App_Data designated exactly for that reason. In case that the file does not exist it will be created. During the creation of interconnection we should choose the type of data source with which we would like to work and the way how we will login to this source (Windows Authentification or SQL Server Authentification). After connecting .NET applications on the data source it is the most important the very first interconnection. There could be a couple of databases running simultaneously on the server. Therefore we have to pass in to the server a few specific arguments like database placement, user id and password in dependence on where and how we want to connect. All this information is stored in so called connection string (ConnectionString). The form for our connection string ought to be found in ServerExplorer after clicking on our data source and its part called "Properties". An important question is where the connection string should be stored. The answer to that is the global configuration file for the whole web application known under the name "web.config". It is an xml file containing settings for the whole application. For storage of all connection strings it has a reserved memory part. When we are creating such an interconnection to send commands to the server we should think of possibility that our connection for any reason may be terminated. Therefore it is a good habit to provide a piece of code handling corresponding exceptions in the blocks of the type try{} or catch{} (Šutka 2010).

2.1 Command class SqlCommand

In order to be able to execute all SQL commands in .NET applications we need to use a suitable command class. For MS SQL we have just for that the class "SqlCommand".

In order to execute commands we need to create an instance of this class and pass information into a newly generated object concerning the type of command, connection type and solely text of the command. By the command text we mean either an SQL query or the name of stored procedure. To execute the command itself it is necessary to call one of the four methods belonging to the object of the class SqlCommand. Before we do that we should think about the output coming out from our command. Provided the command does not return any set of values we should use the method ExecuteNonQuery() e.g. in order to insert data. The only value being here returned is the number of effected lines. In the case that our

SQL command returns a single value we ought to make use of the method ExecuteScalar() e.g. during detection of the number of lines. The method ExecuteReader() is exploited most often. Based on it we may obtain a whole set of results coming from the database. On the other hand the method ExecuteXmlReader() will be in use for eventual reading of output results from other SQL XML commands. By means of the command class and OOP we are able to map individual tables belonging to the database and to implement methods that process various objects. Thereafter it is possible to enter new data into database or erase old data from the database or read it. This technique is typically used for classical three layers applications (Šutka 2010).

3 DATASET

DataSet is a very good set for processing data in the form of XML. It is able not only to represent data but also it can be interconnected directly to an XML file. DataSet allows working with data on the level of a simple transaction processing. That means we are able to return data to its previous consistent state unless we confirm the made data change. To imagine what is DataSet good for and how to work with it we can envision a well known Excel. The same way as Excel also DataSet contains a table DataTable, a row DataRow and a column DataColumn. In order to access its individual items we may do it either by name when it has been created during its generation phase or by an index. When a primary key has been defined it is possible in an elegant manner to search DataSet by means of it. There are a couple of different ways how to create a DataSet e.g. either via an existing program or by means of applying wizards directly from VisualStudio. For visual design there is a need to specify a concrete connection string. In the next step we may simply specify the data e.g. via direct SQL query or via existing procedures or eventually via newly stored procedures. SQL commands may be entered manually or by means of QueryBuilder which allows making commands without too much knowledge of SQL syntax.

DataSet processes table data by means of Table Adapters. Each such table is serviced by such an adapter. The corresponding wizard is able to generate new methods to enter, erase, change or gain data (Vieira 2006). Visual design is very simple and makes possible instantaneous mapping of tables. This is its main advantage against the manual mapping through the command class.

4 LINQ

LINQ (Language INtegrated Query) found its palce in the last version of the application platform of .NET Framework. Physically it is represented by a set of language extension of C# and Visual Basic that support structural data gain from XML sources, relational databases or memory data structures. Virtually understood each application gets or actualizes data in some form (RDBMS, XML, etc.) so that developer's productivity and code reliability are directly proportional to improvements in writing code for accessing data.

Applications accessing data repositories involve an important logic not contained in the programming language itself, but instead it may be represented by text formulations expressed in quotes and is sent for processing to the database. So defined application access to data complicates considerably the corresponding program maintenance and it degrades possibilities of development tools for application changes, since there is no known way to interconnect textual database commands with the application code logic. In this case the development environment is unable to detect shortcomings in database syntax during the program compilation phase. Therefore it usually results in application failure during the code execution. If we had a way to detect such errors already during the code compilation the developer could save his time and effort during the program debugging phase. With support of LINQ the data requirements can be expressed with use of various extensions built-in directly into the programming language in such a way that the compiler in conjunction with development environment is helping the developer to assemble appropriate database commands. Though LINQ does not copy all dialect possibilities of SQL language it affords developers to express themselves straightly in the programming language the most used group of database operations. This is quite in contrary to the previous attempts to interconnect programming languages with database specifications that are oriented on specific databases. LINQ is a universal query system that may be used with almost all data sources. Microsoft incorporated into .NET Framework 3.5 a version of LINQ supporting .NET data types including records and arrays, SQL Server and XML. For instance LINQ for XML allows developers to generate, parse and transform data by using well-known query metaphors instead of using application interfaces for the work with XML such as XPath, XQuery or XSLT. Developers applying them in their design must often solve the dilemma how to use them effectively in large projects. LINQ constructs allow combining data coming from various sources. This way we may interconnect a list of customers, their addresses temporarily stored in the computer memory with corresponding bookkeeping data stored in respective database. Such flexibility along with the possibility to extend the list of supported data sources gives into the hands of developers a set of commands and libraries to read in and update a broad data spectrum (Šutka 2010).

4.1 New level of source data shielding

"**Service Pack 1**" that is available for .NET Framework 3.5 adds to LINQ a new dimension. It contains libraries "ADO.NET Entity Framework" supporting data access based on appropriate entity model. This access technique through data model allows to developers and database professionals to describe relationships among database tables and columns with emphasis on logical entities representing true data meaning rather than their physical structures and placing of tables in used data repository.

When using entity model the very first step is the creation of logical data model. This defines data as a collection of entities and relationships. For instance when we describe a business cycle in entity model we should use in description of tables dealers and employees the entities inheritance, for each dealer is also an employee and therefore the code developer does not need to combine data from SQL tables using a dialect exploiting the command "join". Exactly "Entity Framework", as a higher abstraction level of data source enables us to generate multi-database applications since the change of used database platform does not require any change of the application code except for a change in mapping layer of logical data structure for its physical layer e.g. mapping "ORM".

No doubt "LINQ" is a technology which will massively influence application development that is heavily oriented on data. Simplifying data access coding stored in various types of repositories decreases errors predisposition and consequently it facilities the code debugging and its maintenance. The entity model in LINQ allows applications generation that is independent of existing database providers thanks to its support of object-relational mapping (Šutka 2010).

5 CONCLUSIONS

In this paper we made an overview for data access procedures in databases that are provided in .NET Framework. The work with databases is not simple, but .NET Framework and Visual Studio environment considerably simplifies this job thanks to offered design visualization and due to possibilities of object-relational mapping. In spite of the fact that C# and .NET Framework belong to the beginners in this area they gain more and more fans owing to innovations provided in LINQ. That way the coding of data access procedures is made easier independently on underlying data source. Moreover the whole debugging process is simpler too. Consequently the productivity enhancement can not be made by any better way than it is by shorting the time necessary for writing the corresponding program code.

REFERENCES

Delikát, T. 2007. *Základy projektovania databázových systémov*. Bratislava, Delint, ISBN 978-80-969613-0-6.

McClure, W.B., Beamer, A.G. 2005. *Professional ADO.NET 2 Programming with SQL Server 2005: Oracle and MySQL*. Indianapolis, Wiley Publishing Inc., ISBN 978-0-7645-8437-4.

Oravec, M. 2001. *Malé veľké databázy II*. In: PC Revue, No. 9: 130–131, ISSN 1335-0226.

Rahmel, D. 2000. MSDN Developers Guide, Indianapolis, M&T Books, ISBN 0-7645-4698-8.

Šutka, M. 2010. *Moderné postupy programovania s databázami MS SQL*. Master's Thesis, Dept. of Applied Informatics, Bratislava: FEI STU.

Vieira, R. 2006. *Beginning SQL Server 2005 Programming*. Indianapolis. Wiley Publishing Inc., ISBN 978-0764584343.

Past and future of Human—Computer Interaction

M. Šperka
Faculty of Informatics, Pan-European University, Bratislava, Slovakia

ABSTRACT: The information and communication technologies become human partners in many activities. Innovative concepts, paradigms and applications such as ubiquitous, pervasive or ambient computing bring a new view on the *Human—Computer, Human—Robot and Human—Human Interaction*. The paper summarizes trends of the user interfaces from the perspective of sensory, motor and cognitive abilities of the humans compared with machines, new input/output devices principles, technology and different computing concepts.

1 INTRODUCTION

The aim of this paper is to analyze how the *Human Computer Interaction—HCI* has changed during the evolution of computer technology and computing concepts. The main focus is on the *User Interface—UI* paradigms, concepts, metaphors and interaction styles.

Information and communication technology together with computer controlled machines and robots are the amplifiers of human's sensory, motor and cognitive abilities. Computers are tools for the communication, engineering and office tasks, education and training, controlling home and industry appliances, prototyping and manufacture, entertainment etc.

Sensory, motor and cognitive abilities of computers are approaching human abilities. The computers become our partners in all professional and free time activities. The evolution of computers started with simple tasks, which were programmed and used by experts. Today's computers, which are able to solve complex problems, can be used by people without any technical background.

2 INTERACTION CONCEPTS, METAPHORS AND STYLES

2.1 *Definitions and overview*

Interaction is communication between objects in the form of messages. In a general sense message could be a simple or very complex physical activity—for example push or catch an object. In the HCI exist different models of interaction, each of them is used for a specific purpose. We focus on two main forms of communication—with people and with things. People use language—it is a stream of symbols in the audio or visual form. Human speech can be enhanced with gestures—body language, and a written text with pictures. In this case the communication is parallel. Gestures are time and space configurations of the body and can use real objects.

Objects can be identified by pointing at them, they can be translated, rotated, assembled, destroyed etc. This is a form of the natural interaction with people and things. We can use special things—artefacts, which represent real objects or events in a form of records or representations of reality—drawings, 3D models, photographs, movies, and use them in the same way as real things. For example communicating chatbots—Avatars.

Interaction between subjects is possible when both subjects are familiar with symbols and their meanings and is more efficient, when each subject knows, what the other subject knows about these symbols. In the pre-computer age, things were objects without "intelligence" but now they can be enhanced by embedded microprocessors and are becoming subjects with own cognition—*Things That Think*.

The main issue in the HCI is *usability, accessibility and ergonomics* (Interaction Design Foundation 2014). The usability is the *effectiveness*—that is an accuracy and completeness with which users achieve certain goals, the *efficiency*—a measure how expanded resources help to achieve effectiveness and *satisfaction* with which specified users achieve specified goals in particular environments.

Accessibility deals with the accessibility of a computer to all people regardless of disability or severity of sensory, motor or cognitive impairments. Ergonomics is concerned with an anthropometric and a physiological characteristics of people and their relationship to workspace and environmental parameters.

An important usability issue is *intuitive interface* and *learnability*. Intuitiveness is supported by the application of the *affordance*—one of the main design principles in general. To learn a new tool is easier, when the novice user has some experience using similar tool, for example shifting from the classic typewriter to the word processing software. In this case it is easier when the UI is a *metaphor* of the traditional tool or an *analogy* of the experienced activity.

Metaphor is a set of UI objects and procedures used in an analogical domain. They help users acquire appropriate cognitive model relating to that application, contextualize interface and help with immersion. Some HCI experts doubt about the utility of metaphors (Barr 2003). Others propose to use idioms instead of metaphors (Arvizu 2014).

There exist several metaphor taxonomies in the HCI. For example Lakoff and Johnson (Lakoff 1980) recognize structural, ontological, orientation metaphors and metonymy. Other metaphor classes are: *Metaphors for computation*—computer as calculator, desktop, record player, phone, etc., *office desk top metaphor*—folders, wastebaskets, pen, brush, eraser, ..., *icons and gadget metaphors*—radio button, dashboard of a vehicle, carousel, accordion, *World Wide Web*—W3 is a cyberspace, spider web, global village, highway network, ocean for surfing, library, etc., *game and simulators metaphors*—walking, flying, shooting, first person.

Metaphors determine *interaction styles*—the way of using input/output devices. The main traditional interaction styles are *command line, menu driven, question and answer, form filling, windows (WIMP), direct manipulation style* etc.

3 HCI AND TECHNOLOGY EVOLUTION

3.1 *Prerequisities*

First computers used the batch processing computing. Programmers were often operators or even technicians. Programming was done by the interconnecting wires, later punch tapes and cards, teletypes with command line interaction style. With the introduction of mini computers and CRT alphanumerical and vector graphics displays appeared an interactive computing menu driven and forms filling styles. Progress in memory technology and merging with TV technology contributed to the color raster graphics and *desktop metaphor* style computing with the *Windows, Icons, Menus and Pointers—WIMP*. Pointing, clicking, dragging, cut and paste techniques in the text or image processing became standards and the users with no computing education and skills could use them massively. When the first touch or multi-touch sensitive displays appeared the new style of the direct manipulation appeared. Typical gesture-function-pairs are *tap*—instead of moving the mouse controlled cursor and clicking, the tap uses direct pointing, *double tap, long press, scroll, pan, flick, pinch, zoom and rotate*. This interaction style was invented a long time before the tablets and smartphones era (Krueger 1991).

Integration of computers with telecommunications caused that *computers became also communication tools* and a new concept of distributed and collaborative network computing appeared. They use different communication paradigms, some of them existed long time before the computer era or new, which were enabled by digital technology—*face to face dialogue, conference, chatting room, social network*, etc.

The concept of the *cloud computing*, where the main computing power and data resources are at data centers and applications run on a local desk top or mobile computers existed in a very simple form a long time ago in the era of time sharing and remote terminals.

The visions of intelligent machines, humanoid robots or cyborgs serving humans and communicating with them naturally as shown for example in the movies Knight Rider (Wikipedia/Knight Rider 2014), Star Trek (Wikipedia/StarTrek 2014) or Terminator (Wikipedia/Terminator 2014) are not realistic today, but some functional and natural communication features are possible already now.

What is the prerequisity to implement these ideas? There are several aspects which must be considered:

a. *Miniature, high performing computing machines* with a minimal power consumption and variety of fast communication channels.
b. *Material science and mechatronics* of sensors, displays, actuators and other components for input and output devices.
c. *Ideas, paradigms, concepts, algorithms and software* for executing the above mentioned hardware.

Main technological trends today are: Cognitive wireless sensor networks, embedded, mobile and wireless systems with different wavelengths, modulations and protocols, advanced networking architectures, advanced multimedia systems, ergonomic, usable and natural user interfaces, security, privacy, trust, dependence, reliable and autonomic computing, service and semantic computing, smart agents, intelligent and self-organizing systems and networks, context awareness, social sensing and inference computing.

3.2 *Sensory system and input devices*

First interactive input devices were teletype, light pen, a pair of potentiometers, joystick and mouse. The newer input devices are:

a. *The multi-touch surfaces* massively used in the mobile phones, tablets or information panels.
b. *3D mouse, flock of birds* and other electromagnetic or ultrasound 3D position, orientation and motion trackers.
c. *The exoskeleton, data glove and data suit* provide an input of the position, rotation and bending of fingers, a hand or body and integrated with a haptic feedback provides the touch sense.
d. *The camera* is dedicated to capture still or moving pictures, but can be used in capturing data for the computing objects geometry, position, orientation and motion tracking in 3D space. The depth calculation requires two or more views of the scene, structured light projected by the data projector or markers attached to the objects. Camera can use invisible spectrum—infrared or thermo vision. Merging an infrared laser projector and a monochrome sensor is used in Microsoft Kinect device, which can capture the depth of the scene with objects and provide data for the reconstruction of object geometry and texture. Kinect has also a microphone array, which can localize the position of the sound source in the space.
e. *Sensors* for measuring different physical phenomena for example speed, temperature, pressure, humidity, magnetic field—compass, acceleration, GPS devices, smoke detectors, blood pressure and oxygen content meters, electrocardiograph—EKG or *Brain Computer Interface*—BCI integrated with microprocessors and wireless networks allow remote and continuous monitoring of environment, things and people.
f. *Tangible objects* are 3D artifacts equiped with basic or intelligent sensors and wireless connection. They are elements for creating real object interface and can be manipulated by touching and moving.

3.3 Motor system and output devices

The first data visualization output devices used paper or two colors vector type CRT displays. Later appeared raster and color CRTs. Today flat LCD screens represent the visualization mainstream. The audio output may have one or several channels—stereo, surrounding sound amplifiers with loud speakers and earphones.

a. *LCD monitors and data projectors* render images of different size—from a tiny wrist watch or mobile phones up to large projection panels. Image can be mono or stereoscopic.
b. *Googles and helmets* are devices for creating illusion of the depth and support immersive experience. Different stereoscopic projection methods are exploited—separated images, bicolor anaglyph, passive or active polarization glasses.

   Head-Up Display—HUD is a transparent display that presents data without requiring users to look away from their usual viewpoints. The origin of the name stems from a pilot being able to view information with the head positioned "up" and looking forward, instead of angled down looking at lower instruments. Head or *Helmet Mounted Display*— HMD has two displays that move with the orientation of the user's head.
c. *Complex environments* are used for realistic and immersive simulations. The *Automatic Virtual Environment*—CAVE is a box with 3 to 6 active surfaces, where the mono or stereo image is projected from the outside of the box. Image can change with the time according to external script or user input. Instead of the box, a half dome or sphere can be used. Airplane, ship, car or tank cockpit and dashboard simulators use real control elements with simulated image projected on the windows.
d. *Wearable actuators*, brain and muscle stimulating devices are used in research of direct human machines communication and are basic elements for implementing the idea of cyborgs.

3.4 Cognitive system and Artificial Intelligence

Computers can simulate human cognitive properties—*memory, knowledge representation, reasoning, problem solving, decision making,* etc. Scientists estimate that the brain process about 100 trillion instructions per second, it is 100 teraflops during the visual processing or may be capable of 20 million billions of calculations per second, it is 20 petaflops. One of the world´s fastest supercomputers Sequoia can perform about 6 petaflops. It is not far behind the human brain, but human cognitive processes are not simple logical and numerical computations but complex and unknown activities of billions of neurons.

Animals and human vision, auditory, walking and thinking abilities are the result of million years of evolution. The time when robots will have properties shown in the sci-fi movies are far away, however in many activities machines outperform humans already now.

The future of HCI will depend on the research in the field of Artificial Intelligence, especially machine vision and learning, computational linguistic and knowledge representation. Today the two interaction concepts involving complex cognitive abilities are in focus of research—the *natural language communication* and the *visual and gesture interaction* with the mixed reality and computer vision systems.

3.5 VUI: Voice user interface and natural language communication

Natural language—the result of a long time evolution is the most effective and unique communication form between people. In natural language we are able to describe "almost" everything. One of the first "natural" language pilot projects was the program ELIZA written in 1966 by Joseph Weizenbaum (Weizenbaum 1966). In reality it was a fake, but showed the strong side of this form of interaction. This program simulates Rogerian psychotherapist (Person-Centered Therapy 2014) and provides generic responses. For example, the question to the sentence "My eyes hurt" is "Why do you say your eyes hurt?" Despite the very simple principle, users were surprised and some of them believed in the computer intelligence.

ELISA used typed questions and printed answers. Today exist improved copies of this program, some of them use computer synthetized visual and emotional Avatars.

Language uses abstract signs—words with concrete meanings. They consist of basic written or spoken elements—characters and phonemes. Words can be composed in sentences. With only a few characters or phonemes and using lexical and syntactical rules we are able to create unlimited number of grammatically correct and meaningful sentences. Words or simple sentences can be replaced by abstract or concrete images, but such expressions are ambiguous.

On the other side complex images are hardly expressed by the words. Links between signs and their meaning are not exact and depend on the communicating subject experience and education—each subject has his/her own language of thought called Mentalese (Arbib 1986).

Speech recognition—speech to text subsystem is a multidisciplinary field, merging signal processing, computational linguistics and machine learning. It is a translation of spoken words into text with or without training. Related discipline is speaker identification. The Voice User Interface is not widely used because of the limited reliability, but satisfactory robust language can be achieved in a small special domain.

Speech or voice synthesis and text-to-speech system converts character strings to the audible speech, usually by the concatenating pieces of recorded speech—*phonemes, diphones* or full words, stored in a database. Another method is to simulate a vocal tract. Modern speech synthesizers are able to render *prosody* and *emotional content*.

There are many existing or envisaged applications of the natural language interaction: *Voice dialing*—hands free in car, *direct voice input*—voice input control, *hands-free voice command interface* of different devices and systems—TV set remote control, military airplane cockpits. In the health care speech recognition can be implemented in the *medical documentation process*, where the user dictates—for example diagnosis, into a speech-recognition engine or recorder. The recognized words are displayed and edited. Another examples are SMS or other text reading via cable phone, announcing departures of buses, trains airplanes, screen and book readers for visually impaired persons, children or others with reading difficulty. They can work in connection with real time translation modules. In the future possible candidates for the GUI replacement are in smart phones or tablets, in education—for example learning language, games, office—reading and writing letters and everyday life.

3.6 *VBI: Vision-based user interface*

VBI is a hot issue in HCI research. There exist many publications, tools supporting materials for programming VBI (Davison 2013). The visual communication through the human body language requires using computer vision, graphics and mixed reality systems. It is a multidisciplinary field with following tasks and related disciplines: *Objects and gesture recognition, motion capturing*—Pattern Recognition, Computer Vision, *scene representation*—Geometry Modeling, Artificial Intelligence, *image synthesis, rendering and animation*—Computer Graphics, Virtual Reality, *merging real and virtual worlds*—Augmented or Mixed Reality. The UI goal is to identify, understand and interpret human body language—position and orientation of the skeleton, hands and fingers configurations and movement, face expressions, other object detection and recognition, scene composition. The information about the static and dynamic gestures—gait, working, fighting, dancing is captured with a suitable depth sensing input devices.

Gesture recognition is used in a *sign language interaction* by hearing and speaking impaired persons, *socially assistive robots*—rehabilitation, taking care and patients, *gesture commands for users with motor disabilities*—facial expressions and eye tracking, *virtual reality environment*—simulators, games, remote and *alternative virtual controllers*—for example remote TV set controller or *affective computing*—identifying emotional expressions in education, psychology, criminology etc.

For the computer, the gesture recognition represents the process of information input. The visual output is in the form of computer modeled and animated virtual objects. These objects can be visual clones of existing things, animals or persons. Computer needs a *3D model* of these objects—geometry, surface color, reflection or other property and algorithm of their behavior. *Object's behavior* can be based on some *pre-programmed scenario*—script or based on data from the *motion capture system*—on line sampling in real time or stored samples. Example of a simple sampling movement path is teaching robotic arm with several links to paint or weld complicated shape.

One form of mixed reality interaction is *Synthetic Vision System—SVS* supporting situation awareness—for example in the piloted or airplane. It can use set of locally stored databases—terrain models, obstacles, meteorological data, political and strategic situation, etc., real time image synthesis, GPS and other sensors and software matching simulated with the real world. Remotely operated planes can be used on board cameras or simulated terrain.

4 EMERGING COMPUTING AND INTERACTION CONCEPTS

4.1 *Computing concepts*

In this paper the *computing concept means a way how a computer is used for any goal-oriented activity*. The word computer has a broad meaning and includes networks and robots as well.

There exist various computing concepts, paradigms and styles. Some of them are based on *different technologies*—super, tele, network, quantum, dust or bio computing, *computer organization and architectures*—parallel, collaborative, grid, cloud, green, organic computing. These concepts have impact of the *functionality, performance, reliability or price* and do not influence human factors significantly. From this point of view, closer link with questions like ergonomics, interaction concepts, metaphors and styles are *mobile, ubiquitous, pervasive and ambient computing* as well as *haptic* or *social computing*. All these concepts can be mixed—*jungle computing*.

a. Desktop computing
   This is the main stream in using computers, however being slowly replaced with mobile computing. Desktop computing started with command line and menu driven interaction styles, later replaced with GUI and WIMPs.
b. Mobile and wearable computing
   Mobile communication and personal management tools are becoming smaller and their elements can be integrated to wearable objects—watches, jewelry, cloth, shoes, however today we use this expression for netbooks, tablets and smartphones. Wearable computers may have elements distributed and even implanted in the human body. Existing examples are digital pace maker, cardioverter defibrillator, necklet or collar with *GPS* and transmitter carried by criminals, implanted *RFID* with the authorization and authentication code or payment card or an intelligent vest carried by soldiers and capable to detect injury and transmit this information to the commander.
c. Ubiquitous and pervasive computing
   Ubiquitous computing is a concept where the using computers is possible everywhere and any time including microcomputers embedded in everyday objects such as a wash machine or a pair of glasses. The supporting technologies for ubiquitous computing are network, operating system and middleware for distributed computing, sensors, modern I/O devices, protocols, mobile code, localization and positioning devices.
   Pervasive computing uses the same paradigm as ubiquitous computing. It is closely related with ambient computing. Applications of ubiquitous and pervasive computing vary from domestic systems, such as intelligent houses including home entertainment and health monitoring, complex industry, defense or public sector applications.
   Using existing user interaction metaphors and styles in ubiquitous computing is unacceptable. Possible solutions are Natural and Real User Interfaces metaphors and styles.

d. Dust computing

Dust computing is a limited form of ubiquitous computing and uses miniaturized (mili to nano meters) and simple computing devices or *Micro Electro-Mechanical Systems (MEMS)* without user interfaces distributed on a relatively small area with wireless communication. Dust computing can be used for monitoring pollution, plantations or battle fields etc. Some scientists envisage a systems where billions of miniature, ubiquitous intercommunication devices will be spread worldwide.

e. Skin computing

This concept uses fabrics with light emitting and conductive polymers or organic computer devices. The fabric forms flexible non-planar display surface or is embedded to textile such as cloth or curtains. MEMS device can also be placed onto various surfaces so that a variety of physical world structures can act as networked surfaces of MEMS.

f. Tangible, graspable and clay computing

Graspable and tangible computing uses interfaces in a form of physical artifacts representing and controlling program execution. Such interface directly manipulates its underlying association. Clay interface is an ensemble of MEMS, formed into arbitrary three dimensional shapes as artefact resembling different kind of the physical object. These devices allow embodied interaction and cognition—it is manipulation, creation and sharing objects and acting in a virtual or real environment.

While the dust computing is a pure computing and communication environment, where UI resides in dedicated computer, skin and clay concepts represent real objects interface.

g. Ambient computing

Ambient computing or Intelligence is a term used for the next level of the ubiquitous and pervasive computing. It is an environment—surroundings with the consumer electronics and computers interconnected with telecommunication devices and able to react to the human's presence and activities—*Context Aware Computing*. The response to the human presence and activities can change—*Adaptive Computing* and can anticipate human desires without conscious—explicit mediation—*Anticipatory Computing*. The goal is to support people in their everyday life in a natural way—*Natural User Interface* using information and intelligence hidden in these, interconnected devices—*Things That Think, Internet of Things, Real User Interface*. Important aspect is that the technology is hidden and only the user interface is perceivable. Ambient intelligence is the part of the envisaged concept of service systems with automated adaptation, personalization, context awareness and anticipatory abilities. Possible applications are at homes (Bielikova 2001) or hospitals.

4.2 *Interaction concepts*

a. Symbol based interaction

In GUI real objects are represented by signs—representations of things—icons, symbols. User—interpretant interprets these signs, as they would be real objects. By manipulating—point and click, drag and drop, etc., the interpretant control real objects—matching. One of the Virtual Reality pioneers Jaron Lanier (Lanier 1994) calls this Piercean semiotic model symbolic communication.

b. Post symbolic communication

GUI and windows are perfect for the office work, when it is not necessary to create objects which are hard to be described by words or symbols. On the contrary in applications such as 3D design—architecture, transport, glass or sculpture design, designers are used to make 3D models directly, not from drawings created by the virtual tools represented by the symbols. For them traditional GUI is neither intuitive nor natural. The first prerequisite for post-symbolic communication is *shared virtual reality*. The second one is an *ability to create a virtual world quickly and store it in a common data base*. In this environment it is possible to make up directly shared reality instead of talking about it. Instead of using words for describing things it is possible to make them. For instance, when an architect designs a house, client can make virtual walkthrough in this house. Instead complicated specification for putting a new wall inside, he or she simply puts a new wall and moves it to the requested position. It is not necessary to use drawings and blue prints and the documentation is a 3D model created during the session of the architect with the client. This is a way of communicating without representation.

J. Lanier said "Members of this community could hypothetically communicate by creating rapidly changing content in a shared, objective world. They would create and share content directly, instead of referring to contingencies indirectly with words or other symbolic devices. This is what I call post-symbolic communication" (Lanier 1994).

"While it might at first seem that symbols, abstractions, and categories would be needed to communicate anything substantial, even in this future, that does not appear to be the case. For just one example, instead of abstract categories or platonic ideals, it might be possible to create a concrete, but very large, collection of objects that are to be considered as similar. Such a collection could be held inside a virtual jar, for instance, that is small on the outside but big inside, and could be available as conveniently as a word" (Lanier 1994).

c. Reality and natural user interface

In post—symbolic interaction, the objects are represented by their visual—audio, and in the future maybe touch sensitive clones. They are naturally manipulated but they are still representations and not real things. In *Real User Interface* objects are real objects directly manipulated by the user. The effort to create a user friendly interface under the name "natural" started a quarter of a century ago but the *Natural Interface—NUI* is yet not exactly defined. NUI means that the interaction is natural, rather than that the interface itself is natural. Natural interface is invisible. One example of NUI is *Reality User Interface—RUI* or *Reality-Based Interface—RBI*. This can be wearable computer interacting with real-world objects, where user can tap or show on this object—like on a hyperlink symbol in standard GUI and some event happens in a merged virtual and real world. Another example is using direct-touch and multi-touch technology, however many multi-touch toolkits use traditional GUIs. These NUIs use visual elements—and thus they are GUIs.

In the command level interaction, the users need to know means of input with a strict syntax. GUI enabled easier and more intuitive interaction, but it is based on metaphors. Post-symbolic communication uses visual clones of things. They are not abstract, but still they are virtual.

d. Context Aware Computing

Context Aware Computing is the next level of Ambient Computing. In traditional HCI is the aim to understand the user and the context of use. UI design is focused on supporting anticipated use cases and situations. In Context-Aware Computing more than one context of use are supported. In the HCI, awareness is ability of computers to sense—detect, recognize and classify situations as contexts and react—adapt, change or trigger behavior, based on their environment. Context awareness is the supporting concept and suitable for ubiquitous and wearable computing systems and the Internet of Things. It enables context aware applications such as adapting user interfaces or propose available services, for example automatic setup of the mobile phone during an important meeting—rejecting incoming calls. Context can be physical or human factors related, for example location—absolute, relative, surrounding communicating and computing resources, temperature, emotional state or social environment. It is not easy to determine context based on sensor data, user and task models because this is dependent on the subject too. Well-designed context-awareness supports user-friendly and enjoyable applications. Wrong-designed may cause frustration.

e. Implicit Human Computer Interaction

Implicit HCI Interaction generalizes the concept of Context Aware Computing. Traditional, explicit interaction with computers has visible interface and contradicts the idea of disappearing interface. Explicit and implicit interaction can use different interaction styles including command line, GUI and interaction in the real world.

Implicit HCI is the human interaction with his environment and objects. The aim of interaction is to achieve a specific goal. During this process, the system acquires implicit input from the user and may present implicit output to them.

Implicit input are human actions—behavior with the intention to achieve a goal. The behavior is not primarily regarded as interaction with a computer, but captured, recognized, and interpreted by a computer system as input. Implicit output is the output of a computer that is not directly related to an explicit input, and which is integrated with the environment and the task of the user.

f. Human body as UI, cyborgs and brain computer interface

Today's computer can be very small but the interaction with application software requires input/output devices which can not be tiny. So the idea of human body as an user interface is challenging (Cruz 2014).

A cyborg—cyberorganism is a being with bio-organic, electronic and mechanical parts. In HCI there is a clear distinction between the human and computer functions. In this context, the human with embedded—attached or implanted machine parts is a cyborg. There exist many research projects and pilot applications, where people have machine parts. For example *Ear-Based Interface* uses a small device attached to the ear. Imaginary Interface uses detection of moving fingers with sensors attached to the body.

Electronic media and body artist Stelarc (Wikipedia/Stelarc 2014) has a third robotic arm and has performed projects with distributed neural system connected to the Internet with a possibility of the remote muscles stimulation or controlling his own muscles through neural signals transmitted via Internet. Another artist "eyborg" Neil Harbisson (Wikipedia/ Harbisson 2014), officially recognized as a cyborg by a government was born completely color blind. He uses *Synesthesia*—stimulation of one sensory system to the experiences in a second sensory system in a form of audiovisual device attached to his head. This device, turns color into audible frequencies and is included on his passport photography as an integral part of his identity.

Prosthesis and Brain Computer Interface—extremities, eyes or ears, with digital electronics or embedded processors are building blocks of future cyborg systems. Visions of the *Human—Robots, Cyborg—Human* or *even Brain—Brain* interaction are becoming reality. Moreover, networked cyborgs will have possibility of distributed senses. These facts raise psychological, social and ethical implications.

5 CONCLUSIONS AND FUTURE WORK

Diverse applications prefer different computing and UI concepts, metaphors and interaction styles. Lanier told (Lanier 2004) that "In some cases, as with McLuhan, natural languages disappear and are replaced by a universal mesh of direct brain links". In special cases it might be valid, but in general this statement is more sci-fi vision than real alternative. Natural human language is a perfect communication tool and there is no reason not to use it. The problem is reliability of today's technology.

In some applications, language is not efficient. In this case post-symbolic, Natural or Real Objects Interfaces will be preferred forms. For this reason, reliable and easy to use computer vision systems, advanced, cheap virtual or real autonomous communicating robots—chatbots, avatars will be required. Twenty years ago, the virtual reality was presented as an ultimate user interface design challenge. The same was expected from the Web3D technologies in the Internet, but results of these visions are still not visible. Collaborative environments based on natural interaction need high quality visualization technology—big high resolution monitors on one side or personal wearable displays, mono or stereoscopic rendering, ability to merge image of the real video with rendered virtual worlds in real time. Manipulating real or virtual objects in virtual or real environments requires detecting their position and orientation in 3D space. Despite many existing methods, there is a lack of reliable, mobile, cheap, easy to use and precise devices and algorithms for detecting and measuring spatial objects.

These are tasks not only for user interface professionals, but for wider computer science and engineering as well as mechatronic, automation and telecommunications community. To use efficiently new technology, deeper understanding fundamental problems of interaction semiotics in different application domains and computing concepts is necessary. Usability standards, recommendations and good practices have to be expanded to wider interaction styles. There exist many experimental and pilot concepts, systems, environments, devices and applications using revolutionary principles, but to transfer these ideas and projects to broad spectrum of routine applications is not an easy task. HCI is a multidisciplinary field and to develop new forms of human machines interaction is a challenge not only for IT specialists but also for linguists, psychologists and other human and social science specialists.

REFERENCES

Arbib, M.A., & Hesse, M. 1986. *The Construction of Reality*. Cambridge: Cambridge University Press.
Arvizu, U. (Editor), 2014. Why UX Designers Should Use Idioms Rather Than Metaphors, https://medium.com/ux-ux-human-interfaces/f0e4718f4960.
Barr, P. 2003. *User-Interface Metaphors in Theory and Practice*. MS Thesis. Wellington: Victoria University.
Bielikova, M., Krajcovic, T. 2001. Ambient Intelligence within a Home Environment. *ERCIM News* 47.
Carroll, J.M. 2002. *Human-Computer Interaction in the New Millennium*. New York: ACM Press.
Cruz, L. 2014. *Beyond the Touchscreen: The Human Body as User Interface*. http://newsroom.cisco.com/feature/1275146/Beyond-the-Touchscreen-The-Human-Body-as-User-Interface?utm_medium = rss.
Davison, A. 2013. *Vision-based User Interface Programming in Java*. Kindle Edition.
Harper, R., Rodden, T., Rogers, Y., Sellen, A. 2008. *Being Human: Human-Computer Interaction in the Year 2020*. Cambridge: Microsoft Research.
Interaction Design Foundation, Open educational materials—made by the world's design elite. 2014. http://www.interaction-design.org/.
Krueger, M. 1991. *Artificial Reality II*. Addison-Wesley Publishing Company, Inc.
Lakoff, G., Johnson, M. 1980: *Metaphors We Live By*. Chicago: The University of Chicago Press.
Lanier, J. 1994. Virtual Reality and the Future of Natural and Computer Languages. An abstract of a lecture given at the Columbia University Computer Science Department. http://www.jaronlanier.com/columbia.html.
Lanier, J. 2004. The Next Five Hundred Years of Communication, http://www.jaronlanier.com/lecture.html.
Lanier, J. 2010. You Are Not a Gadget. In *Manifesto*. New York: Alfred A. Kopf.
NN/g Nielsen Norman Group. 2014, http://www.nngroup.com/.
Person-Centered Therapy (Rogerian Therapy). 2014. http://www.goodtherapy.org/person_centered.html.
Rohit T., R., Soumyadeep, P., Chaitanya B., Jyothi, B.V. 2013. The Rise of The Cyborgian Epoch. *International Journal of Computer and Information Technology* 2(2).
Weizenbaum, J. 1966. ELIZA—A Computer Program For the Study of Natural Language Communication Between Man And Machine. In. *Communications of the ACM*, 9(1): 36–45.
Wikipedia. 2014. Stelarc. http://en.wikipedia.org/wiki/Stelarc.
Wikipedia. 2014. Neil Herbisson. http://en.wikipedia.org/wiki/Neil_Harbisson.
Wikipedia. 2014. Knight Rider TV series. http://en.wikipedia.org/wiki/Knight_Rider.
Wikipedia. 2014. Star Trek. http://en.wikipedia.org/wiki/Star_Trek.
Wikipedia. 2014. Terminator. http://en.wikipedia.org/wiki/The_Terminator.

Current Issues of Science and Research in the Global World – Kunova & Dolinsky (Eds)
© *2015 Taylor & Francis Group, London, ISBN 978-1-138-02739-8*

Trends in information visualization for software analysis and development

P. Kapec & P. Drahoš
Faculty of Informatics and Information Technologies, Slovak University of Technology, Bratislava, Slovakia

M. Šperka
Faculty of Informatics, Pan-European University, Bratislava, Slovakia

ABSTRACT: Information visualization is a broad research area with ongoing progress especially in the context of software development and analysis. In recent years many interesting visualization approaches have appeared, but only a few techniques were used in practical applications for software analysis or development. Advanced software visualization is gaining support slowly, but it has already become a part of common software development process. Although there has been rapid progress in this field, many open problems still exist. Technical advances allow us to collect and to process vast amounts of data, but these cause significant issues of scalability in visualization and interaction techniques that have already been addressed. In this paper we present recent trends in context of our software visualization projects that utilize graph representations and various source code visualization techniques.

1 INTRODUCTION

Similarly to other fast developing research fields many open problems still persist in visualization. Problems are caused by various factors and are often driven by fast technical development, new application areas and our better understanding of user's behavior. However, we can identify several factors that complicate the progress in the visualization field (Ward et al., 2010). Today we often deal with terabytes of data that is often dynamic and streamed, unstructured or with errors, thus complicating data processing. New hardware devices like mobile touch devices with HiDPI resolutions, wall-sized displays or immersive technologies for virtual or augmented reality, all of these open possibilities for new types of interaction. GPGPU programming allows to utilize the massive parallel processing power of graphic cards, but algorithm parallelization is not a trivial task. With new interaction options also came new possible application areas for visualizations, however, moving from visualizations for experts, which often combine visual and computation analysis, to visualizations for masses is challenging. A major issue is an evaluation of visualizations where some formalization of evaluation is needed. We do not need to measure only quantitative enhancements, but we need a validation of measurable benefits of visualization for analysis and decision-making. Last, but not least, the psychological and physiological aspects of human visual system and reasoning need to be addressed in the context of visual data mapping and visualization evaluations.

In the following section we briefly introduce information and scientific visualization. The next section describes the software visualization field and we briefly discuss open problems in software visualization. The following sections present an overview of our contribution to the software visualization field: experimental IDEs for the programming language Lua with graphically enhanced code presentation and source code metrics visualization; tools for 3D visualization of software structure, program execution and evolution of software systems.

1.1 Information and scientific visualization

Information and scientific visualization deal with the transformation of data into graphical form and with providing interaction and exploration techniques. The motivation is to help with the understanding of vast amounts of data, complex processes analysis and to help to make software more tangible for software developers. Information visualization deals mostly with abstract data, on the other hand scientific visualization deals with natural phenomena that have inherent spatial information. A good introduction to information and scientific visualization can be found in (Mazza, 2009) (Wright, 2007). For a recent overview of information visualization field see (Liu et al., 2014) (Maalej et al., 2012) and for an overview of scientific visualization field see (Kehrer & Hauser, 2013).

In both fields the visualization process consists of several steps (Ward et al., 2010): *data preparation*, *visual encoding*, *presentation* and *interaction*. Among the various presentation and interaction techniques the most cited visual investigation technique is the visualization seeking mantra: *Overview first*, *zoom* and *filter*, and then *focus* and *details-on demand* (Shneiderman, 1996). According to Keim's classification, information visualization deals with six data types (Keim, 2002): one-dimensional/two-dimensional data, multidimensional data, text and hypertext, hierarchies and graphs, algorithms and software. Although software is categorized as a separate data type, software visualizations often utilize visualization techniques developed for other data types.

1.2 Software visualization

Under software visualization it is often imagined the visualization of program source code. However, software is not only source code, but the term software covers all software artifacts that are created during software development life cycle: data, algorithms, source code, diagrams, documentations, user interfaces, revisions, unit tests etc. and even objects representing software development process: tasks, deadlines, developers, bugs etc. Among the first non-textual representations of programs were pretty printed source code in printed publications, flowcharts, Jackson diagrams or Nassi-Shneiderman diagrams—these techniques can be backtracked to the beginnings of computer science. Since that time programs became more complex and thus specific modeling tools, like the Unified Modeling Language (Rumbaugh et al., 2004), have been developed. However, even UML has limitations as today's complex software systems are usually a set of parallel asynchronous processes implemented in heterogeneous programming environments dealing with huge amounts of data.

The goal of software visualization is to help to understand the structure of software artifacts and their relations, functionality, the concurrency of data and the control flow during program execution. Software visualization aims to "make" the intangible software more tangible allowing programmers to discuss visualizations rather than discussing abstract concepts. Software visualizations are not aimed only at professional software developers, but also at development novices, software project manager and others. Software visualization focuses on three main aspects (Diehl, 2007): *visualizing software structure*, *visualizing behavior* of executing processes and *visualizing the evolution* of software development. Visualization of software structure deals mostly with static aspects and architecture: decomposition into modules, classes, inheritance etc. A recent and detailed survey of static aspect visualizations can be found in (Caserta & Zendra, 2011). Visualization of program behavior deals with algorithm and program state visualization—see (Shaffer et al., 2010) (Jeffery, 2011) for recent surveys. As software changes during development, it is necessary to understand these changes—see survey (Novais et al., 2013).

Although a significant progress in the software visualization field, several open challenges have to be addressed in future (Diehl, 2007), e.g. defining standard visualizations for typical use cases for easier learning how to interpret them; developing automatic focusing in various focus and context techniques; allowing continuous navigation between various levels of abstraction; finding effective ways for displaying structural changes; visualizing state of running programs etc. Modern computer graphics offers means for immersive 3D visualizations, however, such features are not often used in information visualizations. In the following sections we address some of these challenges in our prototype software visualization systems.

2   EXTENDING TEXTUAL REPRESENTATIONS OF SOURCE CODE

Visualization and visual representation of software is closely tied in with interaction. Historically programmers interact with code using text manipulation and express ideas using languages with textual grammars. There have been attempts to introduce languages with alternative grammars that do not rely on text input such as graph based grammars (Bardohl et al., 1999). However, these usually fail to gain enough momentum because of interaction issues and negative impact on productivity of programmers.

In this context we explore the idea of a hybrid programming editor that is capable of both, textual manipulation of existing programming languages and visually enhanced representation of language constructs. The visual enhancements go beyond the common text enhancements found in most programming editors and IDEs that mostly rely only on font style changes, font color and background color manipulation. Our experimental editor TrollEdit tries to organize text into visual blocks that directly represent the underlying grammar of the language.

Interaction with code in TrollEdit is similar to existing text based programming editors, however, users can also directly interact with the code on the block level. Dragging and dropping entire blocks of code is done with a mouse and the editor even provides visual guides where a block can be moved or copied to, so that the code remains syntactically correct. This is achieved by means of incremental analysis of the edited text using the language grammar. The resulting abstract syntactic tree is then used to visualize the result with each node of the tree representing a visual and movable element.

Figure 1 shows a virtual workspace in the TrollEdit IDE containing a fragment of a source code in C language. For illustration the code is displayed as standard text on right, and on the left the same source code is displayed in hybrid textual-graphical form as nested blocks. The arrows pointing outwards from source code actually point to source code comments, which

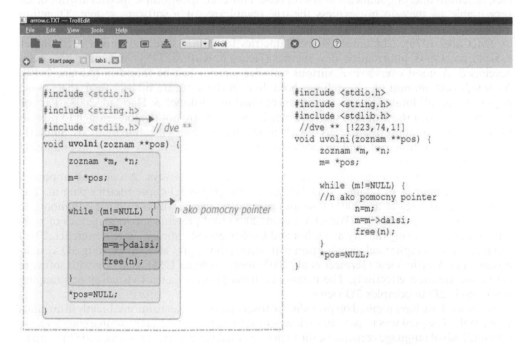

Figure 1.   Source code in C language displayed as standard text (right), and graphically displayed (left).

are unattached so the user can place them freely. Also the source code of a file can be freely placed on the virtual workspace, thus diverting from standard file-based source code view. This approach led to the development of another prototype described in the next section, which is focused on software metrics visualization.

3 INTERACTIVE PRESENTATION OF SOURCE CODE METRICS

Software metrics can provide useful information about the complexity of software design; they can help developers to detect design and implementation errors. Software metrics can notably assist in generating documentation from source code enriched with statistical information that can guide exploring and understanding unfamiliar source code. They are often used to improve the quality of existing software systems and allow comparing different qualities of multiple versions of a software system.

A software metric can be defined as a function that takes various software objects as input and produces a measured, usually quantitative value that describes these software objects. Together with the progress in tools and programming languages for software development, software metrics were also researched. Software metrics are very useful in following development activities (Honglei et al., 2009): *understanding* (e.g. unknown source code or development process), *predicting* (e.g. further development directions), *evaluating* (e.g the current state of development), *monitoring* (e.g. development progress vs. development planning).

Using various software metrics in practice resulted in following criteria for software metrics to be useful (Daskalantonakis, 1992): clarity and precise definition, objectiveness, cost-effectiveness and informative value. As software development is a quite complex process and deals with various heterogeneous software artifacts, three main software metric areas have been identified (Honglei et al., 2009) (Daskalantonakis, 1992): metrics of *software development*, metrics of *software product* and metrics of *software project*. The users of software metrics range from software engineers and software quality managers, to top-level managers and recently also end users.

The basic direct software metrics count the lines of code, number of files, classes, modules, functions and/or comments in source code. However, these simple metrics are not often very usable, as they do not express the true complexity of a software system. Therefore more complex metrics like Halstead metrics (Halstead, 1997), McCabe cyclomatic complexity (McCabe, 1976), Chidamber-Kemerer metrics for object oriented design (Chidamber & Kemerer, 1994), information flow metrics (Henry & Kafura, 1981) and other have been developed. A good overview of various software metrics can be found in (Abran, 2010). Although vast amount of research has been done in the software metrics field, their usage in practice is still limited due to reasons mentioned in (Ordonez & Haddad, 2008). Part of the problem lies in the available software metric tools that are not flexible enough to present various metrics at appropriate levels or lack good interactive visualization methods. Similarly software visualization tools and methods struggle with their use in day-to-day programming practice (Reiss, 2005).

Presentation of software metrics can be done in various ways. A common approach is to present metrics in tabular form like e.g. in the popular Eclipse metrics plug-in. The CodeCrawler (Lanza et al., 2005) tool uses polymetric views, which enhance common tree visualizations. The CodeCity (Wettel & Lanza, 2008) tool presents software metrics of software architecture using the city metaphor and EvoStreets (Steinbrückner & Lewerentz, 2010) enhances this metaphorical visualization with streets, both providing interesting 3D visualizations. The MetricsView (Termeer et al., 2005) tool combines UML diagram with software metric information effectively. The mentioned tools provide metrics visualizations ranging from simple 2D to complex 3D views.

In our work we have focused on providing software metrics visualizations directly in an innovative IDE. The goal was to present various software metrics at different source code levels— from individual language constructs, through individual functions and source code files, up to the total project metrics. This goal is achieved by presenting individual source code artifacts like functions or source code files as individual windows, which are spread on a virtual workspace. This approach is similar to recent innovative IDEs like CodeBubbles (Bragdon et al., 2010) or CodeCanvas (DeLine & Rowan, 2010), which divert from presenting source code based on files. Figure 2 shows the workspace of our prototype showing five windows with source code, each showing source code of a single function, and two windows with different metrics for one selected window.

The windows are connected via links forming a graph-based representation of the software artifacts of a software project. Thanks to the graph-based representation we are able to utilize well-known graph layout algorithms for window placement. The links between windows represent various relations between software artifacts like inheritance, call-graph relations, documentations and metrics. In Figure 2 the largest windows on left-bottom contain code calling three other functions, which are shown in the other connected windows. The metric windows show a bar chart of basic *lines of code* metrics and a pie chart showing count of language constructs. Other metric views can be opened for each window with function source code, depending on user's preferences.

Figure 2. Five windows with source code—links between windows show various relations. Two windows show different software metrics for one function.

4 VISUALIZING STRUCTURE OF SOFTWARE SYSTEMS

Our team has experimented with various approaches for visualization of software structure. In our first prototype (Šperka & Kapec, 2010) we utilized tree visualizations to explore the structure of C# programs focusing on the visualization of object-oriented features. The tree visualization uses a cone-tree like layout in 3D space, which allows exploration of the tree structure from different views by rotating the tree or by navigating a virtual camera. Packages, name-spaces and classes are displayed as boxes with different colors and sizes, depending on the number of elements they contain, and are laid out in circular layers in the cone-tree. Class methods and attributes are displayed as small boxes, colored by their public visibility and placed on the circumference in columns pointing outwards. The visualization tool provides methods for filtering and sorting for better visual analysis of the visualized program. Figure 3 shows the achieved visualizations for a medium sized C# program.

Figure 3 shows the structure of a software system that consists of 17 packages, 97 classes, 439 class methods and 739 class attributes. The first image shows the whole software project—as can be seen, several classes contain much more attributes and methods than the others.

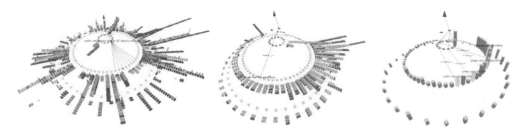

Figure 3. A cone-like tree visualization of the whole software project (left), classes sorted based on number of methods (middle), and displayed without class attributes (right).

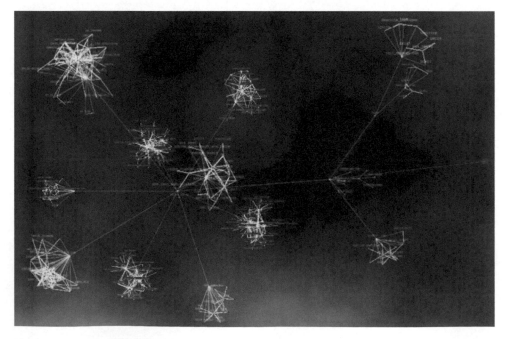

Figure 4. Clusters showing call-graph relations between functions in several software modules. Clusters are interconnected with file-tree.

The middle image in Figure 3 shows the same project but the classes are sorted according to the number of methods and the visualization is made more comprehensible by filtering non-public methods. The right image in Figure 3 shows the result of filtering out class attributes. In our second visualization tool, see Figure 4, we enhanced our approach to support visualizations of graph-based representations of software artifacts, which allows to represent and to visualize heterogeneous software artifacts and their relations. This is similar to our prototype in the previous section, but the visualization is transferred into 3D space and uses 3D primitives to visualize software artifacts. The graph in Figure 4 shows call-graph relations between 11 modules and highly interconnected modules can be identified. The visualization does not show call-relation between functions from different modules to reduce visual clutter.

5   VISUALIZING PROGRAM EXECUTION AND PROGRAM RUNTIME

Program runtime visualization is one of the most difficult tasks in software visualization as it requires solving of two complex problems: 1) monitoring the execution of program instructions and monitoring program state (e.g. class instances, the value of variables etc.), 2) visualization of program state together with data and control flow during program execution. Visualization of program execution can be done in two ways: *live* and *post-mortem*. The live approach directly connects to a running program and as the program executes, the visualization changes. The post-mortem approach makes a recording of the execution of the visualized program and after the program finishes we visualize program execution using the data stored in the recording.

We have developed two unique tools, the first using the live approach and the second the post-mortem approach, for program execution visualization that utilize 3D views. Figure 5 shows a visualization of a running C# program from our first tool (Grznár & Kapec, 2013). The mushroom-like objects represent instances of classes. The cylinders represent individual class methods—as the methods are executed the cylinders grow downwards. This

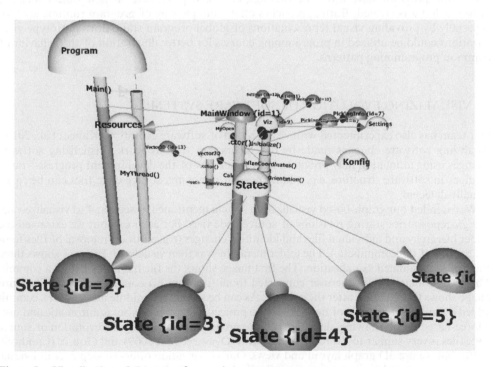

Figure 5.   Visualization of the state of a running program.

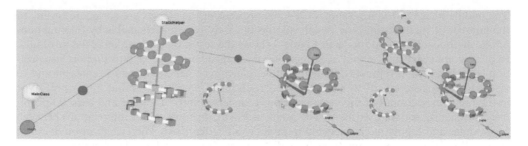

Figure 6.   a) Message sent to class is displayed as a moving blue sphere b) The message before arriving c) The message created a new object.

way the visualization shows both current program state and also the previous execution history. The oriented links show different types of call relations. The live program visualization allows to debug the running program visually—the visualization provides overview of the current program state that is not possible in common debugging tools.

In our second tool we utilize the post-mortem approach (Šperka et al., 2010). The tool monitors the execution of programs implemented in the Java programming language and stores all relevant data into an XML file, which is then used to create the visualization. This allows to playback the recorder program execution in both directions and also to jump to specific execution time, which is not possible in live program visualization/debugging. Figure 6a shows the visualization of a class instance in helix form: the class methods are displayed as orange spheres on the helix, similarly class attributes as green boxes—colors can vary depending on public visibility.

Message passing between objects is displayed as a blue sphere representing the message moving from the invoking object to the receiving object, as shown in Figure 6a. Figure 6b shows another situation in the program state before a message arrives to an object. Figure 6c shows the program state after the message has been processed—a new object instance (a new helix) was created. Both approaches offer a unique way of program runtime analysis especially by providing visual representations of global program state. Both prototype visualizations could be utilized in programming courses for better illustration of the behavior of common programming patterns.

6   VISUALIZING EVOLUTION OF SOFTWARE SYSTEMS

Our team has also experimented with visualization of software evolution (Kapec et al., 2013). Analyzing software changes made by developers is a tedious work. Visualizing software changes using animation may provide a better overview of the development progress—rapid changes in software structure, e.g. adding new, or removing, source code files can be easily visually detected.

We extended our graph-based visualization system mentioned in section 4 to visualize various *git* repositories storing revisions of source code files. For each commit we extracted the folder hierarchy and individual files and identified changes (e.g. addition/removal of files/folders, author of the commit etc.). The folder hierarchy was then visualized. Figure 7 shows three steps of the animated visualization. The first image shows the file hierarchy before a commit. The second image shows the user connected to all files that his commit affected. The third image shows the situation after the commit. As can be seen in the middle image, user's commit added a significant amount of new files to the project. The graph layout is automatic and uses a force-based graph-drawing algorithm. Our approach for visualizing the evolution of source code files is very similar to the tools Codeswarm (Ogawa & Ma, 2009) and Gource (Caudwell, 2010), but we use 3D graph layout and views. Our visualization offers to step over individual commits as opposed to continuous time playback in the previously existing systems.

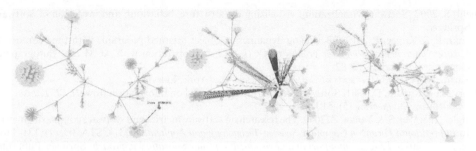

Figure 7.   The visualization of file structure in a software system before, during and after a commit.

7   CONCLUSION

In this paper we presented several prototypes for software visualization. Our prototypes allow new ways of source code presentation and editing in an innovative IDE that allows us to analyze software metrics for source code parts at different code levels. We experimented with software structure visualization utilizing 3D graph visualizations and developed prototypes for program runtime visualizations in 3D. Most of these prototypes are experimental, however, similar approaches are in focus of current research (Polášek et al. 2013).

Future work can be oriented towards various aspects. We are currently focusing on modifying the developed prototypes to be more suitable for applications in praxis, which requires developing suitable interaction techniques and extensive user testing and evaluation. Another goal is to adapt the developed software visualizations for augmented reality, which can bring new challenges in visualization and interaction.

ACKNOWLEDGMENT

This contribution is the partial result of the Research & Development Operational Programme for the project Research of methods for acquisition, analysis and personalized conveying of information and knowledge, ITMS 26240220039, co-funded by the ERDF. This work was supported by the grant VEGA 1/0625/14: Visual object class recognition in video sequences using a linkage of information derived by a semantic local segmentation and a global segmentation of visual saliency. We would like to thank Ivan Šimko, Viliam Kubis and František Nagy for their help in developing the visualization tools presented in this paper.

REFERENCES

Abran, A. 2010. *Software metrics and software metrology.* John Wiley & Sons.
Bardohl, R. & Minas, M. & Schürr, A. & Taentzer, G. 1999. Application of graph transformation to visual languages. *Handbook of graph grammars and computing by graph transformation* (2): 105–180.
Bragdon, A. & Reiss, S. P. & Zeleznik, R. & Karumuri, S. & Cheung, W. & Kaplan, J. & LaViola Jr, J. J. 2010. Code bubbles: rethinking the user interface paradigm of integrated development environments. *In Proceedings of the 32nd ACM/IEEE International Conference on Software Engineering* (1): 455–464.
Caserta, P. & Zendra, O. 2011. Visualization of the static aspects of software: a survey. *IEEE Transactions on Visualization and Computer Graphics* 17(7): 913–933.
Caudwell, A.H. 2010. Gource: visualizing software version control history. In *Proceedings of the ACM international conference companion on Object oriented programming systems languages and applications companion* ACM, 73–74.
Chidamber, S.R., & Kemerer, C.F. 1994. A metrics suite for object oriented design. *IEEE Transactions on Software Engineering*, 20(6), 476–493.
Daskalantonakis, M.K. 1992. A practical view of software measurement and implementation experiences within motorola. *IEEE Transactions on Software Engineering* 18(11): 998–1010.
DeLine, R., & Rowan, K. 2010. Code canvas: zooming towards better development environments. In *Proceedings of the 32nd ACM/IEEE International Conference on Software Engineering* (2): 207–210.

Diehl, S. 2007. Software visualization: visualizing the structure, behaviour, and evolution of software. Springer.

Grznár, F. & Kapec, P. 2013. Visualizing dynamics of object oriented programs with time context. In *Spring conference on computer graphics SCCG 2013* in cooperation with ACM and Eurographics, Comenius University, 75–82.

Halstead, M.H. 1977. *Elements of Software Science*. New York: Elsevier.

Henry, S., & Kafura, D. 1981. Software structure metrics based on information flow. *IEEE Transactions on Software Engineering* (5), 510–518.

Honglei, T. & Wei, S. & Yanan, Z. 2009. The research on software metrics and software complexity metrics. In *International Forum on Computer Science-Technology and Applications*, IFCSTA '09. (1): 131–136.

Jeffery, C.L. 2011. *Program monitoring and visualization: an exploratory approach.* Springer Publishing Company, Incorporated.

Kapec, P. & Paprčka, M. & Pažitnaj, A. & Polák, V. 2013. Exploring 3D GPU-acelerated graph visualization with time-traveling virtual camera. *Journal of Theoretical and Applied Computer Science* 7(2): 16–30.

Kehrer, J. & Hauser, H. 2013. Visualization and visual analysis of multifaceted scientific data: A survey. *IEEE Transactions on Visualization and Computer Graphics* 19(3): 495–513.

Keim, D.A. 2002. Information visualization and visual data mining. *IEEE Transactions on Visualization and Computer Graphics* 8(1): 1–8.

Lanza, M. & Ducasse, S. & Gall, H. & Pinzger, M. 2005. CodeCrawler-an information visualization tool for program comprehension. In *proceedings 27th International Conference on Software Engineering*, ICSE 2005. IEEE, 672–673.

Liu, S. et al. 2014. A survey on information visualization: recent advances and challenges. *The Visual Computer*: 1–21.

Maalej, A. & Rodriguez, N. & Strauss, O. 2012. Survey of multidimensional visualization techniques. *CGVC-VIP'12: Computer Graphics, Visualization, Computer Vision and Image Processing Conference*, 99–107.

Mazza, R. 2009. *Introduction to information visualization.* London: Springer.

McCabe, T.J. 1976. A complexity measure. *IEEE Transactions on Software Engineering* (4): 308–320.

Novais, R.L. & Torres, A. & Mendes, T.S. & Mendonça, M. & Zazworka, N. 2013. Software evolution visualization: A systematic mapping study. *Information and Software Technology* 55(11): 1860–1883.

Ogawa, M. & Ma, K.L. 2009. code_swarm: A design study in organic software visualization. *IEEE Transactions on Visualization and Computer Graphics*, 15(6), 1097–1104.

Ordonez, M.J. & Haddad, H.M. 2008. The state of metrics in software industry. In *Fifth International Conference on Information Technology: New Generations*, ITNG 2008: 453–458.

Polášek, I. et al. 2013. Information and Knowledge Retrieval within Software Projects and their Graphical Representation for Collaborative Programming. *Acta Polytechnica Hungarica*, Vol. 10, No. 2, 173–192.

Reiss, S.P. 2005. The paradox of software visualization. In *3rd IEEE International Workshop on Visualizing Software for Understanding and Analysis*, VISSOFT 2005: 1–5.

Rumbaugh, J. & Jacobson, I. & Booch, G. 2004. *Unified Modeling Language Reference Manual*, The Pearson Higher Education.

Shneiderman, B. 1996. The eyes have it: A task by data type taxonomy for information visualizations. In *IEEE Symposium on Visual Languages*, IEEE, 336–343.

Shaffer, C.A. & Cooper, M.L. & Alon, A.J. & Akbar, M. & Stewart, M. & Ponce, S. & Edwards, S.H. 2010. Algorithm visualization: The state of the field. *ACM Transactions on Computing Education (TOCE)* 10(3):9, 1–22.

Steinbrückner, F. & Lewerentz, C. 2010. Representing development history in software cities. In *Proceedings of the 5th international symposium on Software visualization,* ACM, 193–202.

Šperka, M. & Kapec, P. & Ruttkay-Nedecký, I. 2010. Exploring and Understanding Software Behaviour Using Interactive 3D Visualization. In *8th International Conference on Emerging eLearning Technologies and Applications*, ICETA'2010, 281–287.

Šperka, M. & Kapec, P. 2010. Interactive Visualization of Abstract Data. In: *Science & Military*. Vol. 5, No. 1, 84–90.

Termeer, M. & Lange, C.F. & Telea, A., & Chaudron, M.R. 2005. Visual exploration of combined architectural and metric information. In *Visualizing Software for Understanding and Analysis*, 2005. VISSOFT 2005. 3rd IEEE International Workshop on, IEEE, 1–6.

Ward, M., & Grinstein, G. & Keim, D. 2010. *Interactive data visualization: foundations, techniques, and applications.* AK Peters, Ltd.

Wettel, R., & Lanza, M. 2008. CodeCity: 3D visualization of large-scale software. In *Companion of the 30th international conference on Software engineering,* ACM, 921–922.

Wright, H. 2007. *Introduction to Scientific Visualization.* Berlin: Springer.

Current Issues of Science and Research in the Global World – Kunova & Dolinsky (Eds)
© 2015 Taylor & Francis Group, London, ISBN 978-1-138-02739-8

Model for Management of Business Informatics

Jiří Voříšek
Paneuropean University, Bratislava, Slovakia

Jan Pour
Prague University of Economics, Prague, Czech Republic

ABSTRACT: A number of methodologies (e.g. ITIL, COBIT, ISO 20000, etc.) have been developed over the last two decades but their use, in particular in small and medium sized enterprises is still very limited. In this paper we discuss existing IT management approaches and their limitations with a particular focus on Czech and Slovak organizations. We base our discussion on available literature, our surveys of Czech organizations and our experiences gained from practical assignments. The MBI (Management of Business Informatics) model attempts to overcome the limitations of the existing methodologies and models. It can assist IT executives with solving IT management problems. Customization of the model can take into account various factors that influence enterprise IT management in specific organizations. We describe the basic concepts and features of the MBI model and its future development.

1 INTRODUCTION

This paper describes the characteristics and principles of the MBI (Model for Management of Business Informatics) model. The aim of the MBI model is to offer IT professionals who are responsible for the management of information technology in organizations a comprehensive methodological support based on industry best practices. Application of this model has the potential to increase overall IT effectiveness, improve IT governance and efficiency of IT services, and result in better business performance. MBI model considers important relationships and attributes that are relevant to information systems management and provides a suitable methodological basis for university courses on Business Informatics and IT management.

2 CURRENT ISSUES IN BUSINESS INFORMATICS

Management of business informatics has been the subject of interest of researchers and IT practitioners for a number of decades. Recent increase in the complexity and heterogeneity of the technological infrastructure and enterprise applications, as well as increased number of information services provider options makes the effective management of business informatics challenging. Today, the management of organizations expects improved effectiveness and reduced costs associated with the provision of IT services, forcing IT executives to seek improved methods for the management business informatics. The need to improve the quality of management of business informatics has resulted in numerous approaches; we discuss these approaches and the challenges organizations face in their implementation in the following section.

2.1 *Approaches to Management of Business Informatics*

Increasing demand on the extent and quality of management of business informatics led to the development of various methodologies, models, frameworks, standards and their application in practice. These methodologies include COBIT, ITIL and TOGAF (see IT Governance Institute: Cobit 5. 2012, CMMI Product Team 2010, ISO/IEC 20000). These mostly process-oriented approaches represent current best practice and include recommendations for addressing various IT management issues encountered in practice. Detail comparison of these approaches is available in the literature (Doucek 2007, Schiesser 2011, Voříšek & Pour 2012, Voříšek 2008). However, application of these methodologies in practice is associated with numerous problems. A recent survey of Czech organizations (Pour & Voříšek 2011) indicates that the level of adoption of methods for the management of business informatics is still relatively low. For example, ITIL (the most widely used methodology) is used by 53% of Czech companies, but is fully implemented in only about 6% of these organizations. In the case of COBIT, 53% of respondents indicated that it is not used in their organizations at all, and 12% indicated that this method is only used for strategic management of IT. According to survey respondents, the most important reasons for the low level of utilization of existing methodologies are the complexity of the methodologies, excessive costs associated with their implementation, and considerable demands on the expertize of IT practitioners even in the case of SMEs (Small and Medium-sized Enterprises).

Current trend in business informatics is to strengthen the use strategic applications, i.e., enterprise applications that have the potential to deliver competitive advantage. Such strategic applications include Business Intelligence (BI), Competitive Intelligence (CI), Customer Relationship Management (CRM) (in particular social CRM), and various forms of electronic and mobile business applications. The implementation of these strategic applications is common in multinational corporations, but SMEs face significant problems when attempting to implement such applications caused by limited financial and human resources. According to (Molnár & Střelka 2012) "The use of methods and tools of Competitive Intelligence (CI) is still the domain of large multinational enterprises and Small and Medium-sized Enterprises (SMEs) have little knowledge about what CI is, how to implement CI and what benefits can CI bring them. According to research, conducted in the Czech Republic, main problems of SMEs are lack of financial resources for buying sophisticated and expensive software and shortage of CI experts, due to relatively flat and simple organizational structure."

A very topical issue is the application of metrics in relation to the processes used for the management of business informatics and their integration with methods for IT performance management. According to (Bacal 2012, Maryška 2010, Voříšek 2008 and Wetzstein et al. 2012) performance management of business informatics is one of the most important areas of application of the principles of business informatics and performance management. However, we must differentiate between performance management in organizations that provide IT services (i.e. IT service providers) and performance management in organizations that are not service providers i.e. consumers of IT services (Voříšek 2008, Voříšek & Pour 2012). IT providers tend to focus on effective delivery of a set of IT services that maximize their profit, while IT departments in IT service consumer organizations are focused on ensuring the availability of all IT services needed by the various business units.

Another important research topic concerns the study of the role of IT executives such as Chief Information Officers (CIOs) in ensuring the continuous development and innovation of both IT applications and IT infrastructure, and ensuring harmonization of technology innovations with the business needs of the organization. IT leadership issues, including suitable allocation of responsibility and authority in relation to IT, effective use of human resources, and monitoring and evaluating the effectiveness of IT is extensively covered in the literature (Dohnal 2011, Paladino 2011, Schiesser 2011, Schniederjans et al. 2005 and Voříšek & Pour 2012).

A key question for the management of business informatics concerns the applicability of a single universal model across a range of different types of organization. Many researchers and practitioners argue that each organization needs to develop its own model for the management of business informatics that meets its specific conditions and requirements. Our surveys (see section 2.2) as well as international studies, e.g. Weill and Ross (Weill, P. & Ross 2004), indicate that IT executives tend to develop specific management models that are adapted to the environment that the organization operates in. The factors that influence this environment include industry sector, size of the organization, corporate culture, employees' skills and knowledge, legislative environment, and the availability of IT services from external providers.

The identification and analysis of the above problems and many other issues in the area of management of business informatics form the basis for the formulation of the objectives and principles of the Model of Business Informatics (MBI). An important source of information for the design of the MBI model was the analysis of the results of surveys of Czech organizations during the period of 2010–2012 discussed in the following section.

2.2 *Results of survey of IT management in the Czech Republic*

Business informatics in Czech enterprises and institutions of public administration has undergone a very dynamic development in recent years, in particular in terms of technological innovation. However, international comparisons show that the performance of business informatics in Czech Republic suffers from a number of unresolved problems and unexploited potential (Novotný Voříšek 2011 and Voříšek & Pour 2012). According to international statistics, the Czech Republic is still under the average of EU-27 in the application of e-business, e-government, and the use of electronic information resources (Žák 2013). While there are many factors influencing this situation the quality of management of information systems plays an important role. To substantiate this assertion we have conducted several major surveys involving 1,400 respondents from Czech commercial enterprises and institutions of public administration during the period of 2007–2012. Detailed analysis of the results of these surveys were published in (Doucek et al. 2007, Novotný & Voříšek 2011, Pour & Voříšek 2011 and Voříšek & Pour 2012). We summarize the most important conclusions that were used as inputs into the MBI model below:

- respondents confirmed that the main objective of business informatics and its management must be maximum support for business activities, achieving the highest level of alignment of business needs and functions of business informatics (Business/IT alignment),
- approach to management of business informatics must be adapted to all external and internal factors that affect the organization,
- management of business informatics should follow appropriate standards, methodologies and methods and use suitable metrics to determine the performance and quality of the management processes; necessary supporting documentation and expertise should be provided,
- information systems must be developed and operated at a reasonable cost that corresponds to the importance of IT for the enterprise; focusing on minimizing IT costs may not be the best long-term strategy,
- when evaluating existing applications, and planning new projects it is essential to monitor their potential and actual effects with a primary focus on strategic impact that can deliver competitive advantage for the organization,
- in relation to the previous point, it is useful to identify applications that have strategic importance for the company and are critical for improving competitiveness, and give these applications investment priority,
- innovation in business Informatics should follow developments of the IT market and the state of informatics in organizations of business partners and competitors,
- cooperation between user departments, IT departments and external providers should operate on the basis of service contracts and Service Level Agreements (SLAs),
- from the perspective of ensuring necessary resources, it is essential to choose operational models (outsourcing, cloud computing, etc.) that can deliver cost savings at an acceptable level of risk,
- the required qualifications for users and IT practitioners should be systematically developed,
- organizations should take into account that there will be shortage of IT specialists over the next few years.

3 CONTENT AND PRINCIPLES OF MBI MODEL

Based on the findings of the various surveys we defined the objectives and principles of the MBI model.

3.1 Objectives and principles of MBI model

The primary objective of the MBI model is to provide IT executives in organizations where IT is not core business, a consistent and flexible methodological framework based on best practice for the management of business informatics. The model should assist IT practitioners to:

- document and analyze the existing system for the management of business informatics,
- design and implement a new (improved) management system,
- provide advice and best practice solutions for specific problems in IT management. Examples of problems that MBI provides solutions for include: How to develop information strategy? How to prepare IT budget? What is the structure and content of SLA for application services that are delivered in the form of Software as a Service.

The aim of the model is to enable organizations to improve the performance of their IT, in particular to improve the quality, accessibility, effectiveness and efficiency of IT services, and consequently to improve business performance of the entire organization. To address this requirement several key principles have been defined for the MBI model:

1. the model must allow control of all key features of enterprise information system (Schiesser 2011 and Voříšek & Pour 2012): cover the required functionality, availability, timeliness, accuracy and trustworthiness of required functions and information, compliance with legislations, reliability, user-friendliness, security, flexibility, openness, integrity, standardization, performance, effectiveness;
2. the model must be able to record the responsibilities and authorities of departments and employees in the context of enterprise informatics (Weill, P. & Ross 2004);
3. IT management must be based on a coherent system of metrics that evaluate all important IT services, IT processes and IT resources (Paladino 2011 and Voříšek & Pour 2012);
4. application of the model in practical situations should offer high flexibility. The implementation of individual components (tasks) of the model should be supported without having to implement the entire model. Given the effort involved in the implementation of a comprehensive system for the management of business informatics it is often more effective to address only areas that have been identified as most problematic, or those areas that have the most significant impact on the performance and success of the enterprise;
5. related to the above principle is the ability to effectively deploy the model in organizations operating in different industry sectors and in organization of different size. Implementation of the model in a organization should respect the specific conditions under which the organization operates, including its financial and staff resources. This will allow the application of the model to SMEs that typically have limited financial and human resources.

3.2 Key concepts and content of the MBI model

Any model for management of business informatics must clearly define the types of objects and relationships that the model is based on Figure 1 shows the metamodel that represents the MBI classes of objects and types of relationships between them. The MBI metamodel is drawn using UML class diagram notation 2.0. Task is a key MBI component that represents a basic enterprise IT management unit. Task describes how to proceed in solving a particular IT management problem, for example proposal for sourcing of enterprise IT system, implementation of an IT service, activation of a IT service, implementation of a

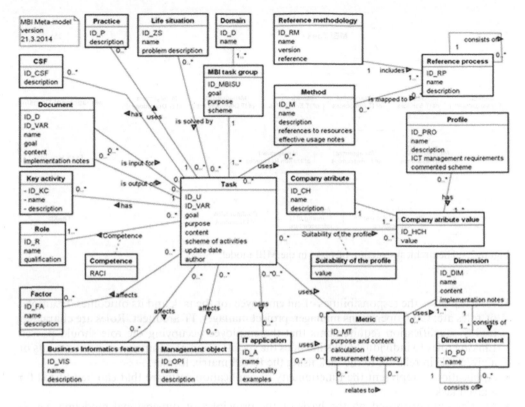

Figure 1.   MBI object classes and relationships.

security audit, etc. The **MBI** model defines a large number of tasks found in **IT** practice and in other **IT** management frameworks. **MBI** presents these tasks as a three-level hierarchy that corresponds to domains of **IT** management as illustrated (Fig. 2).

Task groups are defined within each management domain, for example the domain of Management of IT Resources contains the following tasks: Defining of technological standards, Analysis and planning of software applications, Analysis and planning of IT infrastructure, Configuration management.

Each task is defined by a number of attributes, some of which are directly linked to business performance of the organization and its IT (Fig. 1). A key feature of the **MBI** model are IT processes—schematically expressed management procedures. Processes consist of activities and can be expressed at varying levels of granularity (i.e. only core activities, high-level process diagram, or detailed process diagram) as required for a particular problem under consideration. This enables precise description of the IT process, or on the other hand, can allow for initiative of IT staff as recommended by Method Knowledge Based Process Reengineering (Voříšek 2008). Process tasks are the basis for the specification of management metrics as described in (Pour 2012).

Each task is related to many other **MBI** objects. The most important objects are:

*   Document/Data. Documents include printed or electronic documents that are used as inputs or outputs of tasks. A document may be related to management of business informatics (e.g. a project plan, project objectives, SLA) or provide a solution to particular problem, (e.g. a test scenario, or a tender document used in the process of selecting of a provider of IT services). Data in the MBI model constitute a set of data records that describe some property of an object, for example a results of measurement of a parameter associated with an IT service, e.g. project status within the portfolio of projects, customer information stored in the CRM database, etc.

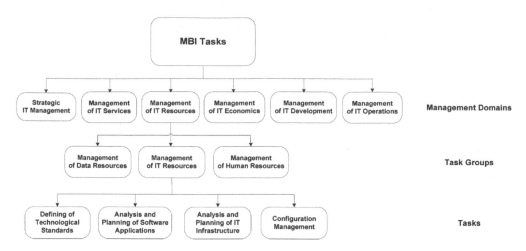

Figure 2.   Hierarchical structure of tasks in the MBI model.

- Role expresses the responsibilities of an employee for the task and its outcomes. Examples of roles are: CIO, operations manager, project manager, IT architect. Roles are characterized by qualification requirements that the employee occupying the role should posses. Linking roles to individual tasks includes the specification of competency requirements of a given role in relation to the task using the RACI matrix format.
- Applications—represent the functionality of application software that can be utilized for a given task.
- Metrics—are expressed on the basis of the principles of dimensional modeling, i.e. as indicators and their analytical dimensions. Metrics determine the key indicators of performance of tasks—KPI (Key Performance Indicator), or KGI (Key Goal Indicator). For example, "IT operating costs" have the dimension of IT services, IT applications, or individual business units. Metrics can provide information such as "What are the operating costs of the service for individual business units."

    Metrics can be regarded as an application of BI in the management of business informatics. The assignment of metrics to individual business processes in the context of tasks ensures that only metrics that are useful in practice are defined. Another essential requirement is that there are data sources that can be used for the measurement and that such data sources are accessible, so that the cost of obtaining data for the monitored metrics does not exceed the benefits obtained by their use.
- Methods—represent the recommended management practices for a particular task.

All defined objects in the MBI model represent "doors" to the model (Fig. 3). This allows the user to access information more directly without navigating the entire model, for example via roles, or some other path that is relevant to the current problem that the user is working on.

Other important components of the MBI model are management factors (Company attributes) that influence the method used to solve a particular task. Management factors include the Size of the organization, Industry sector in which the organization operates, the Type of organization (i.e. private company or public institution) etc. These factors have a significant impact on how a particular task is conducted. For example, consider the task of preparing a Tender for the Supply of IT Services. The approach for solving this task for a public institution has a completely different character than the approach used in the case of a private company, as a publicly funded entity has to follow specific legally mandated tender guidelines. It is therefore apparent that a person responsible for the completion of a given task must identify all the significant factors that impact on the completion of the task and understand how these factors influence the task. The MBI model helps IT managers to

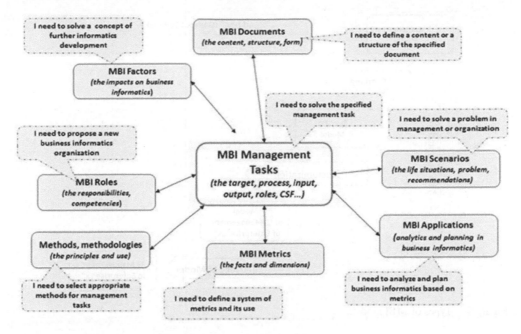

Figure 3. "Doors" into the MBI model for a specific problem.

identify significant factors that impact on each task, and at the same time provides guidance on how to incorporate attributes into specific tasks.

The content of MBI model is available at the site mbi.vse.cz and currently contains data gained from various consulting assignments and derived from literature on this topic.

4 TYPES OF MBI MODELS

As noted in section 2, it is not possible to use a single optimal model of the management of business informatics across all organizations. MBI includes several types of models of business informatics that differ in the level adaptation to specific circumstances of a given organization type.

The most basic and most comprehensive model is the generic reference model. The generic reference model is intended for all organizations and includes generalized, best practice guidelines for the management of business informatics and their specific variants for different types of organizations and for different management factors. Best management practices and practical experiences are captured in the generic reference model in the form of tasks, documents, methods, metrics and other objects as described in section 3.2.

The second type of the MBI model is a specific reference model aimed at organizations that belong to a particular sector of the economy (automotive, banking, public administration, etc.). The content of the specific model is adapted to a particular industry, business, legislative and other conditions that apply to a given sector of the economy. Specific model is created by profiling, i.e. the selection of variants of object occurrences from the generic model that are relevant for a specific sector of the economy as shown in Figure 4.

The third type of model is a model of management of business informatics for a specific organization. This model is created by customizing the specific reference model by taking into account the particular aspects of the organization. Customization includes:

- selection of object occurrences from the specific reference model (i.e. tasks, documents, metrics, etc.) that are relevant for a specific organization. This means that there is a further

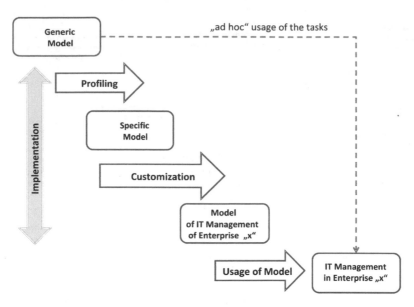

Figure 4.   Types of MBI models.

reduction in the number of occurrences of each object, eliminating the occurrences from the specific reference model that IT management does not regard as necessary for a particular organization and given situation, i.e. those that are not in line with the implementation goals of MBI;

- modification of objects whose contents does not fully meet the specific conditions of the organization. Specific conditions may arise as a result of various factors affecting the organization and its informatics (section 3.2), but also as a result of management efforts to set a different (to MBI model) level of standardization of business and IT processes and its enforcement;
- completion of additional tasks and related objects required by the company not contained in the specific reference model (i.e. capturing specific know-how of the organization);
- definition of the roles of the IT department. The generic reference model assumes a core set of roles for the organization and its IT department. These roles are then assigned responsibilities for individual MBI tasks using the RACI matrix. The roles can be defined differently in a specific organization;
- assigning roles to individual tasks by modifying the RACI matrix for these tasks (determined by the responsibilities and authorities of individual IT roles in the organization— employees with a given role are assigned tasks that they are fully responsible for, and tasks they cooperate on).

The application of the model in an organization involves the use of individual tasks and their implementation by individual actors. The outputs of tasks can be saved back into the model as templates (variants) of documents.

5   APPLICATION OF MBI IN PRACTICE AND ITS FUTURE DEVELOPMENT

IT practitioners have two alternatives for the application of the MBI model. Firstly, IT executives can use the model to create a comprehensive system for the management of business informatics based on the best practices in this area (as illustrated in Fig. 4). In this case, MBI is used to describe all tasks that requires business informatics in the organization, all roles and their responsibilities with respect to tasks, and all other objects that IT management decides to include. MBI then becomes a company regulation that determines who, when and how should solve various IT problems in the organization.

Alternatively, MBI offers IT executives solutions to problems that occur at the strategic, tactical and operational levels. IT executives can find template solutions to current problems using the MBI navigation mechanism based on "access doors", or by searching for specific scenarios and use cases.

Example 1: A user needs to create information strategy for the organization. If the user is only interested in the structure of the information strategy document he can locate the relevant document in the MBI system and use it directly. If the user is interested in the entire process of strategic management he needs to use two tasks groups "Strategic Analysis of Business Informatics" and "Strategic Proposal for Business Informatics".

Example 2: The user needs to prepare SLA for newly introduced services. Using the MBI navigation mechanism the user locates the task "Preparation and Approving of SLAs", follows the recommendations of the task and uses SLA templates stored in the MBI system for inspiration.

Example 3: A user identifies a problem with the performance of enterprise applications—applications have poor availability and slow response time. Using the navigation mechanism the user identifies tasks that impact on performance of enterprise applications, locating the task "Ensuring operational performance and scalability of services".

The authors expect that the second alternative will be used primarily by IT practitioners in SMEs, and that in large enterprises the combination of both alternatives will be used most frequently.

5.1 Implementation of a comprehensive system for management of business informatics

To ensure that the creation of a comprehensive system for the management of informatics using the MBI model brings the desired effect (i.e. IT/business alignment and increase the competitiveness of the company), the MBI implementation must be well managed. MBI authors recommend the following implementation steps:

1. Senior management of the organization should define the objectives of MBI implementation and determine the metrics by which the objectives will be measured. Definition of the objectives will vary depending on the business situation and on the anticipated role of business informatics over the next two to three years.
2. Establish MBI implementation team. The team should have at least three members. The team leader must be the CIO or another IT executive whose responsibility includes the management of business informatics. This executive must possess the necessary competencies, but must also have a natural authority to ensure that the fundamental changes that the implementation of MBI produces can be enforced. The second member of the team should be a business representative that together with the CIO participates in defining IT governance rules and the rules that define the relationship between business and informatics. The third member of the team should be technology oriented, and is responsible for technological aspects of the business model and the definition of technology-oriented tasks. In the case of larger organizations it is appropriate to add a representative of the finance department that will address cost and revenue aspects of MBI implementation.
3. Identify areas (groups of tasks) that business informatics must address including their priorities. The input into this step are the objectives of MBI implementation defined in the first step and its output are the tasks that business informatics must implement including their priority. During this step the implementation team should use the catalog of MBI tasks.
4. MBI implementation plan (stage, schedule, and budget).
5. Select MBI tasks and other objects MBI (documents, metrics, etc.) that suit the list of tasks defined in the previous step.
6. Complete or modify the tasks that the previous step did not successfully resolve. When modifying the tasks the practitioners should comply with the rules that are appropriate

to the description of the process and process maturity. This creates a proposal for a new system of management of business informatics.

7. Top enterprise executives approve the new system of management of business informatics.
8. Provide training to all employees of the organization that have roles described in the MBI model. Every employee is trained in the tasks that relate his role(s).
9. Start using the new system of management of business informatics.

5.2 *Further development of the MBI model*

Given that the lack of proven content of the model could lead to erroneous practices and poor decisions by MBI users it is essential for the development of MBI to be conducted according to rules which ensure its high quality. The MBI Steering Committee ensures the quality of the model. The MBI Steering Committee makes decisions about:

- inclusion of new occurrences of individual objects (tasks, documents, etc.) into MBI. To ensure the quality of MBI content the Steering Committee reviews each new object occurrence by two reviewers. If the reviewers disagree the decisions about acceptance or rejection is made by the steering committee;
- content of the various versions of MBI,
- date of release of a new version of MBI.

The rules governing the creation and modification of a specific model are defined by the CIO and his team and should be part of corporate IT Governance. In general, all tasks and other objects of the generic reference model should be regarded as recommendations that can be modified by the user organization to suit its needs. The authors of the MBI model recommend that when modifying the model, factors affecting the management of business informatics in the organization are taken into account, and that the detail description of the role and maturity of various components and processes are chosen according to the KBPR (Knowledge Based Process Reengineering) method (Voříšek 2008).

6 CONCLUSIONS

The development of the MBI model aims to provide a method for the management of business informatics incorporating all relevant international experiences and at the same time respecting the conditions and customs in individual organizations. MBI model is intended primarily for SMEs, as research indicates that in these organizations the application of available large-scale methodologies (e.g. ITIL, COBIT, etc.) is too expensive and time consuming. An indirect benefit of MBI is that it assists in the clarification of concepts, principles and methods used for the management of business informatics. In its present form MBI is an ongoing project and represents a starting point for further development, aiming to provide a comprehensive platform for the management of business informatics based on precise metrics. This trend towards more effective IT management based on measurable outcomes is evident in recent surveys we have conducted in hundreds of Czech organizations. These studies identify the need for effective methods and tools that enable IT executives to make informed decision based on reliable information.

ACKNOWLEDGEMENT

This paper and books of Business Informatics (Černáková 2014, Farkaš & Ružický 2014, Řepa 2014, Řepa & Voříšek 2014, Voříšek 2014) has been supported, by ESF Operational Programme Education under contract NFP 26140230012 "Quality Education at Pan-European University with International Cooperation".

REFERENCES

Bacal, R. 2012. *Manager's Guide to Performance Management*. New York, McGraw-Hill 2012. ISBN 978-0-07-177225-9.

CMMI Product Team. 2010. *CMMI for Development*, Version 1.3, Software Engineering Institute Carnegie Mellon, 2010. online <http://www.sei.cmu.edu/cmmi/>.

ČSN ISO/IEC 20000-1:2005. 2006. *Informační technologie—Management služeb—Část 1: Specifikace*. Český normalizační institut. Praha, 2006, 1. vyd. (Service Management—Part 1. Specification).

Černáková, I. 2014. *Business Information Systems*. Nitra: Forpress.

Dohnal, J. & Příklenk, O. 2011. *CIO a podpora byznysu*. Praha, Grada, 2011. ISBN 978-80-247-4050-8 (How can CIO support the business).

Doucek, P. & Novotný, O. 2007. *Standardy řízení podnikové informatiky*. E+M Ekonomie a Management. 2007, Iss. 3. ISSN 1212-3609. (Standards of IT management).

Doucek, P., Novotný, O., Pecáková, I. & Voříšek, J. 2007. *Lidské zdroje v ICT—Analýza nabídky a poptávky po IT odbornících v ČR*, Praha, 2007, Professional Publishing, 201 s., ISBN 978-80-86946-51-1 (Human resources in IT—Supply and demand of IT specialists in the CR).

Farkaš, P. & Ružický, E. 2014. Multi—Cloud Hosting IOT Based Big Data Service Platform Issues and One Heuristic Proposal How To Possibly Approach Some of them. In *Current Issues of Science and Research in the Global World*, Vienna Conference. London: CRC press, Taylor & Francis group.

IT Governance Institute: Cobit 5. 2012. *Enabling Processes*. Rolling Meadows, ISACA 2012.

Molnár, Z. & Střelka, J. 2012. *Competitive intelligence v malých a středních podnicích*. (Competitive Intelligence in SME's.) E+M Ekonomie a Management. 2012, Iss. 3, pp. 156–170. ISSN 1212-3609.

Novotný, O. & Voříšek, J. 2011. *Digitální cesta k prosperitě*. (Digital road to competitiveness). Praha: Professional Publishing. ISBN 978-80-7431-047-8.

Paladino, B. 2011. *Innovative Corporate Performance Management: Five Key Principles to Accelerate Results*. Indianopolis, Wiley Publishing, 2011. ISBN: 978-0-470-62773-0.

Pour, J. 2012. *Business intelligence řešení v modelu MBI* (Business Intelligence solution in MBI Model), Systémová integrace 2/2012. ISSN 1210-9479.

Pour, J. & Voříšek, J. 2011. *K výsledkům průzkumu české informatiky*. (Survey of IT management in the Czech enterprises) Systémová integrace, 2011, č. 1, ISSN 1210-9479.

Řepa, V. 2014. Process Management and Modelling. Nitra: Forpress.

Řepa, V. & Voříšek, J. 2014. Role of the Process-Driven Management in Informatics and Vice Versa. In *Current Issues of Science and Research in the Global World*, Vienna Conference. London: CRC press, Taylor & Francis group.

Schiesser, R. 2011. *IT Systems Management*. New York, Prentice Hall 2010. ISBN 978-0-13-702506-0.

Schniederjans, M.J., Hamaker, J.L. & Schniederjans, A.M. 2005. Information Technology Investment: Decision-Making Methodology. New Jersey, World Scientific 2005. ISBN-10: 9812386955.

Voříšek, J. 2008. *Principy a modely řízení podnikové informatiky*. (Principles and models of IT management) Praha: Oeconomia 2008. ISBN: 978-80-245-1440-6.

Voříšek, J. & Pour, J. 2012. *Management podnikové informatiky*, (IT Management). Praha: Professional Publishing, 2012, ISBN 978-80-7431-102-4.

Voříšek, J., Bruckner, T., Buchalcevová, A., Stanovská, I., Chlapek, D. & Řepa, V. 2012. *Tvorba informačních systémů: principy, metodiky, architektury*. Praha: Grada.

Voříšek, J., 2014. *Principles of Development and Operation of Business Information Systems*. Praha: Wolters Kluwer.

Weill, P. & Ross, W.J. 2004. *IT Governance: How top performers manage IT decision rights for superior results*. Working Paper No.326. Boston: Harward Business School Press.

Wetzstein, B., Zengin, A., Kazhamiakin, R., Marconi, A., Marco Pistore, Karastoyanova & D. Leymann 2012. *Preventing KPI Violations in Business Processes based on Decision Tree Learning and Proactive Runtime Adaptation*, Journal of Systems Integration, Vol. 3, No. 1, 2012, ISSN: 1804-2724.

Žák, M. 2013. Konkurenční schopnost České republiky 2011–2012. Praha, Linde 2013, ISBN 978-80-7201-910-6.

Current Issues of Science and Research in the Global World – Kunova & Dolinsky (Eds)
© *2015 Taylor & Francis Group, London, ISBN 978-1-138-02739-8*

Augmented map presentation of cultural heritage sites

Z. Berger Haladová
Comenius University, Bratislava, Slovakia

M. Samuelčík
Comenius University, Bratislava, Slovakia
VIS GRAVIS, s.r.o., Slovakia

I. Varhaníková
Comenius University, Bratislava, Slovakia

ABSTRACT: Augmented reality compared to virtual reality offers a stronger feeling of immersion for the user. In this paper we present an application, which combines augmented reality with presentation of cultural heritage and education. The application named Slovak Augmented Reality consists of a floor map of Slovakia and the Kinect sensor to display and interact with 3D virtual models of cultural heritage objects using gestures. It is extended with the possibility to watch the video of user holding the virtual object which is captured with Kinect device placed under the screen. The user can navigate (scale and rotate around two axes) the virtual object using predefined gestures.

To demonstrate the use of the proposed system we have designed a serious game for identifying Slovak cultural heritage sites and locating them on their geographic position. The Slovak Augmented Reality application can be used for entertainment and as a learning tool in geography or history lessons.

1 INTRODUCTION

Since the first definition by Azuma [Azuma97] in 1997, augmented reality has emerged into most aspects of everyday life. One of the areas, where Augmented Reality (AR) makes a strong contribution, is cultural heritage. There are several possibilities of using AR in the field of cultural heritage. It can be used as an auxiliary tool for archeologist where they can for example virtually restore missing parts of the exhibits, or it can be used to visualize various stages of monuments construction. But mostly, the AR in cultural heritage is used as an enhancement tool of virtual museums tours. Although virtual museum tours are known since the early 90's, the use of AR offers many new possibilities. Touring the virtual museum at home or in a large kiosk in the museum was extended to a walk through the museum enriched by different virtual versions of the exhibits or by their metadata. Augmented tour also allows presenting distant, very small or very large sites. The greatest advantage of this kind of tour lies in the possibility of interaction. Unlike the virtual tours where the participant is mostly passive, using AR induces a greater impression of immersion. Using hand or body gestures enhances this feeling of immersion so the user may gain the impression that he interacts with real exhibits.

There are two most used types of AR applications intended to present cultural heritage: museum/gallery guides (on handheld devices or head-mounted displays) and spatial applications (augmented exhibits [Bimber05, Bimber04] or urban projection mapping). In our application, Slovak Augmented Reality, we decided to create a spatial installation for cultural heritage sites presentation. Our system combines the contexts of a physical map and mirror like displaying of virtual objects.

In the area of AR, several works on interaction with a map were presented, but the authors usually used a small paper map combined with the handheld or head-mounted device [Morrison09]. We decided to create an installation where the user can actually stand on the map on the geographic location of the cultural heritage site, what strengthens the link between the monument and its position.

To display the 3D models we created a mirror-like installation, where the user sees himself standing on the map with the 3D model in the front of him.

Mirror-like installations are popular within the AR community because they allow the user to control his position and his gestures without refocusing from the augmented to the real environment. Previous works on this kind of installations include the Magic mirror [Blum11] (a system for displaying the volume visualization from a CT volume registered on the human body) or the Augmented mirror [Vera11] (an installation where the user can use gestures to navigate virtual characters in the AR environment).

2 SLOVAK AUGMENTED REALITY

2.1 *System*

Our installation consists of a map of Slovakia, a projector, a projection screen (or a display) and a Kinect device (or another RGBD sensor). The map is drawn on a vinyl flooring surface and it shows only the border line of Slovakia. There are no other marks, but the vinyl surface allows us to draw as many markers as we need. In Figure 1 you can see the map with the illustration of the landmark locations. The scheme of the whole installation can be seen in Figure 2.

The initialization of our system is done as follows. The four corners of the map surface are marked on the RGB image acquired from Kinect. We compute the transformation between the four acquired corners and the corners of the model of the map using homography. Then the positions of the heritage sites in 2D (x, y) are computed and the depth is extracted from the initialization frame to achieve the 3D position information.

In the runtime when the user steps on a marked heritage site, we display the 3D model of the corresponding cultural heritage object in front of the user on the projection screen. Then the virtual object can be transformed using defined gestures. To determine the gestures of the user we track the skeleton of the user standing on the map using the Kinect SDK. Our application performs two types of transformations: scaling and rotating along two axes. The corresponding gestures are shown in Figure 3.

The user performs the scaling gesture with both hands. In the starting position he holds them approximately at chest height and width. Enlarging of the 3D object is done by putting

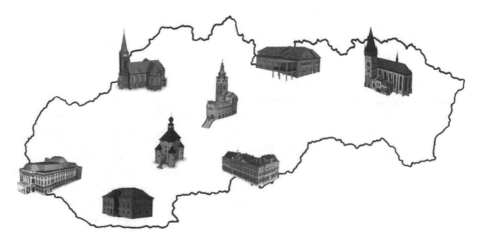

Figure 1.   Map of Slovakia with presented cultural heritage sites displayed.

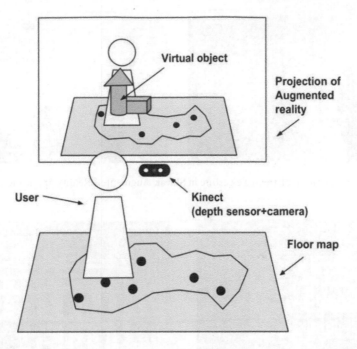

Figure 2. Scheme of installation Slovak augmented reality.

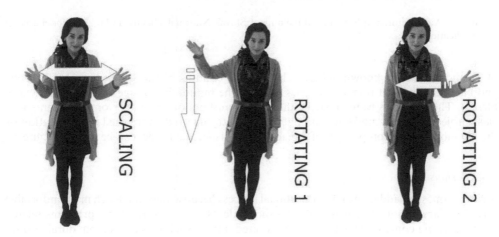

Figure 3. Gestures used while interacting with virtual objects.

his palms further away and shrinking of the object by moving the palms closer together. Rotating around both axes is performed only with one hand. By holding the hand in the head height and moving it downwards the object rotates around the *x-axis* and by holding it in the chest height an moving it from the right to the left object rotates around the *z-axis*. The demonstration of the work with scale gestures can be seen in Figure 4.

Our application runs on the PC (AMD Phenom II X4 3,4Ghz, 4GB Ram, ATI Radeon HD 5700 1GB) with 57 frames per second.

2.2 *Database of 3D models*

Our database of 3D models consists of 8 virtual models of Slovak cultural heritage sites distributed evenly on the map (see Fig. 1). We chose for our application two different kinds

Figure 4. Demonstration of the scale gesture in Slovak Augmented Reality application.

Figure 5. A real photograph of the old building of Slovak National Theatre and its 3D model used in our application.

of 3D models: widely known sights (such as the old building of the Slovak National Theater) and barely known but remarkable ones (such as the mountain cottage "Chata pri Zelenom plese"). These models were created either by the authors of this paper or by students as a class project. Manual modeling and photogrammetry methods were used in the creation of the 3D models. The comparison of a real site with its virtual model can be seen in Figure 5.

2.3 *Serious game*

Serious games are added into the educational process because they can teach in a comfortable way. They are designed primarily so that the students have to solve some problems related to the learning content. The knowledge acquired during such process is more durable since instead of only memorizing the information, the student remembers the particular problem he had to solve with these information and he can use them again.

To extend the use of the presented system we have designed a serious game for identifying the cultural heritage sites of Slovakia and locating them on their exact position. During the game random images of cultural heritage sites are shown and the user has to step on the estimated position on the floor map. His answer is limited by time. There are several possibilities how to customize the settings of this serious game for different age levels or for different educational reasons. For example for younger users we can set different time limitations. Or for teaching a particular historical event images (sites) from limited database connected to the theme could be used.

3 CONCLUSION

In our work we have proposed and built the spatial augmented reality installation called Slovak Augmented Reality. Our installation utilizes the floor map of Slovakia to create a real context for the user exploring the 3D models of Slovak cultural heritage sites. The interaction using gestures is enabled by the usage of the Kinect sensor. Apart from the presentation of the 3D models of cultural heritage we have designed a serious game. The main goal of the game is teaching the users about the cultural heritage sites of Slovakia but it can be easily adapted to current needs of teachers. Our installation was presented several times in the Future Technologies Laboratory at Faculty of Mathematics, Physics and Informatics at Comenius University during the Open days or the conference for pupils Virtual reality without borders 2014.

4 FUTURE WORK

In the future we would like to extend the database of our 3D models in order to be able to presents all UNESCO world heritage sites in Slovakia. Since the map of Slovakia is not very big, we are forced to build a system, where the site is displayed for a particular place only approximately (if there is more than one 3D model for this place—for example Bratislava city). We plan to carry out the usability study focused on pupils. We also plan to extend our system by stereoscopic projection or stereoscopic display (active or passive) combined with stereoscopic rendering of virtual models.

ACKNOWLEDGEMENT

We would like to thank the students for enabling the use of their 3D models in our application. This paper was partly supported by the operational program ASFEU project "COMENIANA—metódy a prostriedky digitalizácie a prezentácie 3D objektov kultúrneho dedičstva", ITMS: 26240220077, which is co-financed from resources of European regional development fund.

REFERENCES

[Azuma97] Azuma, R.: A survey of augmented reality. Presence: Teleoperators and Virtual Environments 6, 4 (1997), 355–385.

[Blum11] Blum, T.: Augmented reality magic mirror using the Kinect. http://campar.in.tum.de/Chair/ Project Kinect Magic Mirror, 2011.

[Bimber05] Bimber, O., Coriand, F., Kleppe, A., Bruns, E., Zollmann, S., Langlotz, T.: Superimposing Pictorial Artwork with Projected Imagery. IEEE MultiMedia 12, 1 (2005), 16–26.

[Bimber04] Bimber, O., Fröhlich, B., Schmalstieg, D., Encarnação, L.M.: The virtual showcase. In SIGGRAPH'06: ACM SIGGRAPH 2006 Courses (USA, 2006), ACM.

[Morrison09] Morrison, A., Oulasvirta, A., Peltonen, P., Lemmela, S., Jacucci, G., Reitmayr, G., Juustila, A.: Like bees around the hive: a comparative study of a mobile augmented reality map. In Proceedings of the SIGCHI Conference on Human Factors in Computing Systems. ACM, 2009, pp. 1889–1898.

[Vera11] Vera, L., Gimeno, J., Coma, I., & Fernández, M.: Augmented mirror: Interactive augmented reality system based on Kinect. In Human-Computer Interaction—INTERACT 2011, vol. 6949 of Lecture Notes in Computer Science. Springer Berlin Heidelberg, 2011, pp. 483–486.

*Faculty of psychology, section man
in the social system of work*

Current issues of psychology and psychologists in the educational system in Slovak Republic

E. Gajdošová
Faculty of Psychology, Pan-European University, Bratislava, Slovakia

ABSTRACT: The new millennium has brought many new social, economical, political, cultural, school and educational changes in Slovakia (changes in the school legislation, in the school system, in the educational process, in the increase of pathological phenomena at schools, etc.) Some of them are closely connected with the tasks, plans and challenges of school psychology and school psychologists at schools and their activities in the education of children. Many of these changes require an important modification of the education and training of school psychologists in Slovakia and modification of the conception of school psychology work. The promotion of health including mental health at schools at the individual, group, system and society level with an emphasis on human and children's rights is the most important task of these professionals.

1 SHORT HISTORY OF SCHOOL PSYCHOLOGY AND ITS PROFESSION

The development of school psychology in the Slovak republic and the strengthening of its position in the educational process begins no sooner than in 1990 after the Gentle revolution in November 1989 was over and the social structure and political situation in Czechoslovakia changed. Here are listed some of the events that took place:

- 1990-the establishment of School psychology association
- 1991-school psychology becomes a separate subject in the pre-graduate studies program for psychologists. Later it is approved as a subject for final state examination and a study program for PHD. study.
- 1992-for the first time, psychologists meet at the ISPA colloquium in Turkey
- 1993-for the first time, school psychology is introduced into the legislation. It is the year when the meeting of ISPA took place in Slovakia (800 psychologists participated). Anton Furman, a school psychologist from Slovakia, later became the president of ISPA.
- 1993–1996 legislation, legal status of school psychologists in the Slovak Republic (Valihorová, Gajdošová, 2009).

The Law of the Slovak National Council No. 279/1993 emphasizes that "the school psychologist is actively involved in the work performed by the school or the institution for special education and provides professional psychological advice to children, their legal guardians and pedagogical employees during the solving of educational problems." Resulting from this law, Slovak schools and school institutions can employ school psychologists for full time, or they can cooperate with school psychologists who work in pedagogical and psychological counselling centers.

The primary objective of the work performed by a school psychologist is to provide psychological and educational services to the pupils, parents, teachers and other pedagogical employees to promote their personal development and mental health using modern forms and methods of identification and diagnostics, consulting and counselling, correctional and preventive work.

School psychologists are required to provide primary prevention and prevention activities in schools and school institutions, with an aim to ensure appropriate social and educational conditions that would prevent or eliminate such negative phenomena as violence, aggression, bullying, intolerance, playing truant, delinquency, drug addition of pupils, but also stress, exhaustion, loss of motivation and interest in learning, or tensions and conflicts among pupils, among teachers and pupils, or teachers and parents and promote personal development and mental health of pupils and teachers.

Since the year 1996, the Organization rules for schools of the Ministry of Education of the Slovak Republic are issued at the beginning of every school year. The headmasters are advised that school psychologists are to be employed in compliance with the Law of the Slovak National Council No. 279/1993 Collection of laws and the Regulation No. 43/1996 Collection of laws, issued by the Ministry of Education of the Slovak Republic, with the aim to humanize schools, to promote the mental health of pupils and to help teachers to cope with the requirements resulting from the educational process.

The government of the Slovak Republic approved the National program for children and youth care, valid from 2008 to 2015, that is based on the European strategy for health improvement and healthy development of children and youth. There are seven priorities. The part "Psychosocial development" is important from the point of view of mental health in schools.

As far as this part is concerned, the general goal of the program is to "ensure appropriate conditions for the instruction and education of children and youth, and to ensure good social background and material conditions for them, and mainly supporting healthy personal development and mental health of pupils." It also emphasizes the fact that the work performed by a school psychologist is important for psychological development and mental health of children. The National program for children and youth care gives the school psychologist the task to "help pupils to decide on their future profession, provide psychological care for gifted and talented pupils, work with pupils who have learning difficulties and behaviour disorders and work with pupils who are integrated into the school, having a physical or mental handicap and to ensure appropriate conditions for their education. School psychologist has to offer his psychological services also to the teachers. He has to support mainly their mental health, personal development, personal relationships, help them solve interpersonal conflicts, give them advice how to deal with burdensome situations, stress and the burn-out syndrome. School psychologist's duties include the supporting of teachers' professional development in psychology, educational process and the optimization of educational process. He has to offer consulting and counselling services to the parents of pupils with the objective to improve their methods of education and child care, if their children have learning problems and behaviour disorders. He also ensures a good cooperation between school and family.

School psychologist performs activities regarding psychological assessment, individual and group psychological counselling, psychotherapy, prevention and intervention care of children and pupils, with strong focus on the process of education and instruction in schools and school institutions. School psychologist offers psychological counselling and consultations to legal guardians of children and to pedagogical employees working in schools or school institutions. He prepares the basic materials for vocational employees of counselling centres (Gajdošová, Herényiová, Valihorová, 2010).

2   THE CURRENT DEVELOPMENT OF SCHOOL PSYCHOLOGY AND THE POSITION OF SCHOOL PSYCHOLOGIST IN SLOVAKIA

The new millennium has brought a lot of new social, economical, political, cultural, school and educational changes. Many of them require the important modification of the school psychologist work and the school psychology activities at schools.

Let me concentrate upon some current issues in Slovakia in the new millennium, which are very closely connected with the development, challenges, tasks and plans dealing with psychology and psychologist in the educational system and activities of psychologists at schools in Slovakia.

1. Current issue: Crisis of Slovak families
   - break up of Slovak marriages: divorce rate in the Slovak republic is 100:47
   - great percentage of divorces of young couples with one or two children (children are educated in incomplete families without one of parents, especially father)
   - school substitutes the role of family (school brings up the child, and provides for academic, social, emotional and moral development of pupils)
   - lack of cooperation between "family and school" and low parents active role in many of school activities
   - increase of parents' aggression towards teachers.
2. Current issue: Crisis of educational process
   - increase of new socio-pathological phenomena in Slovak schools (not only aggression, violence, bullying, truancy, drug addictions, but also non-substance addictions, intolerance, neurosis, depression, emotional emptiness, lack of values, lack of reason for being, suicides of students)
   - large number of elderly teachers at schools (who are no longer able to deal with increasing number of learning and behavior difficulties of pupils and young teachers are very often leaving school setting after 2–5 years of professional life)
   - stress and burn-out syndrome of teachers and other staff at schools
   - lack of lifelong learning of professional staff at schools (giving strong emphasis on current problems of school of the 21st century and on psychology of new generation of pupils and parents)
   - school legislation not always reflects the current problems at schools (with effective resolution of crisis and conflicts)
   - lack of professionals at schools to help with social, mental, behavior and learning difficulties (school psychologists, special teachers, social workers, pedopsychiatrists)
   - low salaries of teachers and school psychologists in comparison with other professions and little possibilities to obtain funds from state budget to cover salaries of young school psychologists
   - school legislation not sufficiently supports school psychologists at schools.
3. Current issue: Crisis of society and social life
   - migration of people, especially young people (the reasons are mainly the better jobs and higher salaries)
   - gender differences
   - deficit of tolerance towards social, physical and mental differences (nationality, ethnicity, religion, social status, personal characteristics, physical differences)
   - social tolerance to unethical behavior of people in various social status, functions, ranks, positions
   - lack of emphasis on moral values
   - depression and anxiety as the prevalent mental diseases of adults and children.

The promotion of health including mental health at the individual, group, system and society level and emphasis on human and children's rights are important tasks of professionals, including psychologists. These problems call for new activities of psychologists in the Slovak schools of the 21st century. They have to be implemented into the curriculum and training of young Slovak psychologists in the educational system/school psychologists and into the psychological services in the schools.

3 NEW CHALLENGES FOR SCHOOL PSYCHOLOGY AND SCHOOL PSYCHOLOGISTS IN SLOVAKIA

Let me emphasize some spheres of challenges for Slovak school psychology and Slovak school psychologists in the new millennium (Gajdošová, 2012):
1st challenge: School psychology—science, research, practice
   To arrange:

   - publicity and promotion of school psychology as a science and school psychologist as its profession in Slovakia (e.g. in conferences, TV, radio, newspapers, journals, billboards)

- international researches and international projects in school psychology (with the help of **EFPA-NEPES** and **ISPA**)
- publishing of school psychology books, textbooks, study materials and monographs (in cooperation with **EFPA** and **ISPA** with the support of **EU** grants)
- school psychology national and international conferences, colloquiums, seminars and training courses for school psychologists (in cooperation with **ESPCT**—European School Psychology Centre for Training and **EFPA** Standing Committee Psychology in Education—**N.E.P.E.S.**).

2nd challenge: Preparation and training of school psychologists
To carry out:

- exchange of students of MA study and doctoral study
- participation of lectors, experts from abroad in study courses, trainings during the school year
- supervised psychological practice for young school psychologists (which is the 3rd level of education for psychologists and one of the requirements to achieve the EuroPsy diploma)
- accredited trainings for supervisors of school psychologists
- lifelong training courses for school psychologists in practice with the help of ESCPT (European School Psychology Centre for Training).

3rd challenge: School legislation
To prepare:

- new legal status of Slovak school psychologist
- professional standards of the school psychologist in the new school legislation
- national program of mental health at the individual, group, system and society level
- innovative care structures for children and teachers with mental health problems
- employ a larger number of school psychologists at Slovak schools.

4th challenge: University curriculum in school psychology
To concentrate upon:

- primary prevention and long-term preventive programmes at schools (e.g. programmes of conflict resolution, school-based mental health, social and emotional learning, multicultural tolerance, effective learning methods, career guidance)
- counselling, consultations, psychotherapy for families and parents of pupils
- counselling and consultations (for teachers and other professionals at school—special teachers, social workers, counsellors)
- crisis management and crisis intervention in schools
- effective forms for school—family cooperation
- group activities (group assessment, group testing, group counselling, group interventions, group prevention)
- multidisciplinary teams (school psychologist, special teacher, counsellor, social worker, clinical psychologist, psychiatrist)
- new pathological phenomena—mobbing, intolerance, addiction to computer and mobiles, gambling, sects, CAN syndrome, depression, suicides of children
- mental health, mental hygiene of learning and teaching process, learning styles of pupils, teaching styles of teachers
- personal management at schools.

5th challenge: School psychologist personality
To develop:

- competences and skills by means of longlife training and education
- competences and skills by self-education
- ethical and moral behavior
- positive personal characteristics.

4 CONCLUSION

The article emphasizes some actual challenges for school psychology as a science and for school psychologist as its profession that emerged from social and educational changes in the national and international context. To deal with these tasks, there is needed the permanent cooperation of psychological institutions and psychological organizations in Slovakia and the narrow cooperation of psychologists of many specializations and subject fields.

REFERENCES

[1] Gajdošová, E., Herényiová, G., Valihorová, M. Školská psychológia. 2010. Bratislava: Stimul.ISBN 978-80-89236-81-7.
[2] Gajdošová, E. 2012. Školský psychológ pre 21. storočie. In: Školský psychológ pre 21. storočie. Zborník z medzinárodnej konferencie. Bratislava: Polymédia, s. 112–116. ISBN 978-80-89453-03-0.
[3] Jimerson, S.R., Oakland, T., Farrell, P. 2006. The Handbook of International School Psychology. Sage Publications Thousand Oaks, California, USA—London, UK.
[4] Merrell, K.W., Ervin, R.A., Oeacock, G.G. 2012. School Psychology for the 21st Century. Foundation and Practices.
[5] Národný program starostlivosti o deti a dorast v Slovenskej republike na roky 2008–2015.
[6] Valihorová, M., Gajdošová, E. 2009. Kapitoly zo školskej psychológie. Banská Bystrica: UMB, ISBN 978-80-8083-817-1.
[7] Zákon NR SR č.279/1993 Z.z. o školských zariadeniach.

Current Issues of Science and Research in the Global World – Kunova & Dolinsky (Eds)
© *2015 Taylor & Francis Group, London, ISBN 978-1-138-02739-8*

Diagnostic of the motivational orientation of managers on the basis of Herzberg's Theory

T. Kollárik
Faculty of Psychology, Pan-European University, Bratislava, Slovak Republic

ABSTRACT: Recent findings confirm that Herzberg's Theory also creates a good platform for the diagnostics of managers' motivation and for the definition of suitability or unsuitability of the current motivational orientation for managers or applicants for a certain position. It fills a gap in the diagnostic toolbox of occupational psychologists, thus contributing to increased success of psychologists in practice.

The problem of worker motivation is one of the main problems in the field of work psychology, which is evidenced by the attention of psychologists and their construct new theories, approaches and various modify current diagnostic methods in various ways. It is especially the "one-dimensional" procedures that leave space for this, which in terms of content and focus of method provides results for good, but simplified interpretation. Often the complexity of motivation as such is forgotten, i.e. structure and content orientation, as well as valid interpretation of results, or even maintaining psychometric qualities of newly developed methods.

Research and experience show that in monitoring motivation and job satisfaction, a number of factors must be captured and respected, which are associated with a specific person, their attitude toward work, and toward the social work environment, relevant to personal criteria. Five aspects are being used most commonly:

1. current level of observed phenomenon (it is now),
2. importance to the individual (it is important),
3. desired level with respect to a given phenomenon (it should be),
4. expectations of the individual for the level of the phenomenon (I would),
5. satisfaction with work factors.

These procedures allow different variations, which Wanous, J.P. and Lawler, E.E. (1972) summarized into nine operational definitions. Their application allows a manager to gain a broader familiarization with work factors that are dependent on the manager's own degree of aspirations, and his expectations of current or potential workplace.

The starting point for using this procedure is Herzberg's two-factor theory. According to Herzberg, employment has an important place in the human realization.

Motivation is generally characterized as a dynamic component of personality, the cause of action, and the source of people's and social groups' behaviour. Therefore, it holds an unquestionably important place in psychology and it can be stated that no psychological direction has been able to avoid it. This despite the fact that there are diametrically opposed positions on the existence of motivation—on the one hand, it is seen as a major factor in human activities and the cause of the activity, on the other hand, it is not considered a necessary factor in explaining people's actions.

Differences are also found when evaluating its structure, when it is differentially attributed an energizing or regulatory function. Similarly manifested are also the presented principles of motivation: a) mental balance—behaviour tends toward an individual's inner mental balance, b) hedonism—focus on finding pleasure and avoiding negative experiences, c) self system— the pursuit of self-actualization, self-realization, self-determination, etc.

Therefore the number of existing motivation theories is not surprising. In terms of work, these can be classified into 3 categories:

1. Theories of needs
   - McGregor's theory of X-Y beliefs,
   - Maslow's hierarchy of needs theory,
   - Alderfer's ERG theory,
   - Murray's theory of needs,
   - Mumford's theory of needs.
2. Specific work motivation theories
   - Argyris' immaturity-maturity theory,
   - Locke and Latham goal-setting theory,
   - McClelland's theory of successful achievement motivation, or achievement motivation, later work in by Atkinson,
   - Lawler's page model,
   - Hackman and Oldham theory of work design and motivation,
   - Warr's "Vitamin Model",
   - Adams' theory of justice,
   - Roe et al. model of work motivation and the quality of working life.
3. Job satisfaction theories
   - Staw disposal theory,
   - Johnson and Holdaway model of job satisfaction,
   - Solomon–Corbitova opponent—process theory of job satisfaction,
   - Wernimont theory of job satisfaction,
   - Model of Vroom,
   - Herzberg's two-factor theory of job satisfaction.

Interest in the matter of motivation in the world of work is firstly conditioned by the nature of motivation in general and secondly by the meaning of work in the life of a person as his main activity in adulthood. The place of motivation with respect to workers is truly held at the top position within the HR strategy of any organization, as can be shown by the following figure.

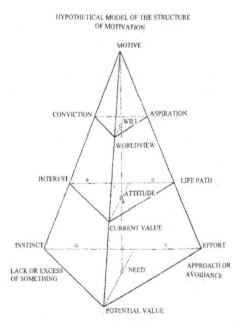

Figure 1.   Hypothetical model of the structure of motivation (In Pardel, T., 1977, p. 72).

People are a key factor in the success of any department or any other organization.

The survival and success of the work group or organization depends on how successful each member is in what he does. This success depends primarily on their knowledge, skills, attitudes and motivations, which are kept at the highest point in the hierarchies of an individual.

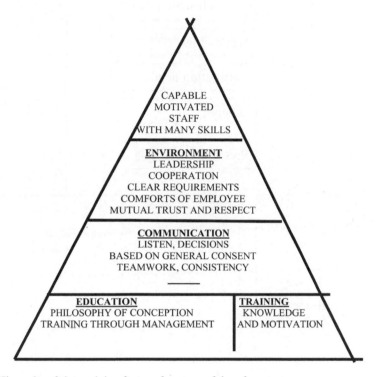

Figure 2.　Hierarchy of determining factors for successful performance.

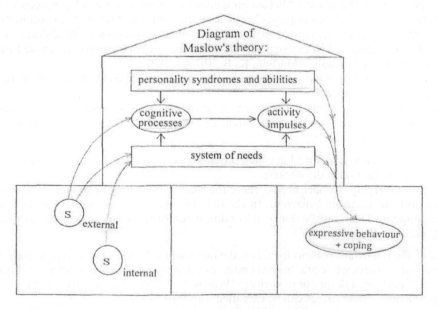

Figure 3.　Diagram of Maslow's theory (Madsen, K. B., 1979, s. 294).

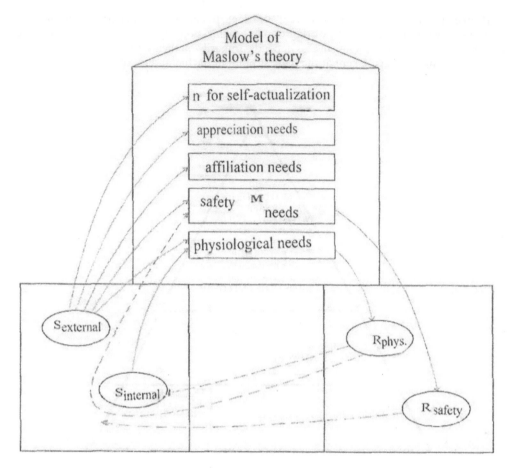

Figure 4. Model of Maslow's theory (Madsen, K. B., 1979, s. 296).

Interpretation of the scheme: "Behaviour consists of expressive and adaptive components. It is determined by "activity impulses", which are in turn determined by a combination of "system of needs", "cognitive processes" and "personality syndromes and abilities." Cognitive needs are determined by external stimuli, while the system of needs is activated by both external and internal stimuli." (Madsen, K. B., 1979, s. 294).

The function and operation of the system of needs in Maslow's hierarchy of needs is expressed by another model.

Interpretation of the scheme: "The system of needs is activated by external and internal stimuli, but only the strongest need in the hierarchy determines the behaviour at a given moment. Dominance is determined by:

1. Hierarchy already determined in which physiological needs are the strongest and the need for self-actualization is the weakest,
2. State of satisfying stronger needs, where the weaker need may be dominant at a certain time and can motivate behaviour. In the diagram, safety needs are the "motivator" for the moment. Their satisfaction is a condition for other needs to become motivators" (Madsen, K. B., 1979, s. 296).

One of the important reasons for this is the fact that motivation has a large impact on the efficiency of workers, while practitioners point out that "people do not do what we think they should do" and they ask questions such as: "Why is that? What are the causes of the performance problems?" These causes can be classified into four groups.

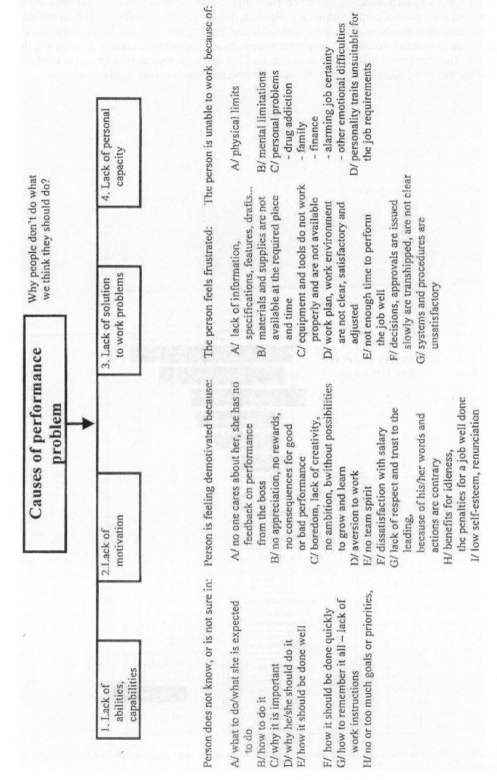

Figure 5. Causes of performance problems.

The position of motivation here is obvious and practice centres the problem into the sphere of personality and its motivational structure (to want), the structure of abilities—capabilities (to know) and their mutual combinations, e.g. classic typology system of orthogonal dimensions.

Of course, this combination is quite schematic and inadequate because, for example, it disregards the impact of conditions—options necessary for the performance in specific conditions. To the theoretical inconsistency are also added methodological problems, particularly in the creation and adoption of an adequate method to measure staff motivation.

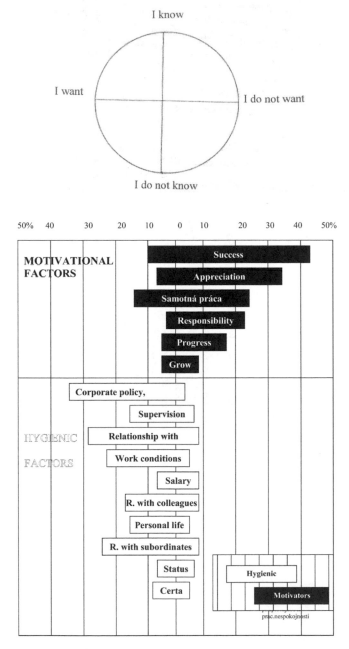

Figure 6.   Motivational and hygienic factors in satisfaction in the workplace, according to Herzberg, F., (1972).

Nowadays a varied range of methods dominates. These are built on the basis of theoretical approaches of individual authors, characterized by different forms, content, approach and qualities. The authors are mostly bound by their attitudes and often use their own method to base their understanding of motivation on, whereby they deviate from following psychometric principles and criteria (standardization procedures, standards, etc.). Then, inevitably, different findings are more or less the work of different theoretical focus of the author, and the quality of the method used. A classic example was the emergence of Herzberg's two-factor theory of motivation, or satisfaction.

Model of diagnostics of managers based on Herzberg's theory.

Herzberg and his colleagues (Maussner, Snyderman), in the fifties of the previous century, stood before the public with a theory that was contrary to the "one factor theories" of worker satisfaction flourishing at the time. Its essence is in distinguishing two groups of factors contributing to satisfaction—dissatisfaction: on one side, it is "motivational factors" that contribute to the satisfaction, and on the other side are "hygienic factors" that contribute to dissatisfaction.

The conclusions of Herzberg and his colleagues were considered a big interference with the perception of workers' sources of satisfaction, while they also provoked a widespread professional dilemma exhibiting a wide variety of views, from highly positive evaluation to the derogatory statements addressed to the authors. One of the most criticized areas was the use of Herzberg's method (a free description of situations leading to satisfaction and dissatisfaction) that directly provoked such differential effect of factors of satisfaction/dissatisfaction. Herzberg's answer to critics was quite fast and, based on results of research conducted in Finland, he presented data in favour of his theory. Still, it was not widely accepted, although it has a firm place in the list of theories of workers' motivation and their satisfaction.

However, it was mostly accepted as explanation of satisfaction or dissatisfaction of workers, less so for its use in diagnostics of workers, particularly in the selection and evaluation of managers. It is precisely in this area where we deemed the approach appropriate and applicable.

In this regard our activities have led to the formation of our own structure of motivational focus, based on the conclusions of Herzberg about different impact on the two groups of motivational factors:

1. Personal focus.		2. Work focus	
1a. Personal profit	1b. Power	2a. Character of the work	2b. Social atmosphere
Success	Authority	Work conditions	Work team
Appreciation	Prestige	Interesting work	Relationships with superiors
Progress	Legal Power	Salary	Relationships with colleagues
Personal growth	Responsibility	Certainty	Relationships with subordinates
Self-realization	Independence	Business/corporate culture	Leadership style

The proposed model presents a motivational style:

1. dominated by personal focus, i.e. personal profit such as success, appreciation, progress, personal growth, self-realization and power, supported by authority, prestige, competence, responsibility, and independence,
2. focused on work factors, it is character of the work, such as working conditions, interesting work, salary, (job) certainty, business/corporate culture and philosophy of focusing on social atmosphere—working team, relationships with superiors, colleagues, subordinates, and leadership style.

This allows for differential diagnosis and identification of managers with different motivational focus:

Manager preferring personality focus	Manager preferring work focus
Emphasizes factors that:	Emphasizes factors that:
– Determine his relationship to the subject of work and those that relate to what a person does,	– Relate to the context of their work to situations and environments in which the work is performed,
– Lead to personal growth, enabling psychological stimulation and self-fulfilment,	– Have a temporary nature, their timeliness is dependent on current events and situations,
– Motivate for better work performance and effort, and lead to job satisfaction based on high relevance of personal needs.	– Contribute minimally to satisfaction from work results.

Behaviour of individual types of managers at work

Manager preferring personal focus

He is motivated by the nature of the task. He exhibits higher tolerance to negative work factors and milder dissatisfaction if there are deficiencies in work factors. He experiences great satisfaction from the results of the work and is able to enjoy the achievements. He holds positive feelings and attitudes toward work and life in general. He benefits professionally from the system of experience. He has a thoughtful and honest system of opinions and convictions, and a clarified philosophy of life. He is motivated by the need for growth, which forms the basis for his psychological stimulation and fulfilment. His current needs—in particular the need for growth, self-realization, self-actualization—are effective in motivating for better work performance and effort, and consequently lead to job satisfaction. Work is seen as a place to succeed, and his work behaviour is a priority determined by internal personality motivators.

Manager preferring focus on working factors

He is motivated more by the nature of his work environment than by his duties. He exhibits increased dissatisfaction with various aspects of work, e.g. salary, working conditions, colleagues. He responds inappropriately to improved work factors—his satisfaction is, however, short lived. He gets very little satisfaction from work results, and consequently shows little interest in the kind and quality of the work he does. He manifests little satisfaction from work results.

He gains nothing professionally from experiences. The only advantage he desires is a more comfortable environment. While he can be stimulated to the activity for a certain time, he does not have his own "generator", he must be continuously stimulated. He only stays motivated for a short time, and only if he receives an external reward.

He sees work as an activity performed in a real environment and under real conditions.

Medium type—situational motivational style

He does not distinguish between personal motivational focus on the task and motivational focus on work conditions and environment. He accepts and adequately copes with shortcomings of work factors, but does not perceive their removal as a source of increasing job satisfaction, nor as an incentive to fix his own motivational structure.

He takes this as a real interference with work conditions, which may or may not have an impact on boosting motivation of team members. He does not see a correlation between the components of two motivational focus types, and due to his own average personal motivation (especially in the sphere of growth, self-actualization, etc.), he does not consider the building and promoting of a motivation style among regular team members important.

He prefers the relationship between the nature of work—team atmosphere—and satisfaction of members, more than their personal motivation. His motivation is relatively unstable, dependent mainly on situational factors affecting the fulfilment or non-fulfilment of a task.

Implications for the organization
Focus on work factors

- People focused on work factors will betray their employer right when their talent is needed the most,
- They are motivated only for a short time and only if they receive an external reward. When an emergency situation arises and the company cannot deal with work factors, these people fail in their work,
- People focused on work factors offer their own motivational characteristics as the scheme to be instilled in their subordinates,
- Holding key positions in the departments of the organizations that manage, they build an atmosphere of excessive interest in the external reward,
- They adversely affect the development and education of future managers of the organization.

Personal focus

- Leads to personal growth, allows psychological stimulation and self-fulfilment,
- Forms a positive motivational structure for work group members, thereby increasing their effectiveness, efficiency and satisfaction,
- Positively and effectively guides the development of managers within the organization, as well as their education in accordance with atmosphere, manifested in the attitudes and behaviour of management,
- By presenting his own motivational structure, he offers his subordinates and other staff an efficient and desirable scheme,
- His motivation is more stable, longer-term and more prosperous for the organization.

Medium type—situational motivational style:

By not presenting a specifically defined motivational style, his motivational presence within the organization is also more registered in the area of fulfilment or non-fulfilment of tasks/ duties. However, this is dependent on a variety of situational factors, acceptance of which brings the effect of his managerial work only with regard to a particular situation. He is governed by his experience, but in the long-term, he does not create an effective system for the development of his subordinates, and does not offer them a favourable motivational scheme for their own application. He often tends to emphasize work factors (eg. improvement of working conditions, pay, etc.), because he assumes that these interventions are most needed and rapidly effective in a given situation. He forgets that the sustainability of these motivational interventions is very short and unstable.

Standardisation of the method

The method is one that goes beyond the current realm of methods in that it is based on a specific theory and the real existence of components being monitored in terms of motivation. It provides valuable information about the current motivational structure of workers from a broader perspective, utilizing the motivational focus mentioned above, based on specific motivational resources. Those can then be defined and evaluated with respect to their holder.

It satisfies a number of psychometric indicators that confirm its objectivity from various aspects:

Matkovčíková N. (2010) presents the complete results of research in motivation and stabilization of young talented managers (N = 160: 84 + 76), similarly, D. Newmanová (2013) analysed data and results of research which confirmed the validity of the method and the convenience of its use (N = 111), V. Blattnerová (2012) used this method to analyse personality determinants of motivation, having brought valuable information not only about the method itself, but also about its qualities (N = 105), J. Šnegoňová (2010) researched quality of working life in the context of life satisfaction of teachers in primary and secondary schools and work-life balance (N = 86), E. Krylov (2006) comparison of the two methods of this kind of focus has confirmed good standardization data and usability not only in research but also in individual psychodiagnostics (N = 41).

Current Issues of Science and Research in the Global World – Kunova & Dolinsky (Eds)
© 2015 Taylor & Francis Group, London, ISBN 978-1-138-02739-8

Author index

Author Index

T - #0277 - 101024 - C0 - 246/174/21 [23] - CB - 9781138027398 - Gloss Lamination